D0948613

Mrs Ann Austin, BSc.
1938–2010

Throughout our 50 years of happy marriage, Ann was most supportive of me in my work.

From our helicopter wedding in 1960 through to her organising and conducting the day tours for the wives of the delegates of the International UAS Conferences, Ann became known and well-liked in several aeronautical circles world-wide.

She did this in addition to bearing and raising our two sons and having her own career in education.

One of her dreams was to see the publication of this book which could not have been written without her encouragement and support.

The realisation of that dream was stolen from her by her death from cancer.

She will be sorely missed by so many.

I dedicate the book to her memory.

Contents

Foreword

The author, Reg Austin, started his career in aeronautics in 1945 at the Bristol Aeroplane Company Ltd, where he served a student apprenticeship and became involved in the design of the Bristol airliners, high-speed fighter projects and helicopters, becoming eventually a senior aerodynamicist in the Helicopter Department. He also learnt to fly gliders and light aircraft.

He moved to become Chief Engineer in the Helicopter Department, Auster Aircraft Limited for three years, but then returned to Bristol to hold various senior positions in Bristol Helicopters.

When Westland Helicopters took over Bristol Helicopters in 1960, Reg was appointed Chief Project Engineer responsible for new Westland designs. Amongst other projects such as compound helicopters and tilt-wing aircraft, he conceived and led the design of the Lynx helicopter and introduced the discipline of Operational Research to the Company.

In 1967 he conceived the idea of the plan-symmetric unmanned helicopter and led the team developing and flight-testing the experimental Mote and Wisp surveillance UAVs. He initiated the work on the Westland Wideye UAV.

In 1980 Reg joined ML Aviation as Flight Systems Manager, covering various experimental projects such as the Kestrel rotary-wing munitions delivery system, the Sea-star naval fixed-wing target and particularly the development and world-wide field trials of the Sprite rotary wing VTOL UAV system.

He was Chairman of the NATO VTOL UAV Engineering Group for some years.

Moving to academia in 1991, Reg has remained busy in the aeronautical field as a Consultant and lecturer, covering all aspects of aircraft design and operation including unmanned air vehicle systems. He is a Professor, Eur Ing, MSc, Fellow of the Royal Aeronautical Society and a member of its Rotorcraft Committee, and Vice Chairman of the Bristol International Unmanned Air Vehicle Systems Conferences which he initiated in 1978.

With such a wide ranging knowledge of all aspects of UAVs he is ideally equipped to write this book. It draws on his extensive first-hand experience of system design, giving examples from his own activities. Rarely do we have the opportunity to learn from engineers who have been pioneers in the application of a new field of technology whilst it is still evolving. This book therefore should be compulsory reading for everyone working (or planning to work) in the field of UAV systems. In particular what might be called non-technical aspects should not be overlooked and Chapter 5 should be very useful in this respect.

David J Walters BSc, CEng, AFRAeS, AinstP
Retired Director of Future Systems, MOD

Acknowledgements

The author wishes to acknowledge the help and advice given to him in the preparation of this book.

The following organisations are thanked for permission to use photographs and data of their products for illustration purposes. AeroVironment Inc., All American Industries, Inc., BAE Systems, Beijing University of Aeronautics and Aerospace, Bell Helicopter Textron Inc., Bernard Hooper Engines, Boeing/Insitu, Cloud Cap Technologies, Controp, Cranfield University, Denel Aerospace, EADS Defence and Security, General Atomics Aeronautical Systems Inc., GFS Projects UK, Honeywell, IAI Malat, Lockheed Martin, Northrop-Grumman, Prox Dynamics, Qinetiq plc., Rafael, RUAG Aerospace, Sandia National Laboratories, Schiebel Elektronische Geraete Gmbh, Selex, Wescam, Yamaha Motor Company.

My thanks are also due to: Colin Coxhead, for data on piston engines; Marilyn Gilmore of dstl, for advice on stealth; Kenneth Munson and Jane's Unmanned Aerial Vehicles and Targets, for data and advice; John Russell, on pilot vision; Arthur Richards of Bristol University, on artificial intelligence; Shephard Press, for information and contacts. I am grateful to The Bristol International UAV Systems Conferences and Jane's Unmanned Aerial Vehicles and Targets for help with the list of abbreviations and acronyms.

My special thanks are offered to Rob Frampton for his advice on autonomy and navigation systems, together with his support in reviewing the draft of this book.

Finally, my Very Special Thanks go to my dear wife, Ann, for her forbearance in accepting the disruption of our social life and for her encouragement, feeding and watering me, during the preparation of this book.

Series Preface

Unmanned air vehicles were once the stuff of rumour and legend, identified in the press as new and mysterious. "New shapes in the sky," declared the headlines at one time in the recent past. Now they seem to be commonplace, on the battlefield at least, where they are seen carrying out surveillance missions and deploying weapons with great accuracy. They are now truly the solution to some of the dull, dirty and dangerous tasks for which they were first proposed.

Apart from military applications there are many jobs to be performed in commercial and government applications in surveillance, monitoring and trouble-shooting in the fields of utilities, maritime rescue, customs and excise and agriculture to name only a few. Some police forces are collaborating with industry to develop systems to replace helicopter surveillance. The main drawback to their application has been the difficulty in obtaining certification to operate in controlled airspace.

There are many classes of unmanned vehicle in existence, and many types within each class, developed by many manufacturers. The classes range from insect like vehicles, through hand launched sub-scale models to full size long endurance types. They are all capable of carrying some form of sensor and of relaying sensor information to the ground. Most are remotely piloted, and some experimental types are being operated on controlled ranges as part of a gradual progression towards full autonomy, where they will be capable of performing missions with minimum human intervention.

This book introduces the classes and types of platforms available, and examines their performance, the design requirements for aerodynamics, structure, propulsion and systems to suit particular roles. Sensor and avionic payloads are discussed as well as the data links and communications suites required to complete the payload. Methods of launch and recovery are presented as an important part of deployment.

This is a good introduction to the subject and it takes the Aerospace Series into the world of remotely piloted unmanned air vehicles, and more importantly to the realm of autonomous systems where great challenges remain to be overcome.

Allan Seabridge

Preface

The development and entry into service of unmanned air vehicle systems has a long, drawn-out history. Unfortunately, the vision of engineers and scientists is seldom matched by that of administrators, regulators or financiers. The availability of UAV systems has also often depended upon maturation of the requisite technology.

UAV systems are now being operated by several military forces and currently, to a more limited extent, by civilian organisations. These latter operations, however, may eventually expand to exceed, in number and diversity, those of the military.

The systematic nature of UAV systems, which is achieved through the combination of many elements and their supporting disciplines, will be emphasised throughout this book. Although the aircraft element is but one part of the coordinated system, it is almost certainly the element which drives the requirements of the other system elements to the greatest extent.

The aircraft itself will have much in common with manned aircraft, but also several differences which are explained. These differences often result from the differences in operational requirements compared with manned aircraft, for example the need to take off from remote, short, unprepared airstrips or to fly for long periods at very high altitudes. The performance of the aircraft is often enhanced by not having to carry the weight of equipment and structure required to accommodate aircrew, and having a lower aerodynamic drag for the same reason. The UAV also often benefits from advantageous scale effects associated with a smaller aircraft.

It is not the author's intention to provide a textbook to expound in detail the many UAV engineering disciplines, which include, of course, aerodynamics, electronics, economics, materials, structures, thermodynamics, etc., but rather to show how the disciplines are integrated into the design, development and deployment of the UAV systems. It is the intention also to explain the manner in which their application may differ from their use in other aerospace and engineered systems.

On occasions, for example in Chapter 3, the author has entered into the theory of the discipline in order to dispel some myths and to bring the reader's attention to the significant aspects without the need for the reader to find his way through a specialist textbook on the subject.

Similarly, the history and evolution of UAV Systems (Chapter 28) is considered only in order to point out how UAS have evolved and where lessons have been, or should have been, learned from that history and the probable way forward.

The systems nature of UAV makes it impractical and undesirable to cover elements in isolation, some aspects therefore appear in more than one chapter, but at different levels of detail. The author intends this to be reinforcement of key aspects. For a more detailed knowledge of the several disciplines, the reader is directed to the specialist works listed in the reference sections.

Examples of several systems and their sub-systems are used to explain the principles involved. The technology continues rapidly to evolve, so that the examples used will not necessarily reflect the latest available when this book reaches publication, but the principles will still apply. The reader is again referred to the reference sections for up-to-date listings of current systems.

In making a comparison between the attributes of the several UAS, where available, the author has used the data supplied by the manufacturers to whom he is grateful for their help. In other cases, where the data has not been made available, the author has derived results from scaling photographs and making calculations of performance using established methods. This is done in order to explain several of the differences between manned aircraft systems and UAV and also between the different configurations of UAV often decided by the specific type of operation that they carry out. The author regrets any misconceptions, should they arise, from this practice but feels that it was necessary to complete the work rather than let it be limited through the lack of authorised data.

The author has been involved with aircraft of all types, both manned and unmanned, during his long career in aviation. It is, however, his time spent working with unmanned rotorcraft and seeing their great versatility, that has contributed most to his understanding of the issues entailed in developing and operating UAV systems.

His 'hands-on' experience with UAV began in 1968 at Westland Helicopters with responsibility for the Mote, Wisp, Pupil and initial Wideye design and development VTOL programmes, and subsequently at M L Aviation, with both fixed- and rotary-wing programmes. The most advanced of all of these programmes was that of the Sprite UAV system at M L Aviation. A number of these systems were deployed around the world, operating by day and night, in all weathers, off-land and off-board ship on many tasks and with a wide range of payloads.

The author was also privileged to have been invited to observe the trials of UAV systems by other manufacturers and to have the opportunity to discuss the activities of many other developers and operators of UAV systems world-wide, especially through contacts made in his involvement over 30 years with the Bristol International UAV Systems Conferences.

However, his leadership of the versatile Sprite programme gave the greatest insight into all aspects of the design, development and deployment of UAV systems, and he makes no apologies for leaning on this experience for many of the examples in the book.

It is the author's hope that this book will be of use not only to future students, designers, developers and operators of UAV systems, but also to procurement and regulatory organisations.

Reg Austin

Units and Abbreviations

Units

All units of measurement throughout this book conform to the *Système Internationale*, with alternative units shown in brackets where appropriate.

Abbreviations: Acronyms

The following terms are not necessarily all to be found in this book, but are terms which are likely to be encountered in general literature or conferences on UAV systems.

AA	Anti-aircraft
AAA	Anti-aircraft artillery
AAIB	Air Accidents Investigation Board
AAM	Air-to-air missile
AAV	Autonomous air vehicle
AC	Alternating current
ACAS	Airborne collision avoidance system/Assistant Chief of the Air Staff
ACGS	Assistant Chief of the General Staff
ACL	Agent communication language/Autonomous control levels
ACNS	Assistant Chief of the Navy Staff
ACS	Airborne control station (system)
ACTD	Advanced Concept Technology Demonstration
ADC	Air data computer
ADF	Automatic direction finder/finding
AEW	Airborne early warning
AFB	Air Force base (US)
AFCS	Automatic flight control system
AFRL	Air Force Research Lab (US)
AFRP	Aramid fibre-reinforced plastics
AFV	Armoured fighting vehicle
AGARD	Advisory Group for Aerospace Research and Development (NATO)
A h	Ampère hour
AHRS	Attitude and heading reference system
AI	Artificial intelligence
AIAA	American Institute of Aeronautics and Aerospace
AIC	Aeronautical Information Circular
AIP	Aeronautical Information Publication

AJ	Anti-jam
AMA	Academy of Model Aviation (US)
AMS	Acquisition Management Systems (UK Ministry of Defence)
ANSP	Air navigation service provider
AoA	Angle of attack
AOA	Aircraft operating authority
AOP	Air observation post
APU	Auxiliary power unit
AR	Aspect ratio
ARAC	Assistant Range Air Controller
ARINC	Aeronautical Radio Inc (US company)
ARPA	Former temporary title of DARPA
ASROC	Anti-submarine rocket
ASTRAEA	Autonomous systems technology related airborne evaluation and assessment
ASV	Autonomous surface vehicle
ASW	Anti-submarine warfare
AT	Aerial target
ATC	Air Traffic Control
ATCO	Air Traffic Control Officer
ATEC	Aircraft Test and Evaluation Centre
ATM	Air Traffic Management
ATR	Automatic target recognition
ATS	Air Traffic Service
AUM	All-up mass
AUVSI	Association for Unmanned Vehicle Systems International (US)
AUW	All-up weight
AV	Air vehicle
Avgas	Aviation gasoline
AVO	Air vehicle operator
BAMS	Broad area maritime surveillance
BDA	Battle damage assessment
BER	Bit error rate
BITE	Built-in test equipment
BLOS	Beyond line-of-sight
BMFA	British Model Flying Association
BVR	Beyond visual range
BW	Bandwidth
BWB	Bundesamt für Wehrtechnik und Beschaffung (Germany)
C2	Command and control
C3I	Command, control, communications and intelligence
C4	Command, control, communications and computers
C4ISTAR	Command, control, communications and computers, intelligence, surveillance, target acquisition and reconnaissance
CA	Clear acquisition (GPS)
CAA	Civil Airworthiness Authority (UK)
C&EA	Customs and Excise Authority
C of A	Certificate of Airworthiness
CAD	Computer-aided design
CAP	Civil Aviation Publication
CAS	Calibrated airspeed

CASA	Civil Aviation Safety Authority (Australia)
CBR	Californian bearing ratio
CC	Collection coordination
CCD	Charge-coupled device
CCI	Command and control interface
CCIR	Comité Consultatif International des Radiocommunications (France)/Commanders Critical Information Requirement (UK/US)
CD	Circular dispersion/Chrominance difference
CDL	Common datalink
CDMQ	Commercially developed, military qualified
CDR	Critical design review
CEAC	Committee for European Airspace Coordination
CEP	Circular error probability
CEPT	European Conference of Postal and Telecommunications administrations
CFAR	Constant false alarm rate
CFD	Computational fluid dynamics
CFRP	Carbon fibre-reinforced plastics
CFT	Certificate for Flight Trials
CG	Centre of gravity
CGF	Computer-generated forces
CIA	Central Intelligence Agency (US)
CKD	Component knocked down
CME	Coronal mass ejection
CMOS	Complementary metal oxide semiconductor
CMT	Cadmium mercury telluride (CdHgTe)
CNC	Computer numerical control
COA	Certificate of Waiver or Authorization
CoM	Centre of mass
COMINT	Communications intelligence
CONOPS	Concepts of operation
CONUS	Continental United States
COTS	Commercial off-the-shelf
CPL	Commercial Pilot Licence
CR	Close range
CRH	Coaxial rotor helicopter
CRT	Cathode ray tube
CS	Control station
CTOL	Conventional take-off and landing
CW	Continuous wave
DA	Danger area
DAP	Director of Aerospace Policy
DARO	Defense Airborne Reconnaissance Office (US)
DARPA	Defense Advanced Research Projects Agency (US)
dB	Decibel
DDD	Dull, dangerous and dirty
DE&S	Defence equipment and support
DEC	Director Equipment Capability (UK Ministry of Defence)
DERA	Defence Evaluation and Research Agency (UK)
DF	Direction-finding
DFCS	Digital Flight Control System

DGA	Délégation Générale des Armements (France)
DGAC	Direction Générale de l'Aviation Civile (France)
DGM	Design gross mass
DGPS	Differential GPS
DLI	Datalink interface
DMSD	Defence Modification and Support Division
DND	Department of National Defence (Canada)
DOA	Design organisation approvals
DoD	Department of Defense (US)
DoF	Degrees of freedom
DOSG	Defence Ordnance Safety Group
DPA	Defence Procurement Agency (UK Ministry of Defence)
DPCM	Digital pulse code modulation
DSA	Detect, sense and avoid
DSTL	Defence Science and Technology Laboratories (UK Ministry of Defence)
DTC	Defence Technology Centre
DTED	Digital terrain elevation data
DUO	Designated UAS operator
EAS	Equivalent airspeed
EASA	European Aviation Safety Agency
ECCM	Electronic counter-countermeasures
ECM	Electronic countermeasures
ECR	Electronic combat reconnaissance
EDA	European Defence Agency
EEPROM	Electronically erasable programmable read-only memory
EER	Effective Earth radius
EHF	Extra-high frequency
EHS	Enhanced surveillance
EISA	Extended industry standard architecture
ELF	Extremely low frequency
ELINT	Electronic intelligence
ELT	Emergency locator transmitter
EMC	Electromagnetic compatibility
EMD	Engineering and manufacturing development
EMI	Electromagnetic Interference
EMP	Electromagnetic pulse
EO	Electro-optical
ERAST	Environment Research Aircraft and Sensor Technology (programme NASA)
ERC	European Radiocommunications Committee
ESM	Electronic support (or surveillance) measures
EUROCAE	European Organisation for Civil Aviation Equipment
EW	Electronic warfare
EXCON	Exercise control
FAA	Federal Aviation Administration (US)
FADEC	Full authority digital engine control
FBW	Fly-by-wire
FCS	Flight control system
FLIR	Forward-looking Infrared
FLOT	Forward line of own troops
FM	Frequency modulation

FoV	Field of view
FPGA	Field-programmable gate array
FPT	Free power turbine
FRTOL	Flight Radio Telephony Operator's Licence
FSAT	Full-scale aerial target
FSEDF	Full-scale engineering development
FSK	Frequency-shift keying
FSRWT	Full-Scale rotary-wing target
FSTA	Future strategic tanker aircraft
FW or F/W	Fixed-wing
FY	Financial year
GCS	Ground control station (or system)
GDT	Ground data terminal
GEN	Generation
GFE	Government-furnished equipment
GFRP	Glass fibre-reinforced plastics
GLCM	Ground-launched cruise missile
GOTS	Government off-the-shelf
GPS	Global Positioning System
GPWS	Ground proximity warning system
GSE	Ground support equipment
GUI	Graphical user interface
HAE	High-altitude endurance
HALE	High-altitude, long-endurance
HF	High-frequency
HFE	Heavy-fuel engine
HIRF	High-intensity radiated field
HITL	Human-in-the-loop
HMI	Human–machine interface
HMMWV	High-mobility multipurpose wheeled vehicle
HTOL	Horizontal take-off and landing
HUD	Head-up display
HUMS	Health and usage monitoring system
HVI	High-value target
Hz	Hertz (cycles per second)
IAI	Israeli Aerospace Industries
IAS	Indicated airspeed
IC	Internal combustion
ICAO	International Civil Aviation Organisation
ICD	Interface cntrol definition (or document)
IDF	Israel Defence Force
IEEE	Institute of Electrical and Electronic Engineers
IEWS	Intelligence, electronic warfare and sensors
IFF	Identification, friend or foe
IFOR	Implementation Force (NATO)
IFR	Instrument flight rules/in-flight refuelling
IGE	In-ground effect
IIRS	Imagery interpretability rating scale
ILS	Integrated logistic support
IMINT	Imagery intelligence

IML	Integration maturity level
IMU	Inertial measurement unit
IN	Inertial navigation
INS	Inertial navigation system
IOC	Initial operating (or operational) capability
IOT&E	Initial operational test and evaluation
IPT	Integrated project team
IR	Infrared
IR&D	Internal research and development
IRLS	Infrared linescan
IRST	Infrared search and tracking
ISA	International standard atmosphere
ISO	International Organisation for Standards
ISR	Intelligence, surveillance and reconnaissance
ISTAR	Intelligence, surveillance, target acquisition and reconnaissance
ITU	International Telecommunication Union
J	Advance ratio
JAA	Joint Aviation Authorities (Europe)/Joint Airworthiness Authority
JAAA	Japan Agriculture Aviation Association
(JARUS)	Joint Authorities for Rulemaking Unmanned Systems
JATO	Jet-assisted take-off
JPO	Joint Project Office (US)
JSIPS	Joint Services Imagery Processing System
JSOW	Joint Stand-off Weapon
JSP	Joint Service Publication
JTIDS	Joint Tactical Information Distribution System
JUEP	Joint UAV Experimentation Programme
JUET	Joint UAV Experimentation Team
JWID	Joint Warrior Interoperability Demonstration
KUR	Key user requirement
LAN	Local area network
lb	st Pounds static thrust
LCD	Liquid crystal display
L/D	Lift-to-drag ratio
LED	Light-emitting diode
LLTV	Low-light television
LMA	Large Model Association
LO	Low observables
LOS	Line-of-sight
LPC	Linear predictive coding/Low-pressure compressor
LR	Long-range
LRE	Launch and recovery element
LRF	Laser range-finder
LRIP	Low-rate initial production
LRU	Line-replaceable unit
MAC	Mean aerodynamics chord
MAD	Magnetic anomaly detection
MAE	Medium-altitude endurance
MALE	Medium-altitude, long-endurance
MARDS	Military Aviation Regulatory Document Set (UK)

MART	Military Aviation Regulatory Team
MAV	Micro-air vehicle
MCE	Mission control element
MDI	Miss-distance indicator
MEMS	Micro-electromechanical system
MER	Multiple ejector rack
MFD	Multifunction display
MIL-STD	Military Standard(s) (US)
MLA ML	Aviation (UK)
MLRS	Multiple launch rocket system
MMI	Man–Machine interface
MMW	Millimetre Wave
MoD	Ministry of Defence (UK)
Mogas	Motor (automobile) gasoline
MoU	Memorandum of understanding
MPCS	Mission planning and control station (or system)
MPI	Message-passing interface
MPO	Mission payload operator
MPS	Mission planning system
MR	Medium-range
MRCOA	Military-registered civil-owned aircraft
MRE	Medium-range endurance
MS&SE	Modelling simulation and synthetic environments
MTBF	Mean time between failures
MTI	Moving target indication
MTOM	Maximum take-off mass
MTOW	Maximum take-off weight
MTTR	Multitarget tracking radar/Mean time to repair
MUAV	Mini-UAV or maritime UAV
NACA	National Advisory Committee for Aeronautics (US)
NAS	Naval Air Station (US)/National Airspace (US)
NASA	National Aeronautics and Space Administration (U)S
NATMC	NATO Air Traffic Management Committee
NATO	North Atlantic Treaty Organisation
NAV	Nano-air vehicle
NBC	Nuclear, biological and chemical (warfare)
NCO	Network-centric operations
NEAT	North European Aerospace Test Range
NEC	Network enabled capability
NFS	Naval fire support
NIIRS	National Imagery Interpretability Rating Scale (US)
NLOS	Non-line-of-sight
NOLO	No onboard live operator (US Navy)
NOTAM	Notice to Airmen
NPV	Naval Patrol Vessel
NRL	Naval Research Laboratory (US)
NTSC	National Television Standards Committee (US)
NULLO	Not utilising live local operator (US Air Force)
OA	Operational analysis
OEI	One engine inoperative

OEM	Original equipment manufacturer
OFCOM	Office of Communications (UK)
OGE	Out-of-ground effect
OLOS	Out-of-line-of-sight
OPA	Optionally piloted aircraft
OPAV	Optionally piloted air vehicle
OPV	Optionally piloted vehicle
OSA	Open system architecture
OSI	Open systems interconnection
OTH	Over the horizon
OUAV	Operational UAV
PACT	Pilot authorisation and control of tasks
PAL	Phase alternation line/Programmable array logic
PC	Personal computer
PCI	Personal computer interface
PCM	Pulse code modulation
PCMO	Prime contract management office
PDR	Preliminary design review
PERT	Programme evaluation and review technique
PIM	Position of intended movement/Previously intended movement
PIP	Product improvement programme
POC	Proof of concept
PPC	Pulse position coded
PPI	Planned position indicator
PPS	Precise positioning service (GPS)
PRF	Pulse repetition frequency
PRI	Pulse repetition interval
PS	Pitot static/Plan-symmetric
PSH	Plan-symmetric helicopter
PVM	Parallel virtual machine
PWO	Principal Warfare Officer
q	Dynamic pressure, pounds per square foot
QPSK	Quadrature phase-shift keyed
R&D	Research and development
RAAF	Royal Australian Air Force
RAC	Range Air Controller
RAM	Random Access Memory/Radar-absorptive material
RAN	Royal Australian Navy
RAST	Recovery, assist, secure and traverse (helicopter)/Radar-augmented sub-target
RATO	Rocket-assisted take-off
RCO	Remote-control operator
RCS	Radar cross-section
RDT&E	Research, development, test and evaluation
Re	Reynolds number
RF	Radio frequency
RFA	Rectangular format array
RFI	Request for information
RFP	Request for proposals
RGT	Remote ground terminal
RISC	Reduced instruction set computer

RL	Ramp-launched
RMIT	Royal Melbourne Institute of Technology
RMS	Reconnaissance management system/Root-mean-square
ROA	Remotely operated aircraft
RON	Research octane number
RPA	Remotely piloted aircraft/Rotorcraft pilot's associate
RPH	Remotely piloted helicopter
rpm	Revolutions per minute
RPV	Remotely piloted vehicle
RSO	Range Safety Officer
RSTA	Reconnaissance, surveillance and target acquisition
R/T	Receiver/transmitter
RTCA	Radio Technical Commission for Aeronautics
RTS	Remote tracking station/Request to send/Release to service
RUAV	Relay UAV
RVT	Remote video terminal
RWR	Radar warning receiver
SAM	Surface-to-air missile
SAMPLE	Survivable autonomous mobile platform, long-endurance
SAR	Synthetic aperture radar/Search and rescue
SATCOM	Satellite communications
SBAC	Society of British Aerospace Companies
SBIR	Small Business Innovative Research (US contract type)
SCS	Shipboard control station (or system)
SCSI	Small computer system interface/Single card serial interface
SE	Synthetic environment
SEA	Systems engineering and assessment
SEAD	Suppression of enemy air defences
SEAS	Systems Engineering for Autonomous Systems (UK DTC)
Sfc	Specific fuel consumption
SFOR	Stabilisation Force (NATO)
SG	Specific gravity
SHF	Super-high-frequency
shp	Shaft horsepower
SIGINT	Signals intelligence
SIL	System integration laboratory
SKASaC	Seeking airborne surveillance and control
S/L or SL	Sea level
SME	Subject matter expert
SMR	Single main rotor
SoS	System of systems
SPIRIT	(Trojan) Special purpose integrated remote intelligence terminal
SPRITE	Signal processing in the element
SPS	Standard positioning service (GPS)
SR	Short-range
SRG	Safety regulation group
SRL	System Readiness Level
SSB	Single sideband
SSR	Secondary surveillance sadar
STANAG	Standardisation NATO Agreement

STK	Satellite toolkit
STOL	Short take-off and landing
SUAVE	Small UAV engine (US)
TAAC	Technical Analysis and Application Center
TAC	Target Air Controller
Tacan	Tactical air navigation
TAS	True airspeed
TBO	Time between overhauls
TC	Type Certificate
TCAS	Traffic (alert and) collision and avoidance system
TCDL	Tactical common datalink
TCS	Tactical control system (US)
TE	Trailing edge
TED	Transferred electron device
TER	Triple ejector rack
TESD	Test and Evaluation Support Division
TFT	Thin-film transistor
TI	Thermal imager
TICM	Thermal imaging common modules
TLC	Through-life costs
TMD	Theatre missile defence
TO	Take-off
Trep	Test report
Treq	Test requirements
TRL	Technology readiness level
TTL	Transistor–transistor logic
TTP	Targeting task performance
TUAV	Tactical unmanned aerial vehicle
TV	Television
TWT	Travelling wave tube
UAS	Unmanned aircraft system
UASCdr	Unmanned Aircraft System Commander
UASSG	Unmanned Aircraft Study Group (ICAO)
UAS-p	UAS pilot
UAV	Unmanned (or uninhabited) aerial vehicle
UAV-p	UAV pilot
UCAR	Unmanned (or uninhabited) combat armed rotorcraft
UCARS	UAV Common automated recovery system (US)
UCAV	Unmanned (or uninhabited) combat air vehicle
UCS	Universal Control Station (NATO)
UGS	Unmanned ground-based system
UGV	Unmanned ground vehicle
UHF	Ultra-high frequency
UMA	Unmanned air vehicle
UNITE	UAV National Industry Team
UNSA	Uninhabited naval strike aircraft
UPC	Unit production cost
USAF	United States Air Force
USD	Unmanned (or uninhabited) Surveillance Drone (NATO)
USMC	United States Marine Corps

USN	United States Navy
UTCS	Universal Target Control Station
UTM	Universal transverse Mercator
UTV	Unmanned (or uninhabited) target vehicle
UV	Ultraviolet
UXB	Unexploded bomb
VCR	Video cassette recorder
VDU	Video (or visual) display unit
VFR	Visual flight rules
V/H	Velocity/height (ratio)
VHF	Very high frequency
VHS	Very high speed
VLA	Very light aircraft/Very large array
VLAR	Vertical launch and recovery
VLF	Very low frequency
VLSI	Very large-scale integration
VME	Virtual memory environment
VoIP	Voice-over internet protocol
VOR	VHF omnidirectional radio range
VTOL	Vertical take-off and landing
VTR	Video tape recorder
VTUAV	Vertical take-off UAV
WAS	Wide area search
WRC	World Radiocommunication Conference
WWI	World War I

1

Introduction to Unmanned Aircraft Systems (UAS)

An over-simplistic view of an unmanned aircraft is that it is an aircraft with its aircrew removed and replaced by a computer system and a radio-link. In reality it is more complex than that, and the aircraft must be properly designed, from the beginning, without aircrew and their accommodation, etc. The aircraft is merely part, albeit an important part, of a total system.

The whole system benefits from its being designed, from the start, as a complete system which, as shown in Figure 1.1, briefly comprises:

a) a control station (CS) which houses the system operators, the interfaces between the operators and the rest of the system;
b) the aircraft carrying the payload which may be of many types;
c) the system of communication between the CS which transmits control inputs to the aircraft and returns payload and other data from the aircraft to the CS (this is usually achieved by radio transmission);
d) support equipment which may include maintenance and transport items.

1.1 Some Applications of UAS

Before looking into UAS in more detail, it is appropriate to list some of the uses to which they are, or may be, put. They are very many, the most obvious being the following:

Civilian uses

Aerial photography	Film, video, still, etc.
Agriculture	Crop monitoring and spraying; herd monitoring and driving
Coastguard	Search and rescue, coastline and sea-lane monitoring
Conservation	Pollution and land monitoring
Customs and Excise	Surveillance for illegal imports
Electricity companies	Powerline inspection
Fire Services and Forestry	Fire detection, incident control
Fisheries	Fisheries protection

Unmanned Aircraft Systems – UAVS Design, Development and Deployment   Reg Austin
© 2010 John Wiley & Sons, Ltd

Gas and oil supply companies   Land survey and pipeline security
Information services   News information and pictures, feature pictures, e.g. wildlife
Lifeboat Institutions   Incident investigation, guidance and control
Local Authorities   Survey, disaster control
Meteorological services   Sampling and analysis of atmosphere for forecasting, etc.
Traffic agencies   Monitoring and control of road traffic
Oil companies   Pipeline security
Ordnance Survey   Aerial photography for mapping
Police Authorities   Search for missing persons, security and incident surveillance
Rivers Authorities   Water course and level monitoring, flood and pollution control
Survey organisations   Geographical, geological and archaeological survey
Water Boards   Reservoir and pipeline monitoring

Military roles

Navy

Shadowing enemy fleets

Decoying missiles by the emission of artificial signatures

Electronic intelligence

Relaying radio signals

Protection of ports from offshore attack

Placement and monitoring of sonar buoys and possibly other forms of anti-submarine warfare

Army

Reconnaissance

Surveillance of enemy activity

Monitoring of nuclear, biological or chemical (NBC) contamination

Electronic intelligence

Target designation and monitoring

Location and destruction of land mines

Air Force

Long-range, high-altitude surveillance

Radar system jamming and destruction

Electronic intelligence

Airfield base security

Airfield damage assessment

Elimination of unexploded bombs

1.2　What are UAS?

An unmanned aircraft system is just that – a system. It must always be considered as such. The system comprises a number of sub-systems which include the aircraft (often referred to as a UAV or unmanned air vehicle), its payloads, the control station(s) (and, often, other remote stations), aircraft launch and recovery sub-systems where applicable, support sub-systems, communication sub-systems, transport sub-systems, etc.

It must also be considered as part of a local or global air transport/aviation environment with its rules, regulations and disciplines.

UAS usually have the same elements as systems based upon manned aircraft, but with the airborne element, i.e. the aircraft being designed from its conception to be operated without an aircrew aboard. The aircrew (as a sub-system), with its interfaces with the aircraft controls and its habitation is replaced by an electronic intelligence and control subsystem.

The other elements, i.e. launch, landing, recovery, communication, support, etc. have their equivalents in both manned and unmanned systems.

Unmanned aircraft must not be confused with model aircraft or with 'drones', as is often done by the media. A radio-controlled model aircraft is used only for sport and must remain within sight of the operator. The operator is usually limited to instructing the aircraft to climb or descend and to turn to the left or to the right.

A drone aircraft will be required to fly out of sight of the operator, but has zero intelligence, merely being launched into a pre-programmed mission on a pre-programmed course and a return to base. It does not communicate and the results of the mission, e.g. photographs, are usually not obtained from it until it is recovered at base.

A UAV, on the other hand, will have some greater or lesser degree of 'automatic intelligence'. It will be able to communicate with its controller and to return payload data such as electro-optic or thermal TV images, together with its primary state information – position, airspeed, heading and altitude. It will also transmit information as to its condition, which is often referred to as 'housekeeping data', covering aspects such as the amount of fuel it has, temperatures of components, e.g. engines or electronics.

If a fault occurs in any of the sub-systems or components, the UAV may be designed automatically to take corrective action and/or alert its operator to the event. In the event, for example, that the radio communication between the operator and the UAV is broken, then the UAV may be programmed to search for the radio beam and re-establish contact or to switch to a different radio frequency band if the radio-link is duplexed.

A more 'intelligent' UAV may have further programmes which enable it to respond in an 'if that happens, do this' manner.

For some systems, attempts are being made to implement on-board decision-making capability using artificial intelligence in order to provide it with an autonomy of operation, as distinct from automatic decision making. This is discussed further in Chapter 27, Section 27.5.

References 1.1 and 1.2 discuss, in more detail, the differences between model aircraft and the several levels of automation of UAS. The definition of UAS also excludes missiles (ballistic or homing).

The development and operation of UAS has rapidly expanded as a technology in the last 30 years and, as with many new technologies, the terminology used has changed frequently during that period.

The initials RPV (remotely piloted vehicle) were originally used for unmanned aircraft, but with the appearance of systems deploying land-based or underwater vehicles, other acronyms or initials have been adopted to clarify the reference to *airborne* vehicle systems. These have, in the past, included UMA (unmanned air vehicle), but the initials UAV (unmanned aerial vehicle) are now generally used to denote the aircraft element of the UAS. However, UAV is sometimes interpreted as 'uninhabited air vehicle' in order to reflect the situation that the overall system is 'manned' in so far as it is not overall exclusively autonomous, but is commanded by a human somewhere in the chain. 'Uninhabited air vehicle' is also seen to be more politically correct!

More recently the term UAS (unmanned aircraft system) has been introduced. All of the terms: air vehicle; UAV; UAV systems and UAS will be seen in this volume, as appropriate, since these were the terms in use during its preparation.

1.2.1   Categories of Systems Based upon Air Vehicle Types

Although all UAV systems have many elements other than the air vehicle, they are usually categorised by the capability or size of the air vehicle that is required to carry out the mission. However, it is possible that one system may employ more than one type of air vehicle to cover different types of mission, and that may pose a problem in its designation. However, these definitions are constantly being changed as technology advances allow a smaller system to take on the roles of the one above. The boundaries, therefore, are often blurred so that the following definitions can only be approximate and subject to change.

The terms currently in use cover a range of systems, from the HALE with an aircraft of 35 m or greater wing span, down to the NAV which may be of only 40 mm span.

They are as follows:

HALE – High altitude long endurance. Over 15 000 m altitude and 24+ hr endurance. They carry out extremely long-range (trans-global) reconnaissance and surveillance and increasingly are being armed. They are usually operated by Air Forces from fixed bases.

MALE – Medium altitude long endurance. 5000–15 000 m altitude and 24 hr endurance. Their roles are similar to the HALE systems but generally operate at somewhat shorter ranges, but still in excess of 500 km. and from fixed bases.

TUAV – Medium Range or Tactical UAV with range of order between 100 and 300 km. These air vehicles are smaller and operated within simpler systems than are HALE or MALE and are operated also by land and naval forces.

Close-Range UAV used by mobile army battle groups, for other military/naval operations and for diverse civilian purposes. They usually operate at ranges of up to about 100 km and have probably the most prolific of uses in both fields, including roles as diverse as reconnaissance, target designation, NBC monitoring, airfield security, ship-to-shore surveillance, power-line inspection, crop-spraying and traffic monitoring, etc.

MUAV or Mini UAV – relates to UAV of below a certain mass (yet to be defined) probably below 20 kg, but not as small as the MAV, capable of being hand-launched and operating at ranges of up to about 30 km. These are, again, used by mobile battle groups and particularly for diverse civilian purposes.

Micro UAV or MAV. The MAV was originally defined as a UAV having a wing-span no greater than 150 mm. This has now been somewhat relaxed but the MAV is principally required for operations in urban environments, particularly within buildings. It is required to fly slowly, and preferably to hover and to 'perch' – i.e. to be able to stop and to sit on a wall or post. To meet this challenge, research is being conducted into some less conventional configurations such as flapping wing aircraft. MAV are generally expected to be launched by hand and therefore winged versions have very low wing loadings which must make them very vulnerable to atmospheric turbulence. All types are likely to have problems in precipitation.

NAV – Nano Air Vehicles. These are proposed to be of the size of sycamore seeds and used in swarms for purposes such as radar confusion or conceivably, if camera, propulsion and control sub-systems can be made small enough, for ultra-short range surveillance.

Some of these categories – possibly up to the TUAV in size – can be fulfilled using rotary wing aircraft, and are often referred to by the term remotely piloted helicopter (RPH) – see below.

RPH, remotely piloted helicopter or VTUAV, vertical take-off UAV. If an air vehicle is capable of vertical take-off it will usually be capable also of a vertical landing, and what can be sometimes of even greater operational importance, hover flight during a mission. Rotary wing aircraft are also less susceptible to air turbulence compared with fixed-wing aircraft of low wing-loading.

UCAV and UCAR. Development is also proceeding towards specialist armed fixed-wing UAV which may launch weapons or even take part in air-to-air combat. These are given the initials UCAV for unmanned combat air vehicle. Armed rotorcraft are also in development and these are known as UCAR for Unmanned Combat Rotorcraft.

However, HALE and MALE UAV and TUAV are increasingly being adapted to carry air-to-ground weapons in order to reduce the reaction time for a strike onto a target discovered by their reconnaissance. Therefore these might also be considered as combat UAV when so equipped. Other terms which may sometimes be seen, but are less commonly used today, were related to the radius of action in operation of the various classes. They are:-

Long-range UAV – replaced by HALE and MALE

Medium-range UAV – replaced by TUAV

Close-range UAV – often referred to as MUAV or midi-UAV.

1.3   Why Unmanned Aircraft?

Unmanned aircraft will only exist if they offer advantage compared with manned aircraft.

An aircraft system is designed from the outset to perform a particular rôle or rôles. The designer must decide the type of aircraft most suited to perform the rôle(s) and, in particular, whether the rôle(s) may be better achieved with a manned or unmanned solution. In other words it is impossible to conclude that UAVs always have an advantage or disadvantage compared with manned aircraft systems. It depends vitally on what the task is. An old military adage (which also applies to civilian use) links the use of UAVs to rôles which are dull, dirty or dangerous (DDD). There is much truth in that but it does not go far enough. To DDD add covert, diplomatic, research and environmentally critical rôles. In addition, the economics of operation are often to the advantage of the UAV.

1.3.1   Dull Rôles

Military and civilian applications such as extended surveillance can be a dulling experience for aircrew, with many hours spent on watch without relief, and can lead to a loss of concentration and therefore loss of mission effectiveness. The UAV, with high resolution colour video, low light level TV, thermal imaging cameras or radar scanning, can be more effective as well as cheaper to operate in such rôles. The ground-based operators can be readily relieved in a shift-work pattern.

1.3.2   Dirty Rôles

Again, applicable to both civilian and military applications, monitoring the environment for nuclear or chemical contamination puts aircrew unnecessarily at risk. Subsequent detoxification of the aircraft is easier in the case of the UAV.

Crop-spraying with toxic chemicals is another dirty role which now is conducted very successfully by UAV.

1.3.3   Dangerous Rôles

For military rôles, where the reconnaissance of heavily defended areas is necessary, the attrition rate of a manned aircraft is likely to exceed that of a UAV. Due to its smaller size and greater stealth, the UAV is more difficult for an enemy air defence system to detect and more difficult to strike with anti-aircraft fire or missiles.

Also, in such operations the concentration of aircrew upon the task may be compromised by the threat of attack. Loss of the asset is damaging, but equally damaging is the loss of trained aircrew and the political ramifications of capture and subsequent propaganda, as seen in the recent conflicts in the Gulf.

The UAV operators are under no personal threat and can concentrate specifically, and therefore more effectively, on the task in hand. The UAV therefore offers a greater probability of mission success without the risk of loss of aircrew resource.

Power-line inspection and forest fire control are examples of applications in the civilian field for which experience sadly has shown that manned aircraft crew can be in significant danger. UAV can carry out such tasks more readily and without risk to personnel.

Operating in extreme weather conditions is often necessary in both military and civilian fields. Operators will be reluctant to risk personnel and the operation, though necessary, may not be carried out. Such reluctance is less likely to apply with a UAV.

1.3.4   Covert Rôles

In both military and civilian policing operations there are rôles where it is imperative not to alert the 'enemy' (other armed forces or criminals) to the fact that they have been detected. Again, the lower detectable signatures of the UAV (see Chapter 7) make this type of rôle more readily achievable.

Also in this category is the covert surveillance which arguably infringes the airspace of foreign countries in an uneasy peacetime. It could be postulated that in examples such as the Gary Powers/U2 aircraft affair of 1960, loss of an aircraft over alien territory could generate less diplomatic embarrassment if no aircrew are involved.

1.3.5   Research Rôles

UAVs are being used in research and development work in the aeronautical field. For test purposes, the use of UAV as small-scale replicas of projected civil or military designs of manned aircraft enables airborne testing to be carried out, under realistic conditions, more cheaply and with less hazard. Testing subsequent modifications can also be effected more cheaply and more quickly than for a larger manned aircraft, and without any need for changes to aircrew accommodation or operation.

Novel configurations may be used to advantage for the UAV. These configurations may not be suitable for containing an aircrew.

1.3.6   Environmentally Critical Rôles

This aspect relates predominantly to civilian rôles. A UAV will usually cause less environmental distur-bance or pollution than a manned aircraft pursuing the same task. It will usually be smaller, of lower mass and consume less power, so producing lower levels of emission and noise. Typical of these are the regular inspection of power-lines where local inhabitants may object to the noise produced and where farm animals may suffer disturbance both from noise and from sighting the low-flying aircraft.

1.3.7   Economic Reasons

Typically, the UAV is smaller than a manned aircraft used in the same rôle, and is usually considerably cheaper in first cost. Operating costs are less since maintenance costs, fuel costs and hangarage costs are all less. The labour costs of operators are usually lower and insurance may be cheaper, though this is dependent upon individual circumstances.

An undoubted economic case to be made for the UAV is in a local surveillance role where the tasks would otherwise be carried out by a light aircraft with one or two aircrew. Here the removal of the aircrew has a great simplifying effect on the design and reduction in cost of the aircraft. Typically, for two aircrew, say a pilot and observer, the space required to accommodate them, their seats, controls and instruments, is of order 1.2 m^3 and frontal area of about 1.5 m^2. An UAV to carry out the same task would require only 0.015 m^3, as a generous estimate, to house an automatic flight control system (AFCS) with sensors and computer, a stabilised high-resolution colour TV camera and radio communication links. The frontal area would be merely 0.04 m^2.

The masses required to be carried by the manned aircraft, together with the structure, windscreen, doors, frames, and glazing, would total at least 230 kg. The equivalent for the UAV would be about 10 kg.

If the control system and surveillance sensor (pilot and observer) and their support systems (seats, displays, controls and air conditioning) are regarded as the 'payload' of the light aircraft, it would carry a penalty of about 220 kg of 'payload' mass compared with the small UAV and have about 35 times the frontal area with proportionately larger body drag.

On the assumption that the disposable load fraction of a light aircraft is typically 40% and of this 10% is fuel, then its gross mass will be typically of order 750 kg. For the UAV, on the same basis, its gross mass will be of order 35 kg. This is borne out in practice.

For missions requiring the carriage of heavier payloads such as freight or armament, then the mass saving, achieved by removing the aircrew, obviously becomes less and less significant.

(a) First Costs

The UAV equipped with surveillance sensors can be typically only 3–4% of the weight, require only 2.5% of the engine power (and 3% fuel consumption) and 25% of the size (wing/rotor span) of the light aircraft.

The cost of the structures and engines within the range of manned aircraft tend to vary proportionally with their weight and power respectively. So one might think that the cost of buying the surveillance UAV would be, say, 3% of the cost of the manned aircraft.

Unfortunately this is not true for the following reasons:

> Very small structures and engines have almost as many components as the larger equivalent, and although the material costs do reduce as the weight, the cost of manufacture does not reduce to the same degree.

> The UAV must have a radio communications system which may not be necessary in the manned aircraft or, at least, would be simpler.

The UAV will probably have a more sophisticated electronic flight control system compared with the manned aircraft and, of course, a day/night surveillance camera system rather than an observer with a pair of binoculars, night vision goggles and digital SLR camera.

In addition, the UAV must have a more sophisticated control station (CS) for interfacing between the operator(s) and the aircraft. The CS may be ground, sea or air based.

So the overall result is not obvious but, depending upon the surveillance requirements, may be of the order of:

UAV	20–40% of manned aircraft cost
UAV control station	20–40% of manned aircraft cost
UAV + UAV control station	40–80% of manned aircraft cost

(b) Operating Costs

These figures are, to some degree, inevitably subjective. They include the following.
Approximate UAV cost as % of manned aircraft cost:

i) Interest on capital employed	40–80%	
ii) Depreciation	30–60% ⎤	Probably about 40%
iii) Hangarage [includes support vehicle and CS]	20%	or less overall
iv) Crew salaries and associated costs*	50%	
v) Fuel costs	5%	
vi) Maintenance	20%	
vii) Insurance**	30% ⎦	

* Aircrew are paid more than ground-based personnel, retire earlier and must pass regular, more rigorous professional and medical examinations.
** This will include aircrew and third party cover and may be initially higher than that suggested here for the UAV System until insurers better understand the risk.

At the other end of the scale, it could be argued that the cost saving of removing the aircrew (the cabin crew would still be required) from a large civil jet transport such as a Boeing 747 would be minimal (although it is understood that some airline circles are thinking along these lines!). The operation would have long-range navigation and control risks and probably would be psychologically unacceptable to passengers and insurers.

In addition, the cost of operating a civilian passenger airline amounts to very much more than just the 'airside' cost. Airside costs include buying, crewing, flying, hangaraging and maintaining the aircraft.

The 'ground' cost, which includes airline publicity, ticketing, check-in, baggage handling, security, policing, fire precautions, customs, air traffic control, facilities maintenance, etc. is the dominant cost which will not be reduced by UAV operation. In fact some of those costs might be increased. Therefore it is unlikely that UAV will ever operate as large passenger transports, though such observations are prone to be proven wrong!

In between these two extreme applications, economic arguments for the use of UAVs are possible, but will depend upon particular circumstances. For example, it is possible that UAV may be considered for the long-range transport of goods in limited circumstances.

1.4   The Systemic Basis of UAS

Technically, a UAV system comprises a number of elements, or sub-systems, of which the aircraft is but one. The technical functional structure of a typical system is shown in Figure 1.1.

There are, of course, other integration facets within the more global system, such as the clearance required to operate within controlled airspace, which are not shown in this figure. These issues are addressed in later chapters.

It is always most important to view each sub-system of the UAV system as an integral part of that system. No one sub-system is more important than another, though some, usually the aircraft, have a greater impact upon the design of the other subsystems in the system than do others. In the early days of UAV development, for example, some unmanned aircraft were designed, and made, with inadequate regard to how payloads would be mounted, the aircraft launched or recovered, communications effected, or the system maintained and transported in conditions under which the system was required to perform. Subsequent attempts to build an operational system around it were either doomed to failure or resulted in unacceptable compromises or unacceptable cost.

1.5   System Composition

The following section outlines the function of each of the major sub-systems. Each of these will be discussed separately in more detail in later chapters, always remembering that they do not exist in isolation, but form part of a total system. Integration of the sub-systems into a total system is addressed in Chapter 17.

1.5.1   Control Station (CS)

Usually based on the ground (GCS), or aboard ship (SCS), though possibly airborne in a 'parent' aircraft (ACS), the control station is the control centre of the operation and the man–machine interface.

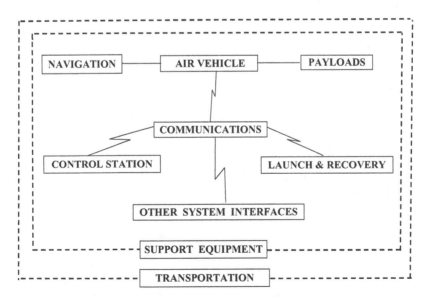

Figure 1.1   UAV system – functional structure

It is also usually, but not always, the centre in which the UAV mission is pre-planned, in which case it may be known as the mission planning and control station (MPCS). Less usually, the mission may be planned from a central command centre and the mission data is sent to the CS for its execution.

From the CS, the operators 'speak' to the aircraft via the communications system up-link in order to direct its flight profile and to operate the various types of mission 'payload' that it carries.

Similarly, via the communications down-link, the aircraft returns information and images to the operators. The information may include data from the payloads, status information on the aircraft's sub-systems (housekeeping data), and position information. The launching and recovery of the aircraft may be controlled from the main CS or from a satellite (subsidiary) CS.

The CS will usually also house the systems for communication with other external systems. These may include means of acquiring weather data, transfer of information from and to other systems in the network, tasking from higher authority and the reporting of information back to that or other authorities.

1.5.2   The Payload

The type and performance of the payloads is driven by the needs of the operational task. These can range from:

(a) relatively simple sub-systems consisting of an unstabilised video camera with a fixed lens having a mass as little as 200 g, through
(b) a video system with a greater range capability, employing a longer focal length lens with zoom facility, gyro-stabilised and with pan and tilt function with a mass of probably 3–4 kg, to
(c) a high-power radar having a mass, with its power supplies, of possibly up to 1000 kg.

Some, more sophisticated, UAV carry a combination of different types of sensors, within a payload module or within a series of modules. The data from these several sensors may be processed and integrated to provide enhanced information, or information which could not be obtained using a single type of sensor.

For example, images from an optical (light) colour video camera, from a thermal (heat) imaging camera and possibly a radar scanner system, may be fused together. Thus the thermal image and radar image may add information hidden to the optical image. The optical colour image will add discrimination, resolution and contrast not available from the reduced contrast of the thermal image or the lower resolution of the radar image. Also, the reduction in performance of one sensor under differing light or atmospheric conditions of precipitation or pollution, may be compensated for by the complementary sensors. The images, or other data, obtained by these systems are processed into a form in which they can be transmitted via the down-link to the control station or other destination as appropriate.

A number of different types of payloads appropriate for carriage by UAV are described in Chapter 8.

1.5.3   The Air Vehicle

The type and performance of the air vehicle/aircraft is principally determined by the needs of the operational mission (see Chapter 2). The task of the aircraft is primarily to carry the mission payload to its point of application, but it also has to carry the subsystems necessary for it to operate. These sub-systems include the communications link, stabilisation and control equipment, power plant and fuel, electrical power supplies; and basic airframe structure and mechanisms needed for the aircraft to be launched, to carry out its mission, and to be recovered.

Other significant determinants in the design of the aircraft configuration are the operational range, airspeed and endurance demanded of it by the mission requirement. The endurance and range requirement will determine the fuel load to be carried. Achievement of a small fuel load and maximised performance will require an efficient propulsion system and optimum airframe aerodynamics.

The speed requirement will determine more fundamentally whether a lighter-than-air aircraft, or a heavier-than-air fixed-wing, rotary-wing, or convertible aircraft configuration, is used. A long endurance and long range mission for military surveillance will predominately require a high-aspect ratio fixed-wing aircraft operating at high altitude. It will be necessary for it to take off from a long paved runway to achieve the high lift-off speed demanded by the high wing-loading required for low aerodynamic drag.

UCAVs (Unmanned Combat Air Vehicles) may be required to operate at high speed. They are likely to have low aspect ratio wings and either take off from a long runway or be air-launched.

The majority of potential civilian uses of UAVs will require the air vehicle to fly at speeds lower than 50 kt (70 km/hr) for much of its mission; and many will need an ability of the aircraft to hover (e.g. for powerline inspection) or greatly benefit from a hover capability (e.g. incident control by police and fire services).

Several military roles will either need, or greatly benefit from, the ability to hover or fly very slowly, e.g. naval decoying, army NBC monitoring and laser target designation and air force base security and the detection and elimination of unexploded bombs (UXB). In addition any application, military or civilian, where operation from off-board ship or from restricted sites is required will probably benefit from a vertical take-off and landing capability in the aircraft.

Provided that the aircraft is not required to have a top speed of more than, say, 150 kt (210 km/hr), then a helicopter configuration offers the most efficient hover and low-speed performance. Also, within its speed range, because of its high rotor-blade loading it is the most insensitive of all aircraft types to air turbulence.

Compounded helicopter configurations add wings and/or a propulsive system to a basic helicopter in order to reduce the thrust required from the rotor and enable the aircraft to achieve higher speeds. The addition of a wing can give a helicopter a speed of over 200 kt. A fully compounded helicopter (with a wing and propulsive system) has reached a speed in excess of 300 kt (550 km/hr), but at considerable cost in reduction of payload and endurance.

'Convertible' aircraft configurations attempt to achieve a viable compromise between the requirement to take off and land vertically and have a long endurance. This is achieved by lifting off with the rotor(s) horizontal, but tilting them into a vertical plane to become propellers for cruise flight with the weight of the aircraft being borne upon wings. These configurations suffer a payload weight penalty compared with either a helicopter or fixed-wing aircraft.

Another rotary-wing configuration of interest is the autogyro, which attempts to dispense with the transmission system of the helicopter in the interest of reducing complexity, but it suffers in that it cannot hover. However, it is able to fly considerably more slowly than can fixed-wing aircraft. These different aircraft configurations are discussed in more detail in Chapter 3.

1.5.4   Navigation Systems

It is necessary for the operators to know, on demand, where the aircraft is at any moment in time. It may also be necessary for the *aircraft* to 'know' where it is if autonomous flight is required of it at any time during the flight. This may be either as part or all of a pre-programmed mission or as an emergency 'return to base' capability after system degradation. For fully autonomous operation, i.e. without any communication between the CS and the air vehicle, sufficient navigation equipment must be carried in the aircraft.

In the past, this meant that the aircraft had to carry a sophisticated, complex, expensive and heavy inertial navigation system (INS), or a less sophisticated INS at lower cost, etc., but which required a frequent positional update from the CS via the communications link. This was achieved by radio tracking or by the recognition of geographical features.

Nowadays, the availability of a global positioning system (GPS) which accesses positional information from a system of earth-satellites, has eased this problem. The GPSs now available are extremely light in weight, compact and quite cheap, and give continuous positional update so that only a very simple form of INS is now normally needed. The accuracy is further improved by the use of differential GPS (DGPS, see Chapter 11).

For nonautonomous operation, i.e. where communication between aircraft and CS is virtually continuous, or where there is a risk of the GPS system being blocked, other means of navigation are possible fall-back options. These methods include:

(a) *Radar tracking.* Here the aircraft is fitted with a transponder which responds to a radar scanner emitting from the CS, so that the aircraft position is seen on the CS radar display in bearing and range.

(b) *Radio tracking.* Here the radio signal carrying data from the aircraft to the CS is tracked in bearing from the CS, whilst its range is determined from the time taken for a coded signal to travel between the aircraft and the CS.

(c) *Direct reckoning.* Here, with the computer-integration of velocity vectors and time elapsed, the aircraft position may be calculated. If the mission is over land and the aircraft carries a TV camera surveying the ground, its position can be confirmed by relating visible geographical features with their known position on a map.

However, in the interests of ease of operation, it is always desirable for the system to be as automatic, if not autonomous, as possible.

1.5.5   Launch, Recovery and Retrieval Equipment

(a) *Launch equipment.* This will be required for those air vehicles which do not have a vertical flight capability, nor have access to a runway of suitable surface and length. This usually takes the form of a ramp along which the aircraft is accelerated on a trolley, propelled by a system of rubber bungees, by compressed air or by rocket, until the aircraft has reached an airspeed at which it can sustain airborne flight.

(b) *Recovery equipment.* This also will usually be required for aircraft without a vertical flight capability, unless they can be brought down onto terrain which will allow a wheeled or skid-borne run-on landing. It usually takes the form of a parachute, installed within the aircraft, and which is deployed at a suitable altitude over the landing zone. In addition, a means of absorbing the impact energy is needed, usually comprising airbags or replaceable frangible material. An alternative form of recovery equipment, sometimes used, is a large net or, alternatively, a carousel apparatus into which the aircraft is flown and caught. An ingenious version of the latter is described in Chapter 12.

(c) *Retrieval equipment.* Unless the aircraft is lightweight enough to be man-portable, a means is required of transporting the aircraft back to its launcher.

1.5.6   Communications

The principal, and probably the most demanding, requirement for the communications system is to provide the data links (up and down) between the CS and the aircraft. The transmission medium is most usually at radio frequency, but possible alternatives may be by light in the form of a laser beam or via optical fibres. The tasks of the data links are usually as follows:

(a) *Uplink* (i.e. from the CS to the aircraft):
   i) Transmit flight path tasking which is then stored in the aircraft automatic flight control system (AFCS).

   ii) Transmit real-time flight control commands to the AFCS when man-in-the-loop flight is needed.
  iii) Transmit control commands to the aircraft-mounted payloads and ancillaries.
  iv) Transmit updated positional information to the aircraft INS/AFCS where relevant.
(b) *Downlink* (i.e. from the aircraft to the CS):
    i) Transmit aircraft positional data to the CS where relevant.
   ii) Transmit payload imagery and/or data to the CS.
  iii) Transmit aircraft housekeeping data, e.g. fuel state, engine temperature, etc. to the CS.

The level of electrical power, complexity of the processing and the antennae design and therefore the complexity, weight and cost of the radio communications will be determined by:

   i) the range of operation of the air vehicle from the transmitting station;
  ii) the sophistication demanded by transmission-down of the payload and housekeeping data;
 iii) the need for security.

1.5.7 Interfaces

All these elements, or sub-systems, work together to achieve the performance of the total system. Although some of them may be able to operate as 'stand-alone' systems in other uses, within the type of system described, as sub-systems they must be able to operate together, and so great attention must be paid to the correct functioning of their interfaces.

For example, although the communications radio sub-system itself forms an interface between the CS and the air vehicle, the elements of it installed in both the CS and air vehicle must operate to the same protocols and each interface with their respective parent sub-systems in a compatible manner.

It is likely that the UAV system may be operated by the services (both military and civilian) in different countries which may require different radio frequencies or security coding. Therefore it should be made possible for different front-end modules to be fitted into the same type of CS and air vehicle when the UAV system is acquired by various different operators. This requires the definition of the common interfaces to be made.

1.5.8 Interfacing with Other Systems

A UAV system exists in order to carry out a task. It is unlikely that the task may 'stand alone'. That is, it may require tasking from a source external to the system and report back to that or other external source.

A typical example is military surveillance where the UAV system may be operating at brigade level, but receive a task directly, or indirectly from corps level to survey a specific area for specific information and to report back to corps and/or other users through a military information network. This network may include information coming from and/or being required by other elements of the military, such as ground-, sea-, or air-based units and space-satellites, or indeed, other UAV systems. The whole then becomes what is known as a 'system of systems' and is known as network-centric operation.

A UAV system (UAS) operating alone is usually known as a 'stove-pipe system'. A representative architecture of a 'system of systems' which may include not only other UASs of similar or different types, but also include other operational elements such as naval vessels, mobile ground units or manned aircraft that provide information or mount attack missions is shown in Figure 1.2.

Similarly, in civilian operations such as fire patrol, the operators in the CS may be tasked from Fire Brigade Headquarters to move the air vehicle to new locations. It will be necessary therefore to

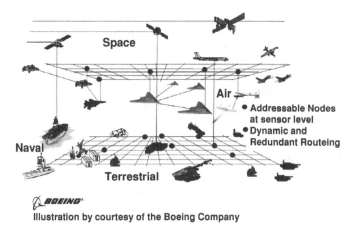

Illustration by courtesy of the Boeing Company

Figure 1.2   Network-centric architecture (Boeing)

provide, probably within the CS, the equipment required to communicate with the external sources and record/display data received and sent.

1.5.9   Support Equipment

Support equipment is one area which can often be underestimated when a UAV system is specified. It ranges from operating and maintenance manuals, through tools and spares to special test equipment and power supplies.

1.5.10   Transportation

A UAV system is often required to be mobile. Therefore transport means must be provided for all the sub-systems discussed above. This may vary from one vehicle required to contain and transport a UAV system using a small, lightweight vertical take-off and landing (VTOL) aircraft which needs no launch, recovery or retrieval equipment and is operated by say, two crew, to a system using a large and heavier ramp-launched aircraft which needs all the sub-systems listed, may have to be dismantled and reassembled between flights, and may require, say, ten crew and six large transport vehicles. Even UAV systems operating from fixed bases may have specific transport requirements.

1.5.11   System Environmental Capability

From the initiation of the concept of the system, it is important to recognise the impact that the environment in which it is to operate will have on the design of all elements of the system, including the provision of an acceptable working environment for the operating and support members of the crew. A system which has been designed with only low-altitude, temperate conditions in mind, will fail in more extreme conditions of altitude, temperature, solar radiation, precipitation and humidity.

It is also necessary to recognise the impact that the UAV system may have on the environment. This can be very significant, though with different accent, in both civilian and military roles. It is therefore necessary to consider all of these aspects carefully at the outset of the system design, and these factors are discussed more fully in Chapter 2, Section 2.4.7, and expanded where relevant in subsequent chapters on design.

References

1.1. Robert Frampton FRAeS, Dstl, UK Ministry of Defence. 'UAV Autonomy'. (*Defence Codex – The Journal for Defence Engineering and Science*, Issue 1, Summer 2008, UK Ministry of Defence.)

1.2. R.A. Frampton and J.M. Keirl Dstl, UK Ministry of Defence. 'Autonomy and its Application to Unmanned Systems'. (Proceedings of the 1st Moving Autonomy Forward Conference, *21–22 June 2006, De Vere Belton Woods Hotel*, Grantham, UK, Muretex Ltd.)

Part One

The Design of UAV Systems

2

Introduction to Design and Selection of the System

The design of most aircraft-based systems, if not of others, will usually be considered to begin in three phases:

a) the conceptual phase,
b) the preliminary design phase,
c) the detail design phase.

Other phases follow after initial manufacture. These include the design of modifications during development and subsequent modifications or improvements whilst the system is in service.

2.1 Conceptual Phase

The raison d'être of most manufacturing is fundamentally to make a profit for the company shareholders. This phase will therefore require the persons involved to have an appreciation of the market trends and either find a 'gap' in the market or see means of producing a product offering better cost–benefit to the customer. Alternatively the product may be one which is thought to open up an entirely new market.

In whichever category the proposed product falls, it is necessary to establish its commercial viability at this early stage. To that end an initial outline design will be made from which the performance and costs of developing, manufacturing and operating the product can be predicted.

a) Is it what the customer needs – (not necessarily what the customer thinks that he wants)?
b) What is the predicted size of the market – i.e. number of units?
c) Will the unit production costs plus mark-up be seen by the customer as value for money?
d) Will the operating costs and system reliability be acceptable to the customer?
e) Will the nonrecurring cost of the programme be recouped in an acceptable time by the return on sales?
f) Are there any forces, political or regulatory, which may prevent sales of the system?

Techniques of operational analysis, cost–benefit and economic studies should be used to answer these questions. Unless a positive conclusion is obtained, the proposed programme should be terminated unless

Unmanned Aircraft Systems – UAVS Design, Development and Deployment   Reg Austin
© 2010 John Wiley & Sons, Ltd

changes could be made to achieve satisfactory answers or the programme is fully supported by external, for example government, funding.

Opportunity may be taken during this phase to carry out wind-tunnel testing of an aircraft model to confirm the theoretical aerodynamic calculations or to determine if any modification to the aircraft shape, etc. is needed. This would expedite the design in the next phase.

It may be decided that the project is only viable if certain new technology is proven. This may apply, of course, to any of the elements of the system, whether it be, for example, in air vehicle control or navigation, or in computation, communications or displays. etc. Therefore a phase of research may be conducted and the decision to proceed or not with the programme will await the outcome.

2.2   Preliminary Design

Given the decision to proceed, the original outline design of the total system will be expanded in more detail. Optimisation trade-offs within the system will be made to maximise the overall performance of the system over its projected operational roles and atmospheric conditions.

A 'mock-up' of the aircraft and operator areas of the control station may be constructed in wood or other easily worked material, to give a better appreciation in three dimensions as to how components will be mounted relative to one another, ease of accessibility for maintenance and operator ergonomics, etc. This facility is becoming less necessary, however, with the availability of 3D computer design programs, though the physical appreciation obtainable from 'real' hardware should not be discarded lightly.

It will be determined which elements of the system will be manufactured 'in house' and which will be procured, at what approximate cost, from alternative external suppliers.

The phase concludes with a comprehensive definition of the design of the complete system with its interfaces and a system specification.

The costing of the remaining phases of the programme and the costs of system operation will have been re-examined in greater detail and the decision to proceed further should be revisited.

So far in the programme, a relatively small number of expert staff and limited facilities will have been employed. Therefore, costs will have been relatively low. It is tempting for the programme management to urge over-hasty completion of this phase, but it could be a false economy.

Careful consideration of options and the addressing of such matters as ease of construction, reliability, maintenance and operation at this stage can save much time and cost in correcting mistakes in the more expensive later phases of the programme.

2.3   Detail Design

At this point the work involved expands and a greater number of staff will be employed on the programme.

There will follow a more detailed analysis of the aerodynamics, dynamics, structures and ancillary systems of the aircraft and of the layout and the mechanical, electronic and environmental systems of the control station and any other sub-systems such as the launch and recovery systems. The detailed design and drawings of parts for production of each element of the system, including ground support and test equipment unless they are 'bought-out' items, will be made and value analysis applied. Specifications for the 'bought-out' items will be prepared and tenders sought.

The jigs and tools required for manufacture will be specified and will be designed unless 'bought-out'. Test Schedules will be drafted for the test phases and initial thoughts applied to the contents of the operating and maintenance manuals.

2.4   Selection of the System

The number of potential roles for UAV systems is legion, especially in the civil field, though with the exception of the use of the Rmax unmanned helicopter for crop spraying, principally in Japan, the wider civilian use of UAS is waiting upon several developments. These are principally the experience gained in their operations by the military, the agreement with airworthiness authorities on regulations for their operation, and possible adaptation of military system production and support facilities.

Although future civilian roles will, in the main, have different priorities from the military, for example economics rather than invulnerability, the main determinants of the system will be the same.

The rôle requirements, as defined by the customer, place demands upon the system which determine the shape, size, performance and costs principally of the air vehicle, but also of the overall UAV system which operates it. Some of the more important parameters involved, beginning with the air vehicle, are briefly discussed below.

2.4.1   Air Vehicle – Payload

The size and mass of the payload and its requirement for electrical power supplies is often the premier determinant of the layout, size and all–up–mass (AUM) of the aircraft. This is perhaps rightly so, as its tasking is the sole reason for the existence of the UAV system. As stated previously, the payload may range in mass from a fraction of a kilogram up to 1000 kg and in volume from a few cubic centimetres to more than a cubic metre, especially in the case of armed air vehicles.

The necessary position of the payload may also be a significant factor in the configuration and layout of the airframe. Imaging payloads for surveillance may require a full hemispheric field of view and others a large surface area for antennae.

Payloads which will be jettisoned must be housed close to the centre of mass of the air vehicle.

2.4.2   Air Vehicle – Endurance

The flight endurance demanded of the air vehicle can range from, say 1 hr for a close-range surveillance system to more than 24 hr for a long-range surveillance or airborne early warning (AEW) system.

The volume and mass of the fuel load to be carried will be a function of the required endurance and the reciprocal of the efficiency of the aircraft's aerodynamics and its powerplant. The mass of the fuel to be carried may be as low as 10% of the aircraft AUM for close-range UAV, but rising to almost 50% for the long-endurance aircraft, thus being a significant driver in determining the AUM of the aircraft.

2.4.3   Air Vehicle – Radius of Action

The radius of action of the aircraft may be limited – by the amount of fuel that it can carry, and the efficiency of its use, its speed or by the power, frequency and sophistication of its communication links.

The data rate requirements of the payload and other aircraft functions will greatly effect the electrical power and frequency range needed for the radio-links. The design and positioning of the radio antennae will reflect these requirements and could have an affect upon the choice of aircraft configuration. The radius of action will also have a significant impact on the choice of navigation equipment affecting both aircraft and control station (these aspects are covered in more detail in Chapters 9 and 11).

2.4.4   Air Vehicle – Speed Range

Driven particularly by the necessary speed of response, this could range typically as follows:

0–100 kt for a close-range surveillance role;

0–150 kt plus for many off-board naval roles;

80–500 kt for long-range surveillance and AEW roles;

100 kt to mach 1 plus for future interception / interdiction roles.

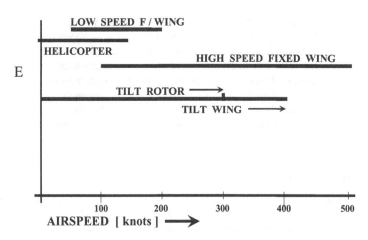

Figure 2.1   Speed ranges of aircraft types

The required speed range will be a dominant factor in determining the configuration and propulsive power of the aircraft.

Figure 2.1 indicates the aircraft configurations most appropriate to the above speed ranges. However, speed generally comes at a cost in terms of fuel consumption and airframe complexity resulting in reduced efficiency of payload and/or range for size, mass and financial cost. A notional value E for relative efficiency is ventured on the vertical axis of the figure.

2.4.5   Air Vehicle – Launch and Recovery

The method for air vehicle launch and recovery, as driven by the operational role, will be significant in determining the aircraft configuration, its structural design and auxiliary equipments. This is discussed further in Chapter 12.

2.4.6   Overall System

At this point, discussion of the demands imposed by the role requirements has been limited to those affecting the air vehicle, since the air vehicle is the element of the system which usually has the greatest effect upon the other elements. For example, the launch and recovery of the aircraft may be achieved:

a) using a wheeled-aircraft from and onto a conventional runway;
b) ramp-launching the aircraft with various alternative means of acceleration and subsequent recovery;
c) without any further equipment for a VTOL aircraft.

The length of the runway required for (a) will depend upon the aircraft acceleration and lift-off speed. The size, power and sophistication of the launcher for (b) will depend upon the aircraft mass and minimum flight speed. In addition, a transport vehicle to retrieve the aircraft may be required, particularly for cases (a) and (b). The various factors outlined above and other factors will influence the sophistication of the communications equipment and of the control station.

The design and integration of the other elements of the system are considered in later chapters, but the following two aspects are common to all elements.

2.4.7   *Environmental Conditions*

It was pointed out in Chapter 1 that:

(i) it is important to recognise, during the system design, the impact that the environment will have on all elements of that system; and
(ii) it is necessary to recognise the impact that the UAV system may have on the environment.

First, a system which has been designed with only temperate conditions in mind, will fail if operated in more extreme conditions of altitude, temperature, solar radiation, precipitation, humidity and other atmospheric conditions. In cold conditions, batteries tire rapidly; electronics performance may vary; some materials become brittle; air and ground vehicles ice up; engines may fail to start; brakes seize, etc. In conditions of extreme heat, engines and actuators may lose power or seize, electronics components may fail, vapour locks appear in fuel lines, etc.. Even worse are areas where large temperature gradients occur between night and day, resulting in changes in system characteristics and considerable problems with condensation.

Extremes of heat or cold, not only whilst the system is functioning, but also whilst standing idle will determine the level of provision of cooling needed for sub-systems such as sensors, communication equipment, etc. or of anti-icing and de-icing systems and systems for the prevention of condensation.

Atmospheric conditions such as wind strength and turbulence in which operation is required can be an important factor in the design of the structure of the ground control station and support equipment as well as the choice of air vehicle configuration. The suitability of all elements of the system must be addressed if it is to be operated in atmospheric 'pollution' such as snow, freezing fog or wind-blown sand particles.

Facilities needed for night operation must be incorporated in the design of both the aircraft and GCS.

With particular concern for the aircraft, it is important to design at the outset knowing the atmospheric conditions in which it is to fly. The altitude, whether say 3000 or 20 000 m, will largely determine the wing or rotor blade area, and the amount of atmospheric turbulence present will have an effect on the configuration to be chosen. An aircraft required to maintain an accurate flight path will not be very acceptable if it is blown around like a leaf in turbulent air conditions.

Second, the other important perspective is the impact that the *system* has on the *environment*. This can be very significant, though with different accent, in both civilian and military roles:

(a) Too high a level of acoustic noise can cause a nuisance in civil operations, whilst it can result in detection of the system in military operations.
(b) Uncontrolled radio frequency transmission can similarly result in interference or detection.
(c) Visual impact of either aircraft or ground-based equipment can be seen as spoiling the environment from the civilian point of view and can lead to vulnerability in the military field.
(d) In military operation, too great an infrared or radar signature, particularly of the aircraft, but also of the ground-based equipment, can lead to detection and annihilation.

It is therefore necessary to consider all these aspects carefully at the outset of the system design. Fortunately, because of their smaller size, UAVs naturally have smaller radar, visual, infrared and acoustic signatures than their larger manned brethren. But, even more significantly, because it is not necessary to compromise the UAV to accommodate aircrew, it is far easier to configure them to have very much lower signatures (see Chapter 7).

The conditions, therefore, in which a UAV system must operate, can have a significant effect upon the design, not only of the air vehicle in determining the power installed, wing area etc., but also upon the other elements of the system.

2.4.8 Maintenance

The frequency and length of time during which a UAV system is nonoperable due to its undergoing maintenance are significant factors in the usefulness and costs of the deployment of the system. These are factors which must be addressed during the initial design of the system, involve control of the system liability to damage, system reliability, component lives, costs and supply, and the time taken for component replacement and routine servicing.

The achievement of adequate accessibility to sub-systems in both airborne and control station elements of the system must be carefully considered in the initial design and may have significance in determining the configuration of those elements. This will be discussed more fully in later chapters.

2.4.9 System Selection as Categories

Chapter 1 listed the eight currently accepted categories of UAV systems as defined by the air vehicle type. In reality, there is a continuum of sizes and shapes of systems, depending upon the operator's needs. However, in order to simplify the approach in the following chapters, it is proposed to consider there being but four categories relating to air vehicle missions. These will be:

a) HALE and MALE systems with the air vehicles operating from runways on established bases away from hostile action, carrying sophisticated payloads over very long distances.
b) Medium-range or tactical systems with air vehicles operating at moderate altitudes, but at moderate to high airspeeds. They may perform reconnaissance, ground attack or air-superiority (UCAV) missions. Typical operating range is of order up to 500 km.
c) Close-range systems in support of land or naval forces operated from the battlefield or from ships. These may also cover most of the civilian roles. They will have an operating range of about 100 km.
d) MAV and NAV which may be hand-launched and of very short range and endurance.

3

Aerodynamics and Airframe Configurations

In Chapter 4 we will look at the characteristics of the aircraft or platform to carry out the four representative categories/roles as defined in Chapter 2. However, before we do so it is appropriate to remind ourselves of some basic aerodynamics and airframe configurations which may be used.

3.1 Lift-induced Drag

Fundamentally, an aircraft remains 'afloat' simply by accelerating an adequate mass of air downwards and, as Newton discovered, the reaction force in the opposite direction opposes the gravitational force which constantly tries to bring the aircraft back to ground. (A swimmer treading water uses the same principle to keep his head above water). This is illustrated in Figure 3.1 for an aircraft travelling with forward velocity V and deflected air velocity u. A disparity of aerodynamic pressures between the upper and lower surfaces of its wings, whether of the fixed or rotating variety, is caused.

The lower pressure on the upper surface and the higher pressure on the lower surface of a wing is merely the 'transfer mechanism' for the reaction force.

The horizontal component of the reaction force is a drag, known as the 'lift-induced drag', which has to be overcome by the propulsion system of the aircraft if it is to maintain airspeed. Therefore it is necessary to minimise the amount of drag caused in the process of creating lift. Figure 3.1 indicates that the best ratio of lift to drag is obtained for small airstream deflection angles φ, where $\tan \varphi = u/V$.

The amount of lift produced, however, is equal to the product of the mass-flow of air entrained and the velocity u that is given to it in the downwards direction, i.e. $PAVu$ N, where the air density ρ is measured in kg/m^3 and the cross-sectional area of the affected airflow A is in m^2.

Hence to create sufficient lift, if u/V is to be smaller for efficiency, the product of the air density ρ and the mass of air being entrained per unit time must be larger.

The amount of air entrained, for a given aircraft velocity, is a function of the frontal area of the wing presented to it but, for efficiency, the incidence of the wing to the air must be kept low in order to retain a small value of φ, as discussed. Also, the ability of the air to remain attached to the upper surface of the wing fails at higher wing incidence, resulting in an increase in parasitic drag (see below). Therefore, especially for the low values of air density at high altitudes, the aircraft must fly fast and/or have a large wing-span b to entrain a large mass of air.

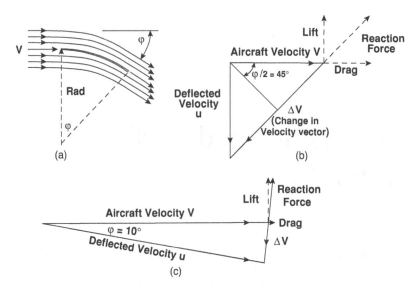

Figure 3.1   Creation of lift (and drag) by air deflection

In summary the induced drag D_i of an aircraft wing varies as the square of the span loading (lift generated, L N divided by span length [b]), the reciprocal of the air density ρ kg/m^2, and the square of the reciprocal of the airspeed V m/s. i.e. from the textbook formula

$$D_i = k_i.(L/b)^2/q\pi \quad \text{or} \quad D_i = k_i.(L/b)^2 / \frac{1}{2}\rho\,\pi\, V^2 \tag{3.1}$$

where k_i is a nondimensional factor, greater than unity, which increases the drag, depending upon the loss of efficiency of the wing due to poor lift distribution, that is away from the ideal elliptical distribution; k_i is typically in the order of 1.1 (see Reference 3.1). L/b is the span loading in N/m and q is the aerodynamic head: $q = \frac{1}{2}\rho V^2$.

Hence calculation of the lift-induced drag can be simply summarised as follows: D_i is equal to the square of the span loading divided by πq and multiplied by a form factor k_i.

The reduction of air density with altitude, shown in Figure 3.2, indicates that at 20 000 m the density of the atmosphere is less than one-tenth of its value at sea level, and therefore a greater volume of air must be accelerated downwards to produce lift at high altitudes compared with the volume required at sea-level. This is why it is desirable to have a wing with a greater span on an aircraft required to operate at high altitudes than for aircraft operating at low altitudes.

The air density shown by Figure 3.2 is for the 'standard atmosphere' where the temperature at sea level is 15°C. In cooler conditions, the air density will be greater over the range shown and, conversely, less dense in warmer air.

3.2   Parasitic Drag

Other factors also create drag on an aircraft. These other origins of drag, which may be collectively grouped as 'parasitic drag', comprise skin friction drag, form drag, interference drag, momentum drag and cooling drag. The origins of these are discussed in detail in most books on aircraft aerodynamics. An extensive treatise on this is given in Reference 3.1, which enables a designer, in the initial stages of a

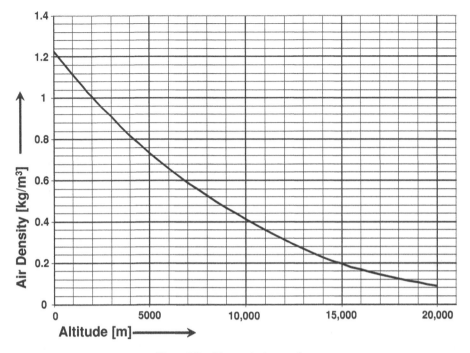

Figure 3.2 The standard atmosphere

project, to make a preliminary estimate of the total drag of an aircraft by calculating the drag contributed by the component parts and summing them.

Suffice it to say that parasitic drag varies, to first order, on an aircraft of defined configuration, with the air density and with the square of the airspeed.

Early in the design of an aircraft, its drag, along with other aerodynamic characteristics, will be measured in a wind tunnel and this will be reduced into a coefficient form by dividing the measured drag by the airspeed, air density and a reference area S, usually the main wing area in fixed-wing aircraft.

That is, the parasitic drag coefficient, $C_{Dp} = D_p / \tfrac{1}{2}\rho V^2 S$, so that the parasitic drag may be estimated for any level flight condition using the expression:

$$D_p = q C_{Dp}.S \qquad (3.2)$$

where S is the wing area and q is the aerodynamic head: $q = \tfrac{1}{2}\rho V^2$.

There is, however, a further term which represents the increased drag which results from a wing being operated at higher incidence. This term is usually small until the wing approaches a stalled condition, when it becomes extremely large. It is caused by an increased skin friction and form drag as the wing incidence increases either to produce more lift or to fly more slowly. The increase generally trends as a function of the square of the lift coefficient C_L, so that the parasitic drag equation then becomes:

$$D_p = \left(C_{Dp} + k_p C_L^2\right) q S. \qquad (3.3)$$

Combining the induced drag and the parasite drag gives the total drag of the aircraft.

Given that for a fixed configuration, the induced drag reduces as the square of the reciprocal of the airspeed, whilst the parasite drag varies with the square of the airspeed, then to obtain a reduction in induced drag, the aircraft must fly faster but, in doing so, the parasite drag increases. Thus there is an intermediate airspeed, where the induced drag equals the parasitic drag and the total drag is a minimum. The power used by the aircraft is equal to the product of total drag and the airspeed, so there is another airspeed at which the power used is a minimum.

There is yet another airspeed, usually faster than either of the former, at which the aircraft is at its most economic in terms of fuel used per distance travelled. All these values are different at different altitudes and they can be a significant determinant in the design of the aircraft, depending upon its operational roles and conditions.

Two basic criteria for flight at any given airspeed are that the wing produces sufficient lift to oppose the aircraft weight and that the thrust of the propulsor (propeller or jet) is equal to, or greater than, the total drag of the aircraft. For a fixed-wing aircraft, if there is a speed below which either of these criteria is not met, then the aircraft cannot sustain flight. This speed is the absolute minimum flight speed. However, it is not practical for the aircraft to attempt flight at this absolute minimum speed since any air turbulence or aircraft manoeuvre can increase the drag and/or reduce the lift, thus causing the aircraft to stall. A margin of speed above this is necessary to define a practical minimum flight speed V_{min}. This important concept of a minimum flight speed will also determine the speed required for the aircraft to take off or be launched.

The lift produced by a wing is given by the equation

$$L = \frac{1}{2}\rho V^2 S C_{\text{L}} \tag{3.4}$$

where C_{L} is a coefficient that determines the ability of the wing of area S to deflect the airstream. This coefficient, itself, is a function of the design of the wing section, the Reynolds number at which it is operating and the wing incidence, increasing in value with incidence and peaking at a value, $C_{\text{L.max}}$, beyond which it sharply reduces.

The value of the absolute minimum flight speed is obtained by rearranging Equation (3.5) as:

$$V = (2L/\rho S \, C_{\text{L.max}})^{1/2}$$

but this provides no margin, as discussed above. A more realistic value of V_{min} can be specified either by allowing a margin in speed or in lift coefficient. The latter approach is adopted by the author.

This results in a value of V_{min} given by:

$$V_{\text{min}} = (2L/\rho S \, C_{\text{lo}})^{1/2} \tag{3.5}$$

where C_{Lo} (operating C_{L}) has been chosen to have a value of about 0.2 less than the $C_{\text{L.max}}$ for the selected aerofoil section. Note that this makes but little allowance for turbulent air or for manoeuvres and so refers to an absolute minimum speed at which the aircraft is able to maintain straight and level flight in smooth air. In that sense it is optimistic.

A typical, moderately cambered, section offers a $C_{\text{L.max}}$ of about 1.2 when used in practical wing construction and without flaps. Hence a value of 1.0 has been adopted.

Few UAV wings incorporate flaps as they represent complication and extra weight and cost. Although their deployment would produce an increase in lift coefficient, it would also add considerable drag. This would demand extra thrust and an increase in fuel-burn, neither effect desirable for take-off or in cruise flight. Their use is normally limited to the landing mode.

Equation (3.6) can be rewritten as:

$$V_{\text{min}} = (2w/\rho C_{\text{Lo}})^{1/2} \quad \text{or} \quad \left[(2/\rho_0)^{1/2} (w/\sigma)^{1/2} \right] \tag{3.6}$$

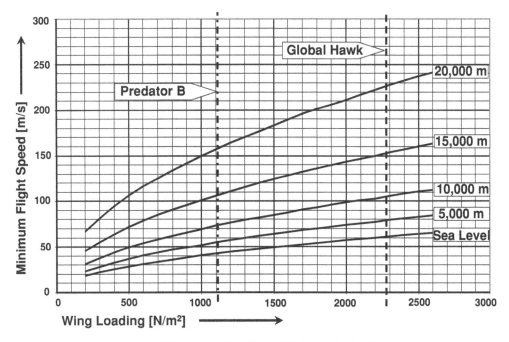

Figure 3.3   The variation of V_{min} with aircraft wing loading

where w is the aircraft wing loading in N/m², ρ_0 is the air density at sea-level standard conditions and σ is the relative air density at altitude.

The variation of V_{min} with aircraft wing loading at several altitudes is shown in Figure 3.3 which covers wing loadings appropriate to the close-range, medium-range and MALE and HALE UAV. Figure 3.4 covers the lower values of V_{min}, resulting from the wing loading relevant to the smaller mini and micro UAV.

3.3   Rotary-wing Aerodynamics

The aerodynamics of rotary-wing aircraft are, by nature, more complex than the aerodynamics of fixed-wing aircraft. Again, the author's intent is to outline some of the essential truths and to refer the reader to more specialised works (see bibliography), including the long-standing work of Gessow and Myers (Reference 3.2) for a well-explained basic understanding and to the work of Simon Newman (Reference 3.3) for a more recent and detailed treatise on the subject.

3.3.1   Lift-induced Drag

The same basic mechanism applies for rotary wings as for fixed wings, the difference merely being that the 'fixed wing' moves on a sensibly linear path in order to encompass the air, whilst the rotary wing, whilst hovering, moves on a circular path. The latter therefore draws in 'new' air from above in order to add energy to it and accelerate it downwards, compared with the former which receives it horizontally to accelerate it downwards.

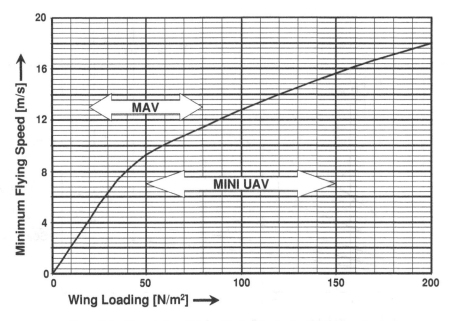

Figure 3.4   The variation of V_{min} with aircraft wing loading (lower values)

At each element of the rotary wing, the aftward-inclined vector of the lift force produces a drag at the element which translates into a torque demand at the rotor hub. For the same reason as with the fixed wing, the rotary wing must induce a large mass of air for efficiency. Therefore the larger the diameter of the circle (or disc) traced out by the rotary wing, the more efficient it is.

Thus the disc-loading p N/m², i.e. lift produced divided by the disc area, may be seen as the equivalent of span loading for the fixed wing. The difference, however, is that in the case of the fixed-wing lift, the lift is produced by the air being deflected over a sensibly single linear vortex so that the induced downwash and drag would theoretically increase to infinity at zero speed if the aircraft were able to fly at that condition. For the rotary wing, lift is produced by the air being deflected downwards over a large area covered by a number of rotating vortices so that the induced velocity and drag of the rotor remains at a finite value at zero airspeed. The velocity induced at the helicopter rotor in the hover is given by:

$$v_i = k_n (p/2\rho)^{1/2} \qquad\qquad (3.7)$$

where k_n is a correction factor to account for the efficiency of the lift distribution and the strength of the tip vortices generated at the rotor blade tips. In practice this can vary between about 1.05 to 1.2, with 1.1 usually being appropriate to rotors of moderate disc loading p.

The induced power in hover flight is then given by $P_i = k_n.Tv_i$, where T is the thrust produced by the helicopter rotor.

In forward flight the rotor is able to entrain a greater mass of air and so, as with the fixed wing, it becomes more lift-efficient and the induced power rapidly reduces with increasing speed.

At a forward speed of about 70 km/hr, the flow pattern through the rotor becomes similar to that passing a fixed wing. Therefore at that speed and above, the same expression may be used to calculate the helicopter-induced power and drag as that used for the fixed-wing i.e. Equation (3.1).

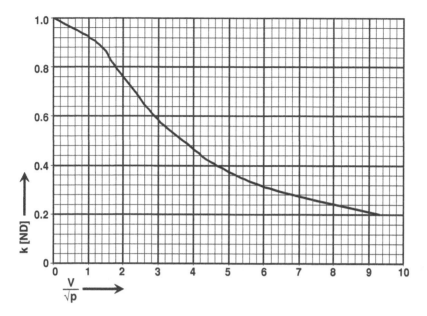

Figure 3.5   The 'Hafner' induced power factor

For airspeeds between zero and 70 km/hr, the airflow through a rotor is more complex and an empirical solution, credited to Raoul Hafner, is normally used. This takes the form of a curve whereby an induced factor k is plotted against a function of aircraft forward speed and disc loading $(V \div \sqrt{p})$, The induced power calculated for the hover is multiplied by the factor k to obtain the induced power obtaining at the intermediate airspeeds. The 'Hafner curve' is shown in Figure 3.5.

3.3.2   Parasitic Drag

Again, the same elements, form drag, friction drag, momentum drag, etc., as discussed for the estimate of fixed-wing aircraft performance, apply to the rotary-wing aircraft. There is a small difference, however, in the accounting.

Due to the more complex 'flight path' of a rotor, its drag is accounted for separately under the heading of 'profile drag' or 'profile power'. The drag of the remaining elements, i.e. fuselage, undercarriage, cooling drag, etc., then comprise the parasitic drag which is calculated in the same way as for a fixed-wing aircraft.

3.3.3   Profile Drag or Power

In hover flight, the profile power required to turn the rotor against the profile drag is computer-calculated by the summation along the length of the rotor blade of the drag of the blade split into elemental sections and multiplied by the local element velocity and radius from the hub. This is referred to as blade element theory. However, for simpler analysis, an average value of drag coefficient δ is used and mathematical integration produces the following expression;

$$\text{Profile power } P_\text{o} = \frac{1}{8} \rho \, \delta A \, V_\text{T}^3 \tag{3.8}$$

where A is the total blade area and V_T is the speed of the rotor tip.

In forward flight, each blade element no longer describes a circular orbit, but a longer, spiral path which becomes increasingly asymmetric as the forward speed increases. The result is an increase of the profile power compared with that of the hover. Power is expended in translating the rotor as well as rotating it. An exact expression of the multiplier to obtain this power is a power series equation in μ, the 'advance ratio' – i.e. the ratio of the forward speed to the rotor tip speed. This multiplier is usually simplified to the close approximation: $(1 + 4.73\,\mu^2)$, so that the profile power in forward flight is calculable as:

$$P_{o\mu} = P_o\,(1 + 4.73\,\mu^2). \tag{3.9}$$

3.4   Response to Air Turbulence

It is highly desirable, if not imperative, to reduce the response of a UAV to turbulent air to a practical minimum. This is primarily in order to more readily maintain payload sensors and beams on to the target. Other equipment, such as navigational sensors, also benefit from a 'smooth ride', but maintaining a course may also be a problem in extreme turbulence.

There are two main causes for an aircraft to have a high response to atmospheric turbulence:

a) if it is designed to have strong aerodynamic stability;
b) if it has large aerodynamic surface areas, coupled with a high aspect ratio of those surfaces, compared with the mass of the aircraft.

If an aircraft is designed to have aerodynamic surfaces whose task is to maintain a steady flight path through a mass of air, by definition, if the air-mass moves relative to spatial coordinates then the aircraft will move with the air-mass. The aircraft will therefore be very responsive to air turbulence (gusts).

To achieve stability with respect to space, it is preferable to design the aircraft to have control surfaces which, together with the aerodynamic shape of the remainder of the airframe, result in overall neutral, or near neutral, aerodynamic stability characteristics. This will ensure minimum disturbance from air turbulence, but the aircraft will then need a system to ensure that it has positive spatial stability to prevent its wandering off course due to other influences such as, for example, payload movement. This will require sensors to measure aircraft attitudes in the three axes of pitch, roll and yaw with speed and altitude and/or height data input. These sensors will be integrated into an automatic flight control and stability system (AFCS) which will control the aircraft in flight as required for the mission.

An aircraft with a large surface area to mass ratio will be disturbed by air gusts far more than an aircraft with a high mass to surface area. Any surface will generate an aerodynamic force though a high aspect ratio lifting surface will generate more than a low aspect ratio wing or fuselage. A high mass, and therefore a high inertia, will reduce the acceleration resulting from an imposed force. The more dense the packaging of an aircraft so will it be less affected by air turbulence. In this respect, UAVs have an advantage compared with manned aircraft as tightly packaged electronics are denser than human aircrew and the room that they require to function. Also, humans come in largely predetermined shapes whereas electronic systems can be tailored to fit.

Unfortunately for fixed-wing (horizontal take-off and landing) aircraft the wing area may be determined by that required to achieve low-speed flight for take-off and/or landing. A compromise may have to be considered between gust reduction and low-speed performance or, if high-lift devices are considered, also with complexity, reliability and cost.

Figure 3.6 shows the results of simple calculations which indicate the degree to which aircraft having different values of wing loading and aspect ratio will respond to a vertical sharp-edged gust of 5 m/s. It

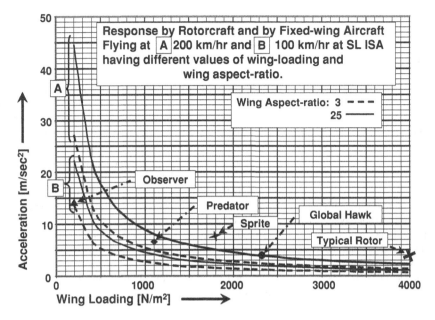

Figure 3.6 Aircraft vertical response to a vertical gust

has to be pointed out that a sharp-edged gust is a rare phenomenon, gusts usually 'ramping-up', and the acceleration shown is the initial response of the aircraft and the response will be damped somewhat as the aircraft moves and by control applied by the AFCS.

Also, of course, the calculations do not take into account that the aircraft, especially whilst flying at the lower speeds, may have stalled before the full acceleration is felt. However, the results are useful in comparative, if not in absolute, terms. It indicates how vulnerable aircraft of low wing loading can be to turbulent conditions, and also the benefits conferred by a low aspect ratio wing in reducing response. Unfortunately, the latter also confers a penalty in producing high induced drag at lower speeds. As ever, in aviation, compromises have to be effected.

An approximate criterion for comparing the unit initial gust response of an aircraft is as follows. The vertical acceleration in response to a 1 m/s vertical gust is approximately given by the expression: acceleration $= K_1 \times K_2 \times V/w_m$, where K_1 and K_2 are constants: $K_1 = \frac{1}{2}\rho a$ and $K_2 = AR \div (AR + 2.4)$. The former constant is a function of the air density ρ, and the two-dimensional aerofoil lift curve slope ($a = 5.73$). The latter is a correction for the aspect ratio (AR) of the aeroplane wing. V (m/s) is the forward speed of the aeroplane or 2/3 of the rotor tip speed of the helicopter. w_m (kg/m^2) is the wing loading of the aeroplane or blade loading of the helicopter expressed in mass per unit area, i.e. the wing loading in N/m^2 divided by the gravitational acceleration g.

Using this expression, the vertical acceleration in response to a 5 m/s vertical gust by various UAV having different values of wing loading and at two representative forward speeds has been estimated and shown in Figure 3.6.

In terms of actual UAV, both mini-UAV shown in Figure 4.23 (see Chapter 4) have wing loading so low and aspect ratio so high, as to be off the scale of Figure 3.6. It is therefore not surprising that the Desert Hawk, at least, has been reported to be limited by wind conditions.

Of the close-range aircraft, the Observer (see Figure 4.17) has one of the lowest values of wing loading at about 200 N/m^2 and yet, as do mini-UAV, has to operate at low altitude where air turbulence is often present. To alleviate this problem, the designers used a low aspect ratio wing of about AR = 3.4

and designed it to have neutral aerodynamic stability. Whilst the latter does not mitigate, may indeed exacerbate, the vertical acceleration, it does prevent large angular responses which would otherwise make it difficult to maintain a stable line-of-sight for an imaging sensor.

Medium-range HTOL UAV tend to have wing loadings in the range 300–600N/m² and MALE and HALE UAV of the order of 1000 and over 2000 N/m² respectively.

Helicopter rotor-blades generally operate at tip-speeds in the order of 200 m/s (720 km/hr) and loadings of over 4000 N/m². This, even whilst the aircraft is hovering, it is in effect equivalent to a HTOL aircraft with a very high wing loading flying at about 500 km/hr (a mean effective speed over the length of a blade) and therefore results in a helicopter rotor having very low response to vertical gusts as shown in Figure 3.6.

A single rotor, however, does have a response to a horizontal gust in the lateral direction, as well as a response, as do all aircraft types, in the vertical and aft directions. Hence 'single-rotor' helicopters, due to their asymmetry and cross-coupling of aerodynamic modes, are more difficult to control manually, so do require more complex flight control systems than a typical HTOL aircraft. However, other rotorcraft configurations do not have this problem (see Section 3.5.2).

Although rotors in forward flight can produce vibrational forces, the forces occur at predictable frequencies and so can be isolated from the airframe using an appropriately tuned airframe-to-rotor suspension.

The helicopter, therefore, with its VTOL and hover capability is well suited to operations at low heights to bring sensors to bear accurately onto small targets.

3.5 Airframe Configurations

The range of airframe configurations available for UAV is as diverse as those used for crewed aircraft, and more since the commercial risk in trying unorthodox solutions is less for the UAV manufacturer. This is principally because the UAV airframes are usually much smaller than crewed aircraft and operators are less likely to have a bias against unorthodox solutions. It is convenient to group configurations into three types appropriate to their method of take-off and landing.

The author prefers to use the acronym HTOL for aircraft which are required to accelerate horizontally along a runway or strip in order to achieve flight speed. In his view the acronym CTOL (conventional take-off and landing) is outdated since VTOL is no longer unconventional. Indeed, many flying organisations and services employ more VTOL aircraft than HTOL aircraft. It also removes the problem in designating fixed-wing aircraft which have a VTOL capability. The following sections will discuss:

a) HTOL or horizontal take-off and landing,
b) VTOL or vertical take-off and landing,
c) hybrids which attempt to combine the attributes of both of these types.

3.5.1 HTOL Configurations

After many years of development in crewed applications, these have reduced to three fundamental types, determined largely by their means of lift/mass balance and by stability and control. They are 'tailplane aft', 'tailplane forward' or 'tailless' types, shown in outline form in Figure 3.7.

All configurations, with the known exception only of the Phoenix (see Figure 4.17), have the power-plant at the rear of the fuselage. This is to free the front of the aircraft for the installation of the payload to have an unobstructed view forward.

From an aerodynamic viewpoint, if a propeller is used, the induced air velocity ahead of the rear-mounted propeller does not increase the friction drag of the fuselage as much as the slipstream would from a front-mounted tractor propeller.

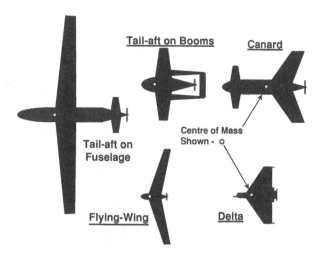

Figure 3.7   HTOL aircraft configurations

(a) Main Wing Forward with Control Surfaces aft

This is accepted as the conventional arrangement and is by far the most ubiquitous. The aircraft centre of mass is forward of the wing centre of lift and this is balanced by a down-load on the tailplane, thus providing aerodynamic speed and attitude stability in the horizontal plane. A vertical fin provides weathercock stability in yaw with wing dihedral giving stability in roll. This established configuration is the datum against which other configurations are compared. Differences within the category are to be distinguished from one another only by how the tail surfaces are carried – i.e. Single tail boom or twin tail booms and by the number of engines used.

Current HALE and MALE, i.e. long-range UAV, all have their tail surfaces carried at the rear of the fuselage. This is probably because the volume of a long fuselage is required to carry the large amount of equipment and fuel load needed on their type of operation.

An indication of the volume demanded is shown in Figure 6.8 (Chapter 6) by an artist's exploded view of a typical long-range UAV.

The twin-boom arrangement is popular for the medium- and close-range UAV as this allows the engine to be mounted as a pusher system just aft of the wing, again freeing the front fuselage for payload installation. It also provides a degree of protection for and from the engine and propeller.

There are also some aerodynamic advantages to be gained with this configuration. A pusher propeller and engine closely behind the aircraft centre of mass reduces the inertia of the aircraft in pitch and yaw. The relative proximity of the propeller to the empennage enhances the control power through the slipstream passing over the elevators and rudders and, with the lower inertia, gives an aircraft that is more responsive to pitch and yaw control. These qualities account for the popularity of this configuration. For examples see the Hunter and Seeker UAV in Figure 4.11.

(b) Canard Configuration (Figure 3.8)

A canard configuration has the horizontal stabiliser, or foreplane, mounted forward of the wing. The aircraft centre of mass is also forward of the wing and the balance is achieved with the foreplane generating positive lift. Aerodynamic stability in the horizontal plane is a result. An advantage of the canard system is that as both planes are generating positive lift, it is aerodynamically more efficient than

This is probably the only example of a UAV that uses the "Canard" "tailplane-first" configuration. At time of press, its status is not exactly known. It would appear to offer a capability between a medium-range and a MALE UAV, combining the facility of runway TOL and catapult / parachute launch and recovery

Leading Particulars currently available:

Wing Span - 6 m	Max Speed - 240 km/hr
Length - 3.2 m	Flight Endurance - 17 hr@130 km/hr
AUM - 180 kg	Service Altitude - 8000 m
Payload - Mass 37 kg	Power-plant - UEL-741 29 kW Rotary.
Payload Type - EO Turret	

Figure 3.8   E.M.I.T. aviation – "Blue Horizon" UAV

the tail-aft configuration. It also has the advantage that, as it is set at higher angles of incidence than the main wing, the foreplane stalls before the main wing. This results only in a small loss of lift and a gentle nose-down pitching motion to a recovery with a small loss of height compared with that following the stall of the tail-aft configuration.

A disadvantage of the canard is that directional stability is less readily achievable since, as the aircraft centre of mass is more rearward, the tail fin (or fins) do not have the leverage that the tail-aft arrangement has. To extend the tail-arm, most canards have the wing swept backwards with fins mounted at the tips.

The most usual propulsive system used in the canard is by aft-mounted engine(s) in turbo-jet or propeller form. An example is the Blue Horizon UAV by E.M.I.T. of Israel (Figure 3.8).

(c) Flying Wing or "Tailless" Configurations (Figure 3.7)

This includes delta-wing aircraft which, as with the above, have an effective 'tail'. The wings have a 'sweep-back' and the tip aerofoils have a greatly reduced incidence compared with the aerofoils of the inner wing. This ensures that, as the aircraft nose rises, the centre of lift of the wing moves rearwards, thus returning the aircraft to its original attitude.

These aircraft suffer in similar manner to the canard in having a reduced effective tail-arm in both pitch and yaw axes, though the rearwards sweep of the wing does add to directional stability.

The argument generally offered in favour of these configurations is that removing the horizontal stabiliser saves the profile drag of that surface. Opponents will point to the poorer lift distribution of the flying wing which can result in negative lift at the tip sections and result in high induced drag.

An example of this configuration is the Boeing-Insitu Scan Eagle shown in Figures 4.18, with technical details in Figure 4.21. The aircraft, with its unique method of recovery, would probably find a tail system (empennage) to be an embarrassment.

(d) Delta-wing Configuration

The delta-wing configuration, such as in the Observer UAV (see Figure 4.17) gives a rugged airframe for skid or parachute landings, without the lighter and more vulnerable tail. It has a lower gust response, due to its lower aspect ratio, than other HTOL aircraft. However, it shares with the flying-wing the criticism of poor lift distribution, resulting in higher induced drag exacerbated by its higher span loading.

The most usual propulsive system used is, as in the canard and flying wing, by aft-mounted engine(s) in turbo-jet or propeller form.

3.5.2   VTOL Configurations

Crewed helicopters are to be seen in many different configurations, largely driven by the means of counter-action of the rotor torque. These are all shown in outline in Figure 3.9.

(a) Single-main-rotor or 'Penny-farthing'

Here the torque of the main rotor, which tends to turn the aircraft body in the opposite rotational direction to the rotor, is counteracted by a smaller, side-thrusting, tail rotor which typically adds about a further 10% onto the main rotor power demands. As discussed earlier, a disadvantage is that the aircraft is extremely asymmetric in all planes which adds to the complication of control and complexity of the algorithms of the flight control system. The tail rotor is relatively fragile and vulnerable to striking ground objects, especially in the smaller size of machine.

These are the most ubiquitous of the crewed rotorcraft since the configuration is most suited to aircraft in the range 600–15 000 kg which currently covers the majority of rotorcraft requirements. The majority

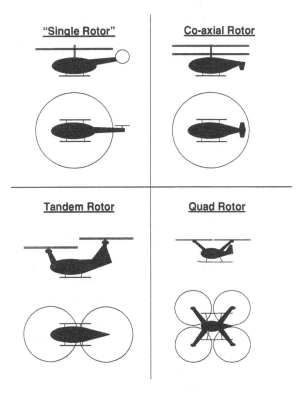

Figure 3.9   Rotorcraft configurations

of VTOL UAV manufacturers have opted for this configuration, possibly because some air vehicles are adaptations of crewed machines and others are adaptations of commercial hobby models. For examples see Figure 4.13.

(b) Tandem Rotor

There is a strong scale effect on the size of helicopter rotors such that the ratio of rotor mass to lift increases strongly with the larger rotor sizes required by the heavier aircraft. Therefore it is more efficient to fit two smaller rotors than one large one to aircraft above a certain AUM. The 'cross-over' point rises as technology, especially in materials, improves. It has increased from a value of about 10 000 kg in the 1960s to about 15 000 kg today.

Until VTOL UAV of this AUM are envisaged the tandem rotor configuration is not appropriate for UAVs, even though the configuration is more symmetric in control than the 'single rotor' and is more power efficient.

The small payload volume of a low AUM aircraft does not require a long fuselage so that the rotors have to be mounted on extended pylons. That is not structurally efficient. For these reasons, the author is not aware of any current tandem rotor UAV examples.

(c) Coaxial Rotor

This configuration, principally from the Russian manufacturer Mil., is in limited use for crewed applications. It is not more generally popular due to its greater height compared with that of the other configurations. It can present disadvantages in maintenance and in hangarage. For UAV application, with much lighter and smaller aircraft, these are no longer disadvantages.

The advantages of the configuration include an almost perfect aerodynamic symmetry, compactness with no vulnerable tail-rotor, efficiency of power and the versatility of providing alternative body designs for different uses, but each using the same power unit, transmission, and control sub-systems. Hence the automatic flight control system (AFCS) algorithms are no more complex than that of a typical HTOL aircraft.

In addition, largely because of its symmetry, its response to air turbulence is the lowest of all of the helicopter configurations, being zero in most modes.

Compromises in design may modify this advantage as is the case of the Sprite UAV. For purposes of stealth, in order to reduce the noise emanating from the rotor, a rotor tip speed lower than the norm was chosen. This had the effect of reducing the blade loading to 180 kg/m^2 and subsequently slightly increasing the response to vertical gusts.

Although the configuration was used in one of the earliest UAV, the USA Gyrodyne DASH of the US Navy, the advantages of the configuration was generally overlooked until the 1970s. Then the Canadian company Canadair used the configuration in their CL 84, as did the British Westland Company in their Wisp and Wideye UAV.

These were followed by M L Aviation's Sprite System of the 1980s (see Figure 4.19). There is now growing interest in VTOL UAVs and in this configuration in particular (see Chapter 27).

Detailed rig testing by M L Aviation, measured the performance of a hovering coaxial rotor compared with the equivalent single rotor. Both had the same total number of blades (four), the same blade loading and tip speed. The rotors were tested over a common range of disc loadings. The coaxial rotors were tested over an appropriate range of rotor spacing. The results showed that, contrary to the previous popular belief, the coaxial configuration used less power in hover flight than did the single-rotor (or penny-farthing) configuration. This is not only because no power is diverted to a torque-control tail rotor, but less power is wasted through 'swirl' energy being left in the rotor downwash. These conclusions have been confirmed by more recently available Russian reporting.

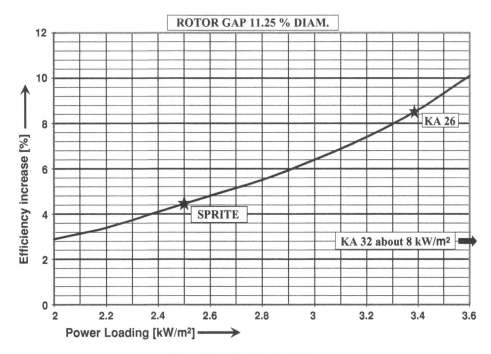

Figure 3.10 Coaxial rotor efficiency

Figure 3.10 shows the incremental percentage improvement in hover efficiency (thrust obtained for the same power) of the contra-rotating coaxial rotor system compared with the equivalent single rotor. The results shown are for a typical coaxial inter-rotor spacing of 11.25% of the rotor diameter. The comparison is more marked for heavier aircraft at higher disc loadings (and therefore higher power loading) for the reason that the induced swirl is greater. However, an advantage within the range of between 3 and 6% is relevant to the power loadings appropriate to helicopter UAV.

In forward flight the shaft between the two rotors will contribute extra drag, but this is offset by the elimination of tail rotor drag.

(d) Tri-rotor

To the author's knowledge there are no tri-rotor rotorcraft currently in existence or planned. The only known representative of this configuration, the manned 'Cierva Air Horse' was flown as a single prototype in the UK in the late 1940s, but the programme was terminated due to mechanical problems. However, the configuration does have some performance and mechanical advantages compared with the quad-rotor configuration and may yet appear.

(e) Quad-rotor

The only current development of this configuration known to the author is by a number of Universities who are attempting to make a simplified VTOL UAV in mini or micro sizes for urban surveillance.

Whilst all of the previously discussed configurations use rotor-head control systems applying both cyclic and collective pitch changes to the rotor blades as the means of aircraft control, the goal of quad-rotor designers is to remove this complication, and also to remove the need for a mechanical transmission system. The idea is to have the rotor blades all fixed in pitch and to achieve thrust changes on each rotor by changing its speed of rotation. Each rotor is individually driven by an electric motor mounted at the rotor head. Thus, for example, for the aircraft to move forward the rotational speed of the two rear rotors would be increased to pitch the aircraft nose-down and direct the resulting thrust vector forwards. At the same time the total thrust must be increased to prevent loss of height and, once established in forward flight, the rotor speeds must again be harmonised.

The control algorithms to achieve this are extremely complicated, taking into account also the changing aerodynamic interference patterns between the rotors. There must be a time-lag in the demanded speed change of each rotor although this becomes less of a problem with the low inertias of a small MAV.

The configuration is naturally more gust-sensitive than the other configurations, and its control response must be expected to be slower. Therefore the achievement of adequate control may be difficult enough in the still air of laboratory conditions, and even more problematic in the turbulent air of urban operations.

Power failures, either of any one of the individual motors or of the power supply, may be considered unlikely, but such an event would spell an immediate uncontrolled descent to earth. The author is, as yet, unaware of any successful practical development of this configuration.

3.5.3   Hybrids

For hover flight, the helicopter has been shown to be the most efficient of the heavier-than-air aircraft. They are limited in cruise speed to the order of 200 kt (370 km/hr) by the stalling of the retreating blade(s). For longer-range missions it is necessary to have the aircraft cruise at higher speed in order to achieve an acceptable response time to the target or area of patrol.

However, the ability to take off and land vertically is a valued asset. From hence comes the wish to have an aircraft which combines the capability of both VTOL and HTOL worlds. Attempts have been made for many years, for crewed aircraft, to achieve this by various devices.

(a) Convertible Rotor Aircraft

One of the most successful methods to date has been to mount a rotor onto each tip of the main wing of a HTOL aircraft. The rotors are horizontal in vertical flight, but tilt forward through 90°, effectively becoming propellers for cruise flight. There are two main variations on the theme, known generally as 'convertible rotor' aircraft.

One, known as the 'tilt rotor', retains the wing fixed horizontally to the fuselage and the rotors, with their pylons, are tilted relatively to the wing (see the first illustration of Figure 3.11). The alternative is for the wing, power-plants and rotors to be constructed as an assembly and for the assembly to be hinged on the upper fuselage. This is known as a 'tilt wing' aircraft and is shown in the second illustration of Figure 3.11.

Within the tilt rotor configuration, there are two options for installation of the power-plant. One is for an engine to be installed in each of the rotor pylons, thus tilting with the rotors; the other is for the engine(s) to be fixed on the wing or within the fuselage.

Similarly, within the tilt wing configuration, two options are possible. The engines are most usually fixed to, and tilt with, the wing and rotors. Alternatively the engine(s) can be fixed in the fuselage, driving the rotors through a system of gears and shafts which pass coaxially through the wing hinge bearings.

Tilting the engines in either tilt rotor or tilt wing requires the engines to be operable over an angular range of at least 90° and leads to some complication in the fuel and oil systems. This complication may be more acceptable, however, than the alternative of the mechanical complexity involved in transferring the drive from a fixed power-plant to a tilting rotor system.

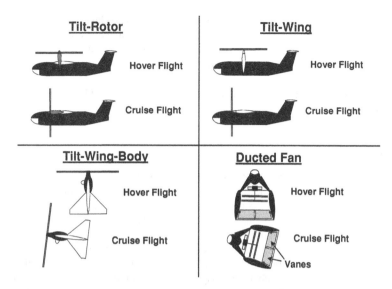

Figure 3.11 Hybrid aircraft configurations

Both these configurations have flown successfully as crewed aircraft, the tilt rotor being the more efficient of the two in hover and the tilt wing being the more efficient in cruise flight. However, as both types include the elements required for both regimes of flight, they have a reduced payload fraction compared with the helicopter and are more expensive in both acquisition and operation (see Reference 3.4 for more details).

To the author's knowledge, only the tilt rotor has, as yet, been adopted in UAV application. See the Sea Eagle UAV, which has the engine fixed in the fuselage, as shown in Figure 4.14.

In the case of crewed aircraft it is desirable to retain the fuselage sensibly horizontal for both flight crew and passengers. No such constraint exists for the UAV. It is therefore theoretically possible to tilt the whole aircraft, as shown in the third illustration of Figure 3.11.

(b) Tilt-wing-body Aircraft

With all convertible rotor aircraft, the critical part of the flight, and which predominantly determines the aerodynamic design of the aircraft, is the transition between full hover flight and full cruise flight. The problems lie in maintaining an attached airflow over the wing and achieving adequate control of the attitude of the aircraft, particularly in the longitudinal pitch angle. The latter is most readily achieved by the use of a helicopter rotor rather than a propeller for propulsion. The use of cyclic pitch can give good pitch and yaw control.

Maintaining flow attachment is a more difficult problem, especially during the conversion from cruise flight back to hover ('transition down') and landing. During 'transition up', the aircraft is accelerating and probably climbing with full thrust on the rotor. In this condition the high air velocity in the slipstream will reduce the otherwise high angle of attack of the wing and reduce the probability of wing stall. Even so, the avoidance of entering a stall will almost certainly require high lift devices, such as Kreuger leading-edge flaps and/or trailing edge flaps on the wing.

In the transition 'down' the velocity of descent will increase the wing angle of attack and will not be mitigated by a high slipstream velocity as the aircraft will be decelerating with low rotor thrust in this manoeuvre. This situation is much worse and may demand a higher disc loading of the rotor to increase the slipstream velocity and a larger wing in order to reduce the wing loading and hence the aerodynamic incidence. The ratio between disc loading of the rotor and wing loading of the wing is known as the 'lift loading ratio'. For a safe transition this may be required to be high, but it also demands more engine power and a larger wing, increasing the mass and cost of the aircraft and reducing its efficiency in both cruise and hover flight modes.

A careful design of the wing with a suitable aerofoil section, probably of low aspect ratio, and appropriate flow attachment devices is the more intelligent solution.

(c) Ducted Fan Aircraft

The ducted fan aircraft, as its name implies, encloses its 'thruster' within a duct. The thruster is called a 'fan' as it will be of constrained diameter and will be of high 'solidity' – i.e. the ratio of blade area to disc area. The fan is most likely to be composed of two contra-rotating elements in order to minimise rotation of the body by a resultant torque. It is unlikely to have either collective pitch or cyclic pitch control available on the blades so that changes in thrust will be obtained by changes in fan rotational speed, and angular control of the body will be by tiltable vanes in the slipstream.

This configuration will inevitably result in a high disc loading of the fan and resulting high efflux velocity of the order 30 m/s (100 km/hr). This may ease the transition into, and back from, cruise flight, but flow separation from the duct is still a concern to be addressed. The greater problem here may be the attitude control of the body as the vanes may lack sufficient force or response to ensure a controllable, stable system.

The Northern Ireland Company, Shorts, were attempting to develop such a configuration in the 1970s. The programme was abandoned, it is understood, through the inability to achieve stable control. The technology available for power-plants and control systems now may make this problem surmountable.

(d) Jet-life Aircraft

A further variant of hybrid aircraft, though not shown in Figure 3.11, is the jet-lift configuration, in which the aircraft is suspended in hover flight on one or more high-velocity jets of air. Other, smaller jets, spaced out on wing tips and front and aft fuselage, are needed for roll and pitch attitude control.

To transit into forward flight, the jet(s) are rotated backwards to provide an element of forward thrust, but retaining a vertical component until a fixed-wing progressively develops lift enough to sustain the aircraft. At this point the jets are effectively horizontal and provide propulsive force only.

This is system has been well proven in the Harrier fighter aircraft, but is very expensive in engine power and fuel consumption. It is not appropriate for an aircraft which is required for other than high-speed missions, yet requiring a vertical take-off and landing.

A variation on this theme is the Selex Damsel-fly, discussed in Chapter 4 and shown in Figure 4.28. It lies, in configuration, between the ducted fan and jet-lift in that it achieves both lift and control from the same four jets of air.

3.6 Summary

It may be concluded that, without the need to carry the weight and volume of a human crew and its support equipment, an unmanned aircraft is able to carry a greater proportion of its total weight in fuel. Without the provision of crew accommodation and outside vision, the fuselage may be shaped for lower drag. The combination of the two will give a UAV a longer endurance than its manned equivalent.

If a long runway is available for the aircraft to operate with a wheeled undercarriage for take-off and landing, then a fixed-wing configuration will offer the greatest speed, altitude and endurance of all configurations. It is the most appropriate for MALE and HALE operations. If no runway is available, such as for a short-range, mobile battlefield system, then a system using a VTOL aircraft has an advantage.

The added complication to achieve VTOL in a VTOL aircraft is outweighed in the fixed-wing aircraft by the need for a low wing-loading for launch and a parachute and impact absorption airbag for its recovery.

As a system, the extra cost of the launcher and transport vehicles and equipment for the fixed-wing aircraft, together with the implied tactical burden of these shows another advantage for the VTOL system.

UAV support to battlefield operations has generally to be conducted at lower altitudes where air turbulence is prevalent. The lower gust-response of the rotorcraft and its versatility in positioning offers further tactical advantage to the VTOL system.

Between these two operational extremes lies a medium-range middle ground where short, improvised landing strips may be available for a fixed wing aircraft. It may, however, need to suffer a burden of rocket-assistance equipment for take-off and a more rugged undercarriage and braking system for landing.

Over the medium ranges, the rotorcraft may have the disadvantage of a lower speed and range. Somewhere there is likely to be a 'watershed' where other factors may determine which offers the better solution or perhaps there is a niche at this point for a hybrid aircraft configuration. These aspects are expanded further in the following chapter.

References

3.1. Dr Ing. Sighard F Hoerner, *Fluid Dynamic Drag*. published by the Author, 1958.
3.2. Alfred Gessow and Garry C. Myers. *Helicopter Aerodynamics*. Macmillan, 1952.
3.3. Simon Newman. *The Foundations of Helicopter Flight*. Edward Arnold, 1994.
3.4. R.G. Austin. *A Comparison of a Configuration*. AGARD Canada, 1968.

4

Characteristics of Aircraft Types

The airborne element of a UAV system is often referred to as the 'platform' for carrying the payload. Although the aircraft is but one element of the system, the author has chosen to address it first because, in his assessment, of all the elements it can have the greatest impact upon the design of the rest of the system. We will now look at the characteristics of the aircraft or platform to carry out the four representative rôles as defined in Chapter 2.

In this chapter a few aircraft types have been selected by the author to typify each platform used in the different rôles. It is not suggested that these have been selected for particular merit in the class. For a wider coverage of platform types the reader is referred to the excellent *Unmanned Vehicles Handbook* updated annually by the Shephard Press – see Reference 4.1.

The attributes and performance of the aircraft are the author's best estimates and do not necessarily reflect those of the actual vehicles in their current configurations. The estimates are provided to explain the design issues and compromises.

4.1   Long-endurance, Long-range Rôle Aircraft

We can now consider rôle (a), i.e. the long endurance roles. These are typified by the Northrop-Grumman Global Hawk, high-altitude, long-endurance UAV and the General Atomics Predator, medium-altitude, long-endurance UAV as shown in Figure 4.1.

Both aircraft types employ a conventional airframe configuration with rear-mounted propulsion units, a turbo-fan for the Global Hawk and a propeller turbine in the case of the Predator B. Each has horizontal and vertical tail surfaces at the rear to provide aerodynamic stability in pitch and yaw respectively.

The task of these UAV types is long-range reconnaissance and they are required to carry a sophisticated (and therefore usually heavy) payload into position over long distances (say 5000 km) and to remain on station for a considerable time, probably 24 hr. They therefore rely upon their communications to be relayed via satellites.

The types of payload are described in more detail in Chapter 8, although as an indication of the payload and communication equipment carried, the following relates to the Global Hawk (Block 20) systems:

Sensors

Synthetic Aperture Radar: 1.0/0.3 m resolution (WAS/Spot)

Electro-optical: NIIRS 6.0/6.5 (WAS/Spot)

Infrared: NIIRS 5.0/5.5 (WAS/Spot)

Unmanned Aircraft Systems – UAVS Design, Development and Deployment   Reg Austin
© 2010 John Wiley & Sons, Ltd

**Global Hawk Block 20 (Tier 2 Plus)
by Northrop-Grumman.**

Wing-span	39.9m
Length	14.5m
MTOM	14,628kg
Max. Endurance	35hr
Max Altitude	19,800m
Payload - mass	1,360kg

Stabilised, high-magnification
Optical and I.R. TV.
Synthetic Aperture Radar

**Predator B
by General Atomics Inc.**

Wing-span	20m
Length	10.6m
MTOM	4,536kg
Max. Endurance	32hr
Ceiling	12,000m
Payload :- mass	230kg

Stabilised, High-mag.
Optical and I.R. TV.
S.A.R.

Figure 4.1 Long-endurance, long-range, HALE and MALE air vehicles (Reproduced by permission of General Aeronautical Systems Inc. and Northrop Grumman)

Communications:

Ku SATCOM Datalink: 1.5, 8.67, 20, 30, 40, 47.9 Mbps

CDL LOS: 137, 274 Mbps

UHF SATCOM/LOS: command and control

INMARSAT: command and control

ATC Voice; secure voice

In addition to the heavy payload, a large amount of fuel must be carried to power the aircraft for its long-endurance missions. This mass of fuel with its containing tank(s) plus fuel pumps, filters, etc., in total principally determines the all-up-mass of the aircraft – more so than the mass of the payload.

It is an escalating effect, however, since the greater the mass of the aircraft, so it will require larger and heavier wings to support it. This, in itself, will require further mass increase and the greater drag caused will require a greater propulsive thrust, more powerful and thus heavier engine(s) which will use more fuel and further increase the aircraft mass – and so on.

It is therefore imperative that the mass of fuel required must be minimised, both by flying the aircraft en route at its most economical speed, i.e. to achieve the maximum distance for the fuel burnt, and by flying on station at the speed for minimum fuel burn. The most economical speed, however, must be high or the transit time to station will be long and the aircraft may be too late to perform a critical surveillance, or other type of, mission.

The design of all aircraft involves a compromise between many factors in order to achieve an optimum result for the mission or missions envisaged. This is no more true than for the design of long-endurance aircraft where a significant task of the designer is to reduce the fuel-burn. Other than to apply pressure to reduce the mass of the payload whilst retaining its mission capability, three main concerns of the airframe designer must be to:

a) keep the aerodynamic drag of the aircraft as low as possible commensurate with the practical installation and operation of the aircraft systems such as the payload, power-plant, radio antennae, etc.;
b) use the latest practical structural technology to obtain the highest possible ratio of disposable load to aircraft gross mass – this is also known as the 'disposable load fraction'.
c) install a reliable power-plant which provides an adequate level of power, yet is light in weight and is fuel efficient, particularly under the conditions at which the aircraft will spend the majority of its time operating.

4.1.1   Low Aerodynamic Drag

In addition to paying fundamental care in the airframe design to achieve low drag by careful shaping using established and innovative technology, parasitic drag can be kept low by limiting the aircraft to fly at low speeds. However, this is unacceptable if a short transit time is necessary. Also, of course, the induced drag becomes high in low-speed flight.

At high altitude the parasitic drag will be reduced as the air density is low. However the induced drag becomes high in low density air unless the aircraft has a low value of span loading, i.e. a longer wing span than is usual for its gross mass.

To obtain long range, therefore, the designer is driven to design an aircraft which will cruise at high altitude and have a long wing in order to reduce the induced drag at high altitude. The wing area must not be greater than that necessary for take-off at a reasonable speed and length of run, and an acceptable minimum flight speed at altitude; otherwise the parasitic drag will be increased.

This results in a very slender wing of aspect ratio perhaps in the range 20–25 which then gives a structural design challenge to achieve it without incurring excess weight. (The aspect ratio of a wing is the ratio of the wing span to the mean chord of the wing. This is often better derived by dividing the square of the wing span by the wing area, i.e. b^2/S).

Long-endurance UAV, and particularly HALE UAV therefore, are characterised by high aspect ratio wings. This is shown graphically in Figure 4.2 where the silhouette of a Global Hawk (A) HALE UAV, with wing aspect ratio (AR) of 25, is compared in plan view with a Boeing 747-200 airliner, with a wing aspect ratio of merely 7.

Of even greater significance, for high-altitude–long-range operation, is the comparison of their span loadings of 3.23 kN/m and 28.29 kN/m respectively. (Span loading is the weight of the aircraft divided by its wing span).

Both Global Hawk UAV and Boeing 747 wings are optimised for their respective tasks. The latter typically cruises at half the altitude of the UAV and has an endurance of little more than one-quarter of that of the UAV.

4.1.2   High Disposable Load Fraction

This type of aircraft is not required to be particularly manoeuvrable and may be designed to sustain lower levels of acceleration than, for example, combat aircraft. It must, however, be capable of sustaining loads imposed by high-altitude air turbulence and from landing.

In addition to careful structural design, advantage may be taken of advanced materials in both metallic and plastic composite form commensurate, of course, with cost and serviceability. See Chapter 6, Section 6.2.

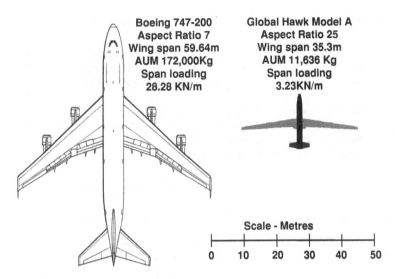

Figure 4.2   Comparison of wing aspect ratios

A further advantage of a UAV is that general fuselage pressurisation is not necessary, thus reducing both steady and flight-cycle fatigue loads compared with a passenger airliner. Protection of certain payloads against low pressure and temperature may still be necessary. The typical disposable load fraction for this type can be in excess of 60%. This disposable load has then to be shared between the payload and the fuel to be carried.

4.1.3   Aspects of Power-plant Selection

Power-plant selection is covered more fully in Chapter 6, Section 6.5, as applicable to all types of UAV. Aspects relating particularly to the HALE and MALE UAV are:

a) The achievement of a beneficial trade-off between engine mass and fuel consumption to obtain the lowest mass of combined power-plant installation and the fuel required for the long-endurance mission. The lightest engine may not result in the lightest overall package if the light engine uses more fuel.
b) To ensure that the power-plant gives a satisfactory performance at altitude. The power or thrust available from the engine will be reduced and it is necessary to ensure that this remains adequate to achieve the required aircraft performance. The specific fuel consumption (sfc) can be expected to remain sensibly constant up to about 11 000 m as the effect of the reduced air density on combustion efficiency is compensated by the reduced air temperature. Above that altitude, however, the air temperature remains constant whilst the air density continues to reduce and the sfc will progressively worsen. It is necessary to ensure that, in operating at greater altitudes, the increase in sfc does not negate the reduction of required power achieved through reduced airframe drag.
c) Other issues to be addressed, of course, include the need to prevent icing of the air intake(s) and the effect on fuel metering of the increase in fuel viscosity at the lower temperatures. The latter is of particular importance if 'heavy' fuels are to be used.

4.1.4   Representative Performance of a HALE UAV

The author is not party to classified detailed data for the Global Hawk UAV and it would not be appropriate to present them here anyway. Therefore the following figures are merely indicative of the performance characteristics of the generic type and its power-plant using published geometric and mass data pertaining to the Global Hawk A model rather than the more developed Block 20 system aircraft shown in Figure 4.1.

The airframe particulars of the A model used in the analysis are as follows:

Wing-span	35.3 m
Length	13.5 m
MTOM	11 636 kg
Payload – mass	608 kg

Generalised standard methods of calculating aircraft performance have been used in the analysis, e.g. Reference 4.2. The calculations are based upon the International Standard Atmosphere (ISA) where the temperature and pressure at sea level are 15°C and 1.013 mb (14.7 psi) respectively.

An estimate of the aircraft drag when operating at its maximum weight (114 kN) at various airspeeds and altitudes is shown in Figure 4.3., together with the expected thrust available from the Rolls-Royce, North America AE 3007 turbofan engine. The speed for minimum drag occurs at about the respective minimum safe airspeed at each altitude level, and both increase with increasing altitude.

If the wing area was to be increased to allow the aircraft to fly more slowly, the rise in drag at the lower speed (added to which would be the extra drag of the enlarged wing area) would deter operation at those lower speeds.

The necessity to have a low value of wing-span loading for operation at high altitude is indicated more expressively in Figures 4.4 and 4.5. Looking first at Figure 4.4, the estimated drag of the aircraft is shown at 10 000 m by the full curve and at 20000 m by the dashed curve. The estimated drag is shown for the actual aircraft with a wing aspect ratio (AR) of 25, but also for a hypothetical aircraft configured with the same wing area but an AR of only half, i.e. 12.5. (Note that this is still quite a high AR compared with the majority of aircraft). The 'estimated' thrust of the AE 3007 engine is also shown for the same two altitudes.

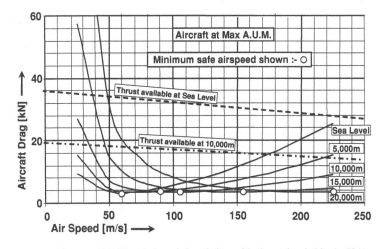

Figure 4.3   HALE UAV variation of aircraft drag with airspeed and altitude (ISA)

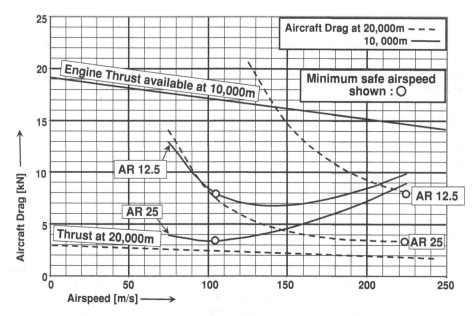

Figure 4.4   HALE UAV: effect of wing aspect ratio on aircraft drag at high altitudes (aircraft at gross mass)

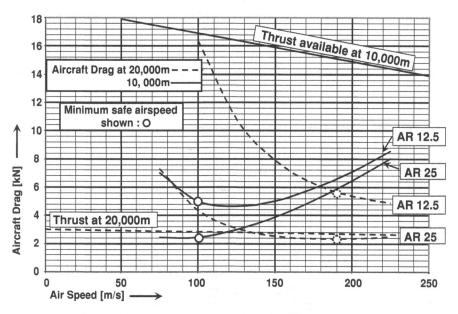

Figure 4.5   HALE UAV: effect of wing aspect ratio on aircraft drag at high altitudes (aircraft mass with half fuel)

Even at the altitude of 10 000 m, an altitude at which most modern long-haul airliners operate, the reduction in aircraft drag due to the higher AR is immediately obvious. At the respective speeds for minimum drag, the aircraft with the longer wing has half the drag of the other. The difference is even greater at the higher altitude, although even the AR25 aircraft may not achieve that altitude through having inadequate thrust.

This analysis is repeated in Figure 4.5 for the aircraft operating at a point at which it will have consumed about one-half of its fuel, resulting in a reduced weight of about 82 kN. The drag of both configurations is seen to be reduced, but the AR25 aircraft can now achieve the altitude level on the thrust available, whilst the AR12.5 aircraft would require a larger engine producing twice the thrust and, presumably, being about twice the weight and having about twice the fuel consumption at that altitude.

Although the larger engine would be throttled back at lower altitudes, it would still use considerably more fuel than the smaller engine as turbine engines are inefficient at reduced power. The extra mass of engine and fuel plus system could amount to 800 and 6000 kg respectively. Although the shorter wing would have less mass, this might amount to a saving of only about 300 kg.

To carry this extra mass would require a larger airframe requiring more power, using more fuel and so entering a vicious circle of spiralling size, mass and cost. This indicates most vividly the reason for the HALE UAV needing wings of ultra-high aspect ratio.

Assessment of the power usage at different altitudes and at different speeds for the representative HALE UAV gives the results shown graphically in Figure 4.6 for the aircraft at its maximum weight and in Figure 4.7 for its weight with half of its fuel used.

As discussed in Section 3.2, the speed for minimum power usage occurs when the induced power and the parasitic power are equal. The figures show that the airspeed at which minimum power is required by the aircraft, at both values of AUM, is at its lowest level at low altitude, and is the speed at which the aircraft uses, approximately, the minimum amount of fuel per hour. (This may be varied very slightly by the trend of engine specific fuel consumption with power.) It is, therefore, the speed at which the aircraft can remain airborne for the longest time and is known as the 'loiter speed'.

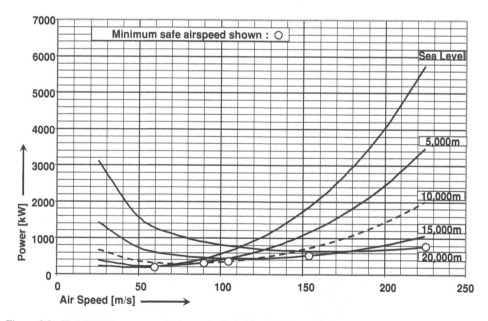

Figure 4.6   HALE UAV: power required to maintain height. Variation with airspeed and altitude – 114 kN AUW, ISA conditions

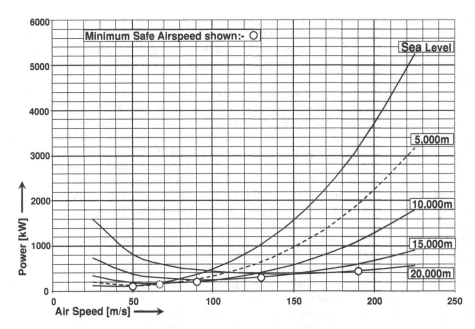

Figure 4.7   HALE UAV: power required to maintain height. Variation with airspeed and altitude – 81.8 kN AUW, ISA conditions

Other considerations may require the aircraft to loiter at a somewhat higher speed. These include a margin well above the stalling speed, especially in gusty conditions and for reasons of aircraft stability and response. An aircraft usually is more stable when operating on the rising sector of the power–speed curve.

The speed for minimum power occurs at very low airspeeds, so that the aircraft is not moving very far for the amount of fuel used. The airspeed for maximum distance travelled for the mass of fuel burned, the 'maximum-range speed' or 'most economical cruise speed' (V_{mec}), is obtained by taking the tangent from the origin to the curve on a graph showing the rate of fuel burned against airspeed. To close order, this is true of a graph showing the variation with airspeed of aircraft power required to maintain height, i.e. as in Figures 4.6 and 4.7.

If the aircraft is flying into a headwind, then the tangent will be taken, not from the graph origin, but from the airspeed, on the baseline, equal to the headwind. Therefore, when an aircraft is flying into a headwind, its V_{mec} airspeed is always faster than when it is flying in still air.

Note that not only is V_{mec} obtained at higher speeds at the greater altitudes, but the level of power (and rate of fuel burn) is less. Thus flying at altitude offers a higher transit speed and longer range, so best satisfying the requirements of the typical HALE UAV.

The table of Figure 4.8 summarises the airspeeds for minimum safe flight, for minimum power and for the most economical cruise, for a range of altitudes and for two values of the AUM of our representative HALE UAV. It indicates that the mission profile must be to climb immediately after take-off, to cruise at the maximum altitude of which it is capable at its initial AUM and continue to increase its flight altitude as the fuel-burn reduces its AUM. This flight profile will, however, be tempered with a knowledge of wind speed and direction at specific altitudes.

It will be recognised, therefore, that the design of such a platform requires the designer to strike a delicate balance in the selection of the several aircraft parameters in order to achieve the extreme range and endurance needed for the mission and to strive continuously to obtain saving in sub-system masses.

Aircraft at Max Gross Weight 25,600lb (114kN)   **Aircraft at Weight with ½ Fuel 18,350lb (81.8kN)**

TAS [m/s] / Alt. [m]	Vmin	Vmec	Vmin power
Sea Level	60	70	50
5,000	80	75	60
10,000	105	100	75
15,000	153	150	120
20,000	225	225	175

TAS [m/s] / Alt. [m]	Vmin	Vmec	Vmin power
Sea Level	50	55	50
5,000	68	70	50
10,000	90	90	65
15,000	130	140	100
20,000	190	200	150

Where: TAS is True Air Speed
Vmin is the minimum speed at which the aircraft can normally safely fly.
Vmec is the speed for max range (most economical cruise) in still air.
Vmin power is the speed at which the aircraft is using minimum power.

Figure 4.8   HALE UAV: V_{min}, V_{mec} and $V_{min\ power}$. Variation with weight and altitude

Turning now to the MALE UAV, we may take the Predator B as typifying the type. This is less ambitious in terms of mission range. The table of Figure 4.9 compares the general characteristics of the Predator B (MALE) with the Global Hawk A (HALE). Note that both aircraft are capable of carrying payloads externally, for example weapons, but this increases the aerodynamic drag. In that condition

(Comparison of Leading Particulars of Predator and Global-Hawk UAV)

	Predator B	Global Hawk A
All-Up-Mass [kg]	4,536	11,636
Payload [kg]	360	608
Fuel Load [kg]	1,360	6,590
Wing Loading [kN/m²]	1.108	2.283
Span Loading [kN/m]	1.450	3.234
Wing Aspect Ratio	16	25
Vmin @ S.L. ISA [kt] #	80	120
Altitude Ceiling [m]	12,000	20,000
Loiter Speed [kt] ^	150?	340
Cruise Speed [kt] ^^	230	450?
Max.Flight Endurance [hr]$	32	42
Range [km]	3,400?	5,500
Endurance on station [hr]	24	36

at Maximum Take-off Mass.
^ at Operating Altitude
^^ at Cruise Altitude
$ at Loiter Speed

Figure 4.9   MALE and HALE comparison (Reproduced by permission of General Aeronautical Systems Inc)

both the range and the endurance are reduced (see Section 4.1.5). The numbers quoted here relate to the aircraft when carrying internal payloads only.

The major driver of the size and other characteristics of both types is the combination of the loiter endurance and transit range required by the missions. A greater overall endurance requires more fuel. More fuel requires a larger aircraft to carry it and thus begins a vicious spiral of size growth as previously described.

The aircraft with reduced demands for endurance will be smaller, lighter and carry less fuel as a fraction of its all-up mass. As a result it can also carry a greater mass of payload as a fraction of AUM.

To obtain its longer range, the HALE UAV is obliged to cruise at higher altitudes and at greater speed. It is therefore equipped with a propulsive unit, a turbo-fan engine, which is appropriate to that speed. The MALE UAV, flying more slowly, is best suited by a turbo-propeller engine which is the more efficient at lower speeds.

Operation at lower altitude does not require such a high aspect ratio wing. Therefore a lower wing loading is possible without undue mass penalty. This then gives the aircraft a greater manoeuvrability and, together with the low-speed thrust of the turbo-propeller engine, offers a much shorter take-off run. With the greater availability of shorter air-strips, the MALE UAV system is tactically more versatile. It is also cheaper and thus more affordable.

The longer-range capability of the HALE system is undoubtedly needed but, until far more efficient propulsive units are invented, it will always be pushing technology towards a limit and its cost up. The MALE systems are therefore likely to be seen in greater numbers than the HALE system.

4.1.5   Long-range Strike UAS

The original Predator A was designed purely to perform a reconnaissance mission at long range. However, it soon became obvious that, when it identified priority mobile targets, by the time that strike facilities had been brought to the area, the targets had long since disappeared. An immediate strike action was necessary to take advantage of the long-range target detection. Predator B was therefore developed from the original design to carry a limited weapon load of two Hellfire missiles, as seen in Figure 4.10.

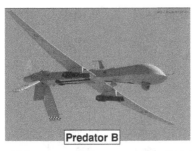

Wing Span	20m
AUM	4536kg
Cruise Speed	230kt.
Armament	2 Hellfire missiles

Wing span	20m
AUM	5090kg
Cruise Speed	260kt
Armament	4 Hellfire missiles and two 500lb bombs

Figure 4.10   Armed MALE UAV (Reproduced by permission of General Aeronautical Systems Inc)

Operations in Iraq and Pakistan showed the success of this concept. The aircraft was then developed to carry yet more weapons and given the name Reaper. This is also shown in the same figure.

To date only fixed-wing aircraft have been used as the platform in HALE and MALE systems. Development is currently under way of a rotorcraft to perform the MALE role. This is the Boeing/Frontier Systems Hummingbird which is treated in detail in Chapter 27. The US Special Operations Command (SOCOM) plans to acquire 20 Hummingbird systems from 2012 for surveillance and strike applications for which it is expected to have a flight endurance of up to 30 hr.

4.2   Medium-range, Tactical Aircraft

There is a plethora of different types in operation and under development in both fixed-wing and rotary-wing configurations, and these are part of systems principally conducting reconnaissance and artillery fire control duties. The fixed-wing aircraft in this category generally have wheeled undercarriages to take off from, and land onto, runways or airstrips, sometimes with rocket assistance for take-off and with arrester-wires to reduce landing run distance. Exceptionally the Ranger, see Section 4.2.1, has the option of a ramp-assisted take-off. VTOL aircraft in this category are often designed for off-board ship operation and this includes operations such as fleet shadowing and mine detection and destruction. The distinction between medium-range tactical systems and MALE systems, however, is becoming increasingly blurred.

It is a known fact in the aeronautical world that an aircraft does not achieve its ultimate efficiency until it has been in service for a while and been 'stretched'. Even in the day of computer aided design, a new type of aircraft will contain components which are found to be 'over-strong' (i.e. have over-large reserve factors) when measured by tests. By strengthening those parts which are found to have realised just the designed factors, the aircraft can be extended in weight to carry more payload and/or more fuel. The power or thrust of the engines may also need to be increased to maintain or enhance performance.

Thus the aircraft, which began life with a medium-range capability, may soon become extended in service ceiling and in endurance, assisted by improved communications, to move towards MALE performance. An example is the Hunter series of UAV described below.

4.2.1   Fixed-wing Aircraft

Typical of these are:

a) the Hunter RQ-5A UAV by IAI, Malat and Northrop Grumman, USA;
b) the Seeker II UAV by Denel Aerospace Systems, South Africa;
c) the Ranger UAV by RUAG Aerospace, Switzerland;
d) the Shadow 600 UAV by AAI Corp., USA.

They are illustrated in Figures 4.11 and 4.12, with further technical data shown in Figure 4.15.

The majority of medium-range aircraft, as in the representative types discussed here, use an airframe configuration with the surveillance payload in the nose of the fuselage, or in a 'ball-turret' beneath the forward fuselage, balanced by a power-plant with a pusher propeller at the rear. The fuel tank is mounted, near the centre of mass, between the two. The tail surfaces, for aerodynamic stabilisation and control, are mounted on twin tail-booms.

The Hunter, in course of development, has acquired a second engine in the nose which precludes a nose-mounted camera. IR and optical TV systems are mounted in the under-belly rotatable turret.

Although Hunter A models are still widely used in the medium-range role, development through the B and E models has extended the aircraft endurance and altitude capability. This development is illustrated

IAI Malat – Hunter Heavy Tactical

All-Up-Mass	885kg
Power (Heavy Fuel)	2 x 50kW
Speed	200km/hr
Radius of Action	250km
Flight Endurance	21hr
Payload	Mass 100kg

Optical & IR TV combined
SAR, COMINT & ESM
Comms. Relay, NBC Monitor
Customer-furnished payloads

Denel Aerospace - Seeker II

All-Up-Mass	275kg
Power	38kW
Speed	220km/hr
Radius of Action	250km
Flight Endurance	10hr
Payload	Mass 50kg

Optical & IR TV
Electronic Surveillance

Figure 4.11 Medium-range UAV: Hunter and Seeker (Reproduced by permission of Denel Aerospace). *Source*: Israeli Aircraft Industries

in the following table:

Hunter model	AUM (kg)	Wing-span (m)	Span loading (N/m)	Endurance (hr)	Cruise speed (km/hr)	Service ceiling (m)
RQ5A	727	8.84	807	12	202	4,600
MQ5B	816	10.44	767	15	222	6,100
MQ5C	998	16.6	590	30	222	7,620

Note the progressive reduction in span loading required to enhance endurance and altitude performance.

RUAG Ranger

All-Up-Mass	285kg
Power	31.5kW
Speed	240km/hr
Radius of Action	180km
Flight Endurance	9hr
Payload	Mass 45kg

Optical & IR TV
Laser Target Designator

AAI Shadow 600

All-Up-Mass	266kg
Power	39kW
Speed	190km/hr
Radius of Action	200km
Flight Endurance	14hr
Payload	Mass 41kg

Optical & IR TV
Customer Specified

Figure 4.12 Medium-range UAV: Ranger and Shadow (Reproduced by permission of RUAG Aerospace, HQ and All American Industries Inc.)

The operating range of the aircraft, however, has not been extended, limited by its relatively slow cruise speed and its communication system range. The latter is still 125 or 200 km using a second aircraft to act as radio relay. The use of satellites for radio relay is probably not worthwhile with the speed-limited range. A strike capability has, however, been added to the C model with its ability to carry missiles beneath the wings.

4.2.2   VTOL (Rotary-winged) Aircraft

Until the current millennium, relatively little development of VTOL UAV systems took place. This may be thought surprising in view of the advantages that VTOL systems bring to the medium-range and, especially, close-range operations. Perhaps this was because there are far fewer organisations having experience of rotorcraft technology than those with fixed-wing experience, especially within the smaller organisations from where most UAV systems originated. However, their worth is now being realised and a few examples are now to be seen.

In the medium-range category these are represented by:

a) The Northrop-Grumman Firescout, which utilises the dynamic components from a four-seat passenger helicopter within a new airframe.
b) The Schiebel Camcopter, which is an aircraft specifically designed as a UAV.
c) The Textron-Bell Sea Eagle, tilt-rotor aircraft, which uses the technology from military and civilian passenger aircraft in the design of a smaller UAV aircraft. Although this aircraft has been operated in various trials, further development is currently on hold.
d) The Beijing Seagull – a coaxial rotor helicopter a little larger than the Camcopter.

They are illustrated in Figures 4.13 and 4.14.

A summary of technical data for all the medium-range UAV considered here is shown in Figure 4.15, and affords some interesting comparisons.

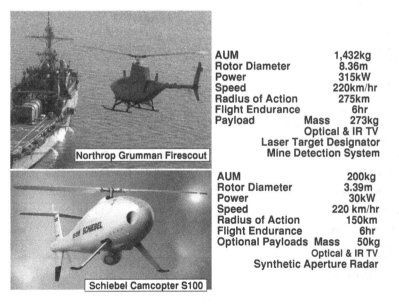

AUM		1,432kg
Rotor Diameter		8.36m
Power		315kW
Speed		220km/hr
Radius of Action		275km
Flight Endurance		6hr
Payload	Mass	273kg
	Optical & IR TV	
	Laser Target Designator	
	Mine Detection System	

Northrop Grumman Firescout

AUM		200kg
Rotor Diameter		3.39m
Power		30kW
Speed		220 km/hr
Radius of Action		150km
Flight Endurance		6hr
Optional Payloads	Mass	50kg
	Optical & IR TV	
	Synthetic Aperture Radar	

Schiebel Camcopter S100

Figure 4.13   Medium-range VTOL UAV systems: Firescout (Reproduced by permission of Northrop Grumman) and Camcopter (Reproduced by permission of Schiebel)

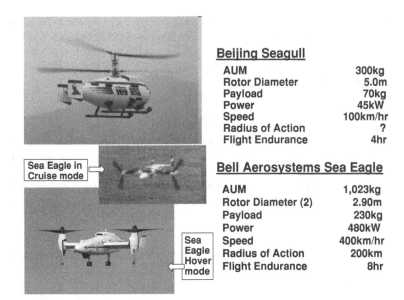

Beijing Seagull

AUM	300kg
Rotor Diameter	5.0m
Payload	70kg
Power	45kW
Speed	100km/hr
Radius of Action	?
Flight Endurance	4hr

Bell Aerosystems Sea Eagle

AUM	1,023kg
Rotor Diameter (2)	2.90m
Payload	230kg
Power	480kW
Speed	400km/hr
Radius of Action	200km
Flight Endurance	8hr

Figure 4.14   Medium-range VTOL UAV systems: Seagull (Reproduced by permission of Beijing UAA) and Sea Eagle (Reproduced by permission of Bell Helicopter Textron Inc.)

UAV Type Data	Hunter RQ-5A	Seeker II	Ranger	Shadow	Firescout	Sea-gull	Cam-copter	Sea Eagle
AUM [kg]	885	275	285	266	1,432	300	200	1,023
Wing Span [m] Rotor Diam. [m]	10.5	7.0	5.71	6.83	8.36	5.0	3.39	3.1 2x2.9
Wing Aspect Ratio	7.7	6.35	8.5	10.29				
Wing Area [m²] Disc Area [m²]	14.28	7.7	8.5	4.5	54.89	19.6	9.03	13.21
Span Loading [N/m] Disc Loading [N/m²]	827	385	472	382	256	150	217	3,240 760
Wing Loading [N/m²] Blade Loading [N/m²]	608	350	317	580	?	?	?	?
Installed Power [kW]	2 x 50	38	31.5	39	315	45	30	480
Power Loading [N/kW]	87	71	85.6	66.9	44.6	65.4	65.4	20.9
Cruise Speed [km/hr]	202	220	240	190	220	100?	220	400
Loiter Speed [km/hr]	140#	115#	128	140	140#	60#	100#	140
Flight Endurance [hr]	21	10	9	14	6	4	6	8
Radius of Action [km]	250	?	180	200	275	?	150	200

Estimated at S.L. ISA

Figure 4.15   Medium-range UAV technical data. (Reproduced by permission of All American Industries Inc.)

The amount of engine power installed per unit of aircraft mass is similar for all the aircraft with piston engines, irrespective of their being HTOL or VTOL aircraft. The gas-turbine-powered aircraft, i.e. Firescout and Sea Eagle have more power installed, partly because both use a higher disc loading (especially in the case of the Sea Eagle) but also because the turbine engines deliver more power for their mass.

With the exception of the tilt-rotor Sea Eagle and the Seagull, all types have a similar cruise speed of about 200 km/hr. The Sea Eagle has twice the cruise speed of the others, as is expected and has power to match. The actual speed of the Seagull is not confirmed, but it may well be slower than the other aircraft since it is the only one which is configured to accommodate an optional single pilot, making it less compact and having greater aerodynamic drag than the more dedicated UAV.

With the exception of the Ranger, all the HTOL aircraft offer longer flight endurance than the VTOL aircraft. This may be due as much to the difference in their operating roles as to their fuel efficiencies. The radii of action of all aircraft are not dissimilar and may be determined more by similar communications limitations as for any other reason.

4.3   Close-range/Battlefield Aircraft

This type of system with its multitude of rôles, military, paramilitary and civilian, (see Part 3 for details) many of which are carried out at low altitude and require a rapid response time, probably poses the greatest challenge to the designer.

The low-altitude military operation, usually' over enemy territory, requires that the system, particularly the aircraft, remain invulnerable to enemy countermeasures in order, not only to survive, but to press home the mission upon which so much depends. Flying at low altitudes most frequently means that the flight is in turbulent air, yet a stable platform is necessary to maintain sensors accurately aligned with the ground targets.

Added to those problems is the fact that, unlike the previous two categories of UAV systems, both for civilian roles and military roles, these systems are required to be fully mobile.

This mobility includes the GCS and the facility of aircraft launch and recovery. The systems operate within a very restricted area and often from wild terrain, so frequently no metalled runway or airstrip is available to them. Other means of launch and recovery therefore must be devised.

Unlike the aircraft within the system categories of MAV and NAV, they are too heavy to be hand-launched.

It is convenient to sub-divide this category into two sub-types.

a) those systems which use aircraft that depend upon additional equipment to enable their launch and/or recovery, i.e. non-VTOL;
b) those systems which use aircraft that have a VTOL capability.

4.3.1   Non-VTOL Aircraft Systems

The design characteristics of aircraft are often, if not usually, driven by the compromise between take-off and flight performance.

If a long runway is available along which the aircraft can be accelerated at a moderate rate of acceleration to flight speed, then the thrust/power installed in the aircraft need be little more than that required, in any case, for flight manoeuvres. Also, the wing area may be merely that dictated by normal flight manoeuvres. Hence no great premium of thrust/power or wing area has to be added to the aircraft purely for take-off.

No such facility is available for the battlefield aircraft. Its 'runway' has to be carried as part of the UAV system and usually takes the form of a ramp, mounted atop of a transport vehicle, along which the aircraft is accelerated to flight speed. These launch systems are considered in more detail in Chapter12, but suffice it to say here that the need to achieve a difficult compromise is presented to the system designer. Too long a ramp is ungainly and difficult to transport, but too short a ramp length requires a high value of acceleration be imposed, not only upon the aircraft, but upon its often delicate and expensive payloads and sensor systems. The ruggedisation of airframe and payloads can add considerably to their mass and cost.

This compromise can be ameliorated to some extent by reducing the aircraft minimum flight speed needed to reliably sustain flight as the aircraft leaves the end of the ramp. This requires either an increase of wing area or wing flaps, both of these not only adding to aircraft weight and cost, but increasing the aircraft aerodynamic drag, power required and fuel consumption in cruise flight.

The problem, of course, does not end with the design problems of the launch. The aircraft, now airborne, must be recovered at the completion of its mission. But there is no convenient runway awaiting its return, nor is it feasible to align it to decelerate back along the ramp. Two alternative recovery methods are generally employed.

The most ubiquitous is the deployment of a parachute from the aircraft and, to cushion its impact on landing, an airbag is deployed. Both of these sub-systems, together with their operating mechanisms, must be carried within the airframe, further adding to its mass, cost and volume.

Non-VTOL systems are represented here by the IAI Pioneer, BAE Systems Phoenix, the smaller Qinetiq/Cranfield Observer and Boeing/Insitu Scan Eagle UAV systems.

The IAI Pioneer continues with the ubiquitous pusher-propeller, twin-boom configuration which is the most popular for the medium and close-range UAV systems. A three-view drawing of the Pioneer UAV is shown in Figure 4.16 as representative of this configuration.

As with the medium-range aircraft described in Section 4.2.1, the configuration offers a compact fuselage with the option of alternative payloads and electronics in the nose, aft-mounted engine and pusher propeller which distances the power-plant and its ignition system from the electronics and provides an uninterrupted view forwards for the payload. The two booms provide some protection for personnel from the propeller. For recovery, a parachute can be mounted above the fuel tank and aircraft centre of mass. The main challenge in the structure of the configuration is to achieve sufficient stiffness in the twin booms to prevent torsional and vertical oscillation of the empennage.

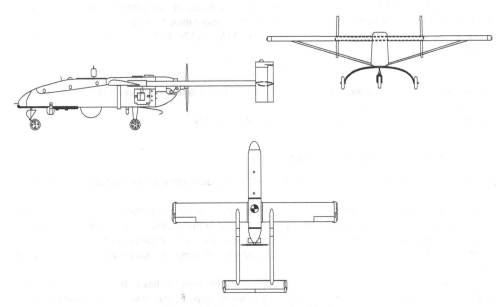

Figure 4.16   Pioneer three-view drawing

Observer	
All-Up-Mass	36kg
Wing span	2.42m
Wing area	1.73m²
Engine power	5.25kW
Wing loading	184N/m²
Span loading	120N/m
Cruise speed	125km/hr
Loiter speed	110km/hr
Mission radius	25km
Endurance	2 hours

Phoenix	
All-Up-Mass	177kg
Wing span	5.5m
Wing Area	3.48m²
Engine power	19kW
Wing loading	500N/m²
Span loading	316N/m
Cruise speed	158km/hr
Loiter speed	126km/hr
Mission Radius	50km
Endurance	4 hours

Figure 4.17   Close-range UAV systems: Observer and Phoenix (Reproduced by permission of Cranfield Aerospace Ltd)

Close-range systems adopting other airframe configurations are shown in Figures 4.17 and 4.18.

The Phoenix system started operation with the British Army about 1990 after an extended development phase. The prime contractor of the system was GEC Avionics who put the emphasis on the aircraft carrying a separate payload and avionics pod slung beneath the fuselage under the aircraft centre of mass. This presumably was the reason for installing the power-plant, unusually, at the front of the aircraft in order to achieve a longitudinal balance. The system saw extensive service in the Balkans and Gulf, but is no longer in service.

The Observer offers a simpler and more rugged airframe, tailored to improve its spatial stability in air turbulence by designing it, as far as is possible, to have neutral aerodynamic stability and stabilising it

All-Up-Mass	18kg
Wing span	3.10m
Wing area	0.62m²
Engine power	(23cc) 1.1kW
Maximum speed	120km/hr
Cruise speed	90km/hr
Endurance	15 hours

Interchangeable payloads:-
Optical & IR video, Mini SAR

Figure 4.18   Boeing/Insitu Scan Eagle. *Source*: Boeing-Insitu Inc; Cloud Cap Technologies; Insitu

electronically in space coordinates. It is fitted with an ingenious surveillance payload which uses three miniature TV cameras which look at increasing elevations along the major axis of the aircraft to provide contiguous coverage. This achieves a large image footprint with high resolution and is claimed to reduce operator workload to a minimum. More details of this system are presented in Reference 4.3.

The Scan Eagle system uses an innovative sky-hook recovery method, but this adds a further vehicle and equipment to the system. However, the system has been successfully introduced into service with several military and naval operators, including the US Army, US Navy, Australian, Canadian and Singaporean Forces. The system has amassed more than 200 000 operating hours within its first five years of deployment.

The 'flying wing' configuration was presumably chosen in view of the demands of the recovery system as an empennage might have fouled the sky-hook. For more details see Chapter 12.

4.3.2   VTOL Aircraft Systems

As previously recorded, little development of VTOL UAV systems took place until relatively recently, following the demise of the 1960s' Gyrodyne Dash system. The close-range systems were particularly neglected.

There are two exceptions: the M L Aviation Sprite system (Figure 4.19) which began life in 1980 and the Yamaha R Max developed from about 1997. These are taken as the main examples of the class since both have seen extensive operation, though in quite different applications.

The EADS Scorpio 30 is one of a number of VTOL systems now in development and, although information on it is scant, it is shown, together with the R Max, in Figure 4.20.

M L Aviation Sprite

The Sprite system was designed from the outset to meet both civilian and military design and airworthiness requirements. To this end critical systems in both GCS and aircraft have back-up systems and include, in

All-Up-Mass	36kg
Rotor span	1.60m
Engine power	2 x 5.25kW
Maximum speed	126km/hr
Loiter speed	0 - 60km/hr
Max. Endurance	3 hours

A wide range of Interchangeable payloads include:-
Colour TV Camera,
Low Light Level TV Camera,
Thermal Imaging TV Camera,
NBC Monitor + LLLTV,
Laser Target Designator + LLLTV,
Radar Confusion Transmission + TV
all with full hemisphere field of regard.
1 or 2 (optional) 500W rotor-driven alternators supply electrical power.

As the launch and recovery of the aircraft is via an automatic vertical take-off and landing, the whole system is contained within and operated by two operators from a single all-terrain vehicle.

Figure 4.19   ML Aviation Sprite close-range VTOL UAV

Yamaha R Max

All-up-Mass	N/A
Rotor Diameter	3.13m
Engine Power	15.4 kW
Payload	Mass 7.4kg + 16kg
	Spray Equipment and Fluid

EADS (France) Scorpio 30

All-up-Mass	38kg
Rotor Diameter	2.20m
Max Speed	50km/hr
Endurance	2hr.
Payload	Mass Unknown
	Optical and I.R. TV

Figure 4.20   Yamaha R Max and EADS Scorpio 30 VTOL UAV. (Reproduced by permission of Schiebel Elektronische and Yamaha Motor Company, Japan). *Source:* EADS - France

the aircraft, two independent power-plants with their fuel supply. The aircraft can hover on one engine. Communication between the GCS and aircraft is obtained via two parallel radio systems of widely separated frequency bands. A logic system aboard the aircraft selects the frequency with the better signal-to-noise ratio. Both GCS installation and aircraft are of modular construction which enables ease of build and maintenance and, particularly with the aircraft, gives a very compact solution. It also enables the aircraft to be transported within the one GCS vehicle.

The Sprite aircraft is designed to have neutral aerodynamic stability and relies upon the AFCS to provide positive spatial stability. It has demonstrated extreme steadiness when operating in turbulent air.

The Sprite UAV also offers extremely low detectable signatures (covered in more detail in Chapter 7).

Yamaha R Max

Unlike the Sprite, which is not designed to transport a dispensable load, other than minor ordnance, the R Max was expressly designed for spraying crops with fluid. It can carry 30 kg of fluid and spray gear and is over $2^1/_2$ times the gross mass of Sprite. It is not designed to be covert or to fly out to distances, but to fly efficiently at low speeds over local fields. Therefore it uses a large-diameter rotor with a lower disc loading than Sprite. The R Max has found a very profitable niche market, over 1500 aircraft being in operation.

4.3.3   Comparison of Close-range Systems

As discussed in Chapter 3, the HTOL aircraft is usually more efficient than the helicopter in cruising flight since the helicopter carries a penalty for its VTOL and hover capability. However, ramp-launched aircraft also carry a penalty for their launch and recovery method.

It is instructive to see how these aircraft compare, with each type carrying a penalty for their different methods of launch and recovery. The available technical data is shown in Figure 4.21 for a small number of close-range UAV, both HTOL and VTOL.

UAV Type / Data	Pioneer	Phoenix	Obser-ver	Scan Eagle	Sprite A	Sprite B	R Max
AUM [kg]	203	209	36	18	36	36	?
Span / Diameter [m]	5.11	5.5	2.42	3.10	1.60	1.60	3.11
Wing area [m²] / Blade area [m²]	3.05	3.48	1.73	0.62	0.2	0.2	?
Wing Loading [N/m²] / Blade Loading [N/m²]	653	589	204	285	1766	1766	?
Span Loading {N/m] / Disc Loading [N/m²]	390	373	146	57	176	176	121
Installed Power [kW]	20	19	5.25	1.1	5.25x2	5.25x2	15.4
Power Loading [N/kW]	100	108	67.3	160	67.3**	67.3**	59.9
Take-off speed [km/hr]	127*	110*	65*	80*	(Blade 324)	(Blade 324)	?
Take-off C_L	1.0	1.0	1.0	1.0	0.5	0.5	?
Max. Speed [km/hr]	158	158	130	120	126	216*	?
Max Endurance Speed / Max Range Speed [km/hr]	130* / 150*	? / ?	72* / 85*	? / ?	72 / 108	100* / 153*	? / ?

* Estimated ** Power restricted to one engine only

Figure 4.21 Close-range UAV technical data

Wing Loading

In order to keep the length of the launch ramp to a manageable length and the acceleration imposed upon the aircraft (in particular on sensitive equipment) to an acceptable level, the ramp-launched (RL) aircraft must be able to leave the ramp top and sustain flight at an airspeed which is considerably lower than that of aircraft which have the advantage of using a runway or airstrip. This means that the RL aircraft must have a large wing area for its weight, i.e. a low wing loading. A RL aircraft has a wing loading typically one-tenth of a HALE aircraft, and half that of a medium-range aircraft, and therefore it carries the penalty of a very high level of friction drag at cruise speeds.

These trends are reflected in the data of Figure 4.21 where a comparison shows the take-off speed varying with the wing-loading for the four fixed-wing aircraft. The Phoenix has the highest wing loading of the three ramp-launched aircraft and requires the longest launch ramp. It does therefore offer the highest cruise speed. The Scan Eagle would then offer the next highest speed, but it is power-limited.

Propeller Efficiency

As the RL aircraft leaves the ramp it is at its most vulnerable to air turbulence when its airspeed is at a small margin above its stalling speed. Any side-gust may cause a lateral roll, side-slip and ground impact. An up-gust may cause a nose-up attitude, an increase in lift and drag and, if not an immediate stall, a reduction in airspeed into a regime where the aircraft drag increases as the airspeed reduces, resulting in an inevitable stall.

To combat this the propeller must be designed to give its maximum thrust to rapidly accelerate the aircraft from a low airspeed. Therefore unless the extra complexity of a variable-pitch propeller is accepted, the usual fixed-pitch propeller, optimised for acceleration away from the ramp top, will suffer poor efficiency in cruise flight, thus increasing the power required and the cruise fuel consumption. In addition, the propeller diameter of the RL aircraft is likely to be restricted for ramp clearance, thus exacerbating the problem.

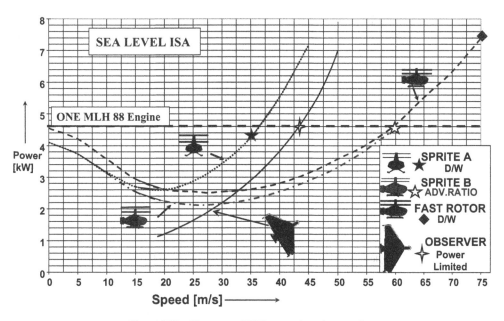

Figure 4.22   Close range UAVL: power/speed comparison

Landing Equipment

Both Phoenix and Observer, typifying such systems, are forced to carry a parachute and airbag to reduce the impact loads on touch-down. These and their means of deployment add to the mass, volume and cost of the aircraft. The Scan Eagle aircraft largely avoids these penalties by being arrested onto a vertical cable, but the system has to bear the extra cost and operation of a further mast-deploying vehicle.

Comparison of Ramp-launched and VTOL Aircraft

The performances of Observer and Sprite A and Sprite B in terms of power/speed are shown in Figure 4.22 and compared. The choice of these aircraft for representative comparison of RL and VTOL aircraft is particularly relevant since they are of the same all-up mass, have the same mass of payload and use the same engines. (The MLH 88 twin-cylinder, two-cycle units were developed specifically by M L Aviation for Sprite and were later made available for projects sponsored by the UK Royal Aircraft Establishment).

The figure shows that Sprite A, with the higher drag, axisymmetric, fuselage has a maximum speed limited to 35 m/s (126 km/hr) by its drag-to-weight ratio.

The author's calculations show that Observer should offer a maximum speed of about 43 m/s (155 km/hr). However, the manufacturers rate it at only 125 km/hr – similar to Sprite A. This may be because the author has been optimistic in assuming its propeller efficiency. The alternative Sprite configuration, the B model with the more streamlined, but less stealthy fuselage, has an estimated maximum speed of 60 m/s (216 km/hr) limited by its advance ratio, i.e. stalling of the retreating blades. However, if the rotor speed were to be increased and some of the power from the second engine used, 75 m/s (270 km/hr) could readily be achieved but at a penalty of a small increase in rotor noise.

It is also of interest to note that comparing the masses of the two airframes with their equipment, the 'extras' required in both aircraft for launch and recovery are essentially equal. The combined mass of the parachute and inflatable bag installed in Observer for its recovery is equal to the combined mass of the Sprite transmission system plus its bonus of the second power-plant installation which gives it its one-engine-failed hover capability. The latter is an excellent safety feature for hovering over urban areas.

If the costs of the two systems are compared subjectively, an assessment can be made using the established method of costing sub-systems as a function of mass and complexity. On the basis that the cost of the more extensive airframe of the RL aircraft equates to the cost of the rotors plus simple airframe of the VTOL aircraft, this cost is used as a 'cost unit'. The remaining sub-systems may then be assessed as follows:

Sub-system	RL aircraft	VTOL aircraft
Airframe plus rotors	1	1
Parachute and airbag	1	0
Undercarriage	0	$^1/_2$
AFCS and actuators	1	1
Communications	1	1
Power-plant(s) and electrics	$^1/_2$	1
Transmission	0	1
Payload	1–3	1–3
Launch ramp on vehicle	2	0
Control Station	5	5
System cost with 1 UAV	$12^1/_2$–$14^1/_2$	$11^1/_2$–$13^1/_2$
System cost with 2 UAV	$17^1/_2$–$19^1/_2$	$17^1/_2$–$19^1/_2$

The author has concluded that a properly configured VTOL rotorcraft system is better suited to the close-range/battlefield scenario than is a ramp-launched HTOL aircraft system:

(a) it is less vulnerable to enemy attack in the air (see Chapter 7) and less vulnerable on the ground as it is more mobile with far less ground equipment and personnel deployed (see also Chapter 26);
(b) VTOL will cost no more as a system to procure and will cost less in operation through lower cost of personnel;
(c) it has the advantage of lower response to air turbulence and the versatility of hover and low-speed flight.

4.4   MUAV Types

The concept of the mini unmanned air vehicle system was that the system could be back-packed, assembled and deployed by no more than two persons. MUAV are prolific largely because they are seen by many people as simple adaptations of model aircraft. Provided that the aircraft has a mass less than 10 kg (originally 5 kg), in some circumstances they can be flown at will under the very restricted rules for model aircraft.

It is often thought that with the addition of a simple video camera system to a model aircraft airframe, controlled via model aircraft radio equipment, a cheap MUAV system can be produced. Little can be further from the truth, but the idea has generated a large number of poor, unreliable 'systems' from entrepreneurs where no consideration is given to proper system integration. Some of them have learned the hard way but, in the process, have often given the MUAV a bad reputation. Few, if any, model aircraft equipments undergo the rigorous design and testing regime necessary to achieve the reliability required by realistic UAV operation. For this reason, the two examples selected for discussion in this volume are from established manufacturers with aeronautical experience and are shown, with their data, in Figure 4.23.

Lockheed Martin Desert Hawk III

All-up-Mass	3.86kg
Mass empty	2.95kg
Wing Span	1.32m
Wing Area	0.323m²
Wing loading	120N/m²
Power- Electric motor	?kW
Cruise speed	92km/hr
Endurance	90min
Operating Range	up to 15km
Payload Mass/Vol.	0.91kg/4720cc
Payload	Optical, LLL or IR TV

Bluebird Skylite

All-up-Mass	6.0kg
Wing Span	2.4m
Wing Area*	0.8m²
Wing loading	74N/m²
Power- Electric motor	?kW
Cruise speed	75km/hr
Endurance	1.5 hr
Operating Range	10km
Payload Mass	1.2kg
Payload	Optical or IR TV

* estimated

Figure 4.23   Mini-UAV systems. (Reproduced by permission of Lockheed Martin Corp). *Source*: Rafael Defense

MUAV were originally expected to be hand-launched and controlled via a laptop computer with display showing video images and navigation and housekeeping data. To this end, the aircraft could not, realistically, have an AUM exceeding about 6 kg (for hand-launching) with the total system grossing about 30 kg distributed between two packs. Initially, the aircraft were powered by small petrol- or diesel-powered engines and this required the fuel supply to be included in the backpack(s).

More recently, with the development of improved battery technology and lightweight electric motors, this power source makes back-packing more realistically possible. On grounds of safety, the carriage of a supply of rechargeable batteries is preferable, to inflammable fuels. Battery performance in low temperatures might, however, be a cause for concern. Referring to Figure 4.23 both the Desert Hawk and Skylite are in use with military forces.

It is a tribute to recent development in electronic technologies that a thermal imaging surveillance payload, aircraft stability sensors, flight control computation, GPS navigation equipment, and radio communication equipment can be battery-powered and yet leave sufficient battery power to provide propulsion to maintain the aircraft airborne for an hour within a total platform mass of little more than 3 kg in the case of the Desert Hawk.

The design of any aircraft is a compromise between several aims, and for one which has to be disassembled to be carried by back-pack, this is especially so. Just one compromise here is between the wing size which is practical to be carried, the wing area and span. A larger wing area is necessary for low-speed flight for ease of launch and surveillance at low altitudes, yet will increase the aircraft vulnerability to air turbulence. A larger span is desirable to achieve low propulsive power requirements at low speed yet is limited by practical packaging.

The author calculates that the minimum flying airspeed for the Hawk and Skylite is 50 and 40 km/hr respectively. This is not readily achievable, without herculean effort, by a hand-launch; hence both systems employ other means of boosting the aircraft to above the minimum airspeed.

The Hawk operators hitch a 100-metre-long bungee to the aircraft and tension it between them before the aircraft is released. This would appear to be somewhat hazardous for the second operator (not shown in the illustration), but seems to work in practice. The Skylite system employs a mechanically tensioned, bungee-powered, foldable catapult to launch the heavier aircraft. Both systems are reportedly achieving operational success though the Hawk operators report that the aircraft, with its low wing loading, is limited to operation in moderate wind conditions. The Skylite, surprisingly with its even lower wing loading, is claimed to be 'weather-proof''.

No VTOL aircraft seem to have been successfully developed in this category and certainly none have entered service.

4.5   MAV and NAV Types

Both of these, until recently, largely remained the province of academia. Now commercial companies such as Aerovironment and Prox Dynamics are developing systems, but as yet, neither are believed to have achieved full operational status.

4.5.1   MAV

The concept of the micro Air Vehicle system is that it is a personal system capable of immediate deployment by one person and operated via an iPod or similar device. The original definition of an MAV was that the air vehicle should be of no more than 150 mm (6 inches) span/diameter, but this has been relaxed somewhat of late. Although small, it is required to carry a surveillance camera, means of control and of image transmission. Its use is seen primarily as urban and indoor surveillance (see Chapters 24 and 26).

The development of MAV systems may adopt one of four forms of airframe:

a) fixed-wing
b) rotary-wing
c) flapping wing (ornithopter)
d) ducted lift-fan.

Fixed-wing

The majority of MAV are, understandably, of fixed-wing configuration since it is seen that these are easier to construct and Figure 4.24 shows two types from well-known aircraft companies. Note that although both are denoted by their manufacturers as MAV, they considerably exceed the size definition of MAV. This is not surprising as the development of an aircraft to carry a useful electro-optical payload with its communications, control, power source and propulsion sub-systems, and yet achieve meaningful flight endurance, within such a small size, is still a formidable task.

Scale effects with reducing size are discussed in detail in Chapter 6. The trend with reducing size is generally to the advantage of structural and mechanical elements, but to the disadvantage of aerodynamic performance. With the very small MAV, the better understanding of the aerodynamics involved at the very low Reynolds numbers is becoming critical, and new technical approaches may be necessary to achieve further miniaturisation.

Also pictured in the figure are four fixed-wing MAV typical of projects being carried out in universities around the world. The small mass and low wing loading of fixed-wing MAV makes them very vulnerable to air turbulence and heavy precipitation. Without a brief hover capability the fixed-wing MAV is likely to have but a limited use.

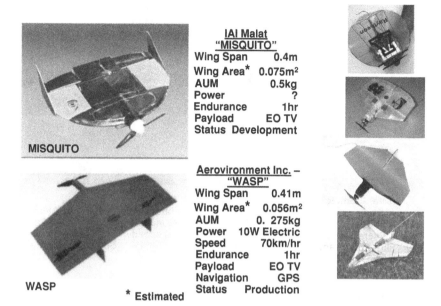

Figure 4.24 Micro air vehicle systems (Reproduced by permission of AeroVironment Inc)

Rotary-wing, Ornithopter and Ducted Lift-fan

If the declared aim of indoor flying and to 'perch and stare' is to be achieved (see Chapter 24), however, this category of UAV must be addressed more actively. Although this work is known to be pursued by a few individuals, little tangible success had been reported until recently.

As earlier suggested, the scale effect of size should benefit the mechanisms of these types and the oscillatory aerodynamics of the former two types should give better lift generation (Warren effect) than fixed aerofoils. The higher wing loadings of all types promise a reduced vulnerability to air turbulence and heavy precipitation compared with their fixed-wing equivalents.

The development of micro-sized rotors and mechanisms for the rotary-wing and ducted lift-fan MAV should present no great problems, but the oscillatory mechanism to provide wing flapping frequencies in the ornithopter of 20 Hz or faster may be more of a challenge. The efficiency of the duct of the fan may be questioned at the very small scale due to the difficulty in maintaining ultra-small fan-to-duct clearance and the greater friction drag of the duct at very low Reynolds numbers. The duct surface area may also make the aircraft vulnerable to urban turbulence.

It also may be that the successful outcome of all types has been awaiting the development of miniature attitude sensors and control systems. That these are now appearing is demonstrated by the successful appearance of the rotary-wing NAV by Prox Dynamics, as reported below.

4.5.2 NAV

Nano air vehicles originally sponsored by DARPA, and now by other organisations, are predicted to achieve an aircraft with dimensions of less than 5 cm in any direction, have an AUM of less than 10 g, including a payload of 2 g. These dimensions, themselves, do not imply nano-scale which relates to matter of a few millionths of a millimetre. The use of 'nano' in the UAV context is that these small aircraft

Prox-Dynamics "Pico-flyer"   **Lockheed-Martin / DARPA**
                                                        "Maple-Seed"

 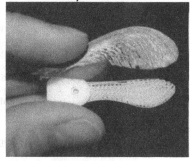

Rotor Diameter	60mm	Objectives:-	
AUM	3.3gm	Flight Endurance	2 min.
Battery	1 x 3.7V, 30mAh	Camera mass	2g
Camera System	?	Forward speed	10m/s
Flight Endurance	1 min	Power	Solid rocket in tip
Radio Link	900 MHz	Cost	<20$
Forward speed	10m/s?		

Figure 4.25 Nano air vehicles. (Reproduced by permission of Proxdynamics)

will require the embodiment of nanotechnology within the subsystems such as computers, sensors, communications, structures, electric motors and batteries, etc. which, with their precise integration, will be small and powerful enough to provide an aircraft capable of a realistic mission within the defined size and mass constraints.

Their future use is yet to be fully determined, but may include flying into and around the interior of buildings and natural structures, such as caves, to provide information as to the position of the structure's contents and condition.

We are looking significantly into the future for their operation but research and development of the possibilities has begun. However, the same concern regarding their operation in wind conditions and precipitation applies as for micro air vehicles, but even more critically. Two examples of NAVs are shown in Figure 4.25.

One is the Prox Dynamics Picoflyer which is the smallest of an existing range by the Company. The range includes the Nanoflyer having a rotor diameter of 85 mm with a 10 min. endurance and the largest, the Microflyer, having an AUM of 7.8 g, a rotor diameter of 128 mm and an endurance of 12 min.

These have been more recently complemented by Prox Dynamics' Hornet range of NAV demonstrated at the 24th Bristol UAV Systems Conference, and in particular the Hornet 3 having an AUM of 15 g, a rotor diameter of 100 mm and achieving a flight of 25 min in benign conditions out of doors.

The other example shown in Figure 4.25 is being developed by Lockheed-Martin and DARPA. It is not believed yet to have flown, but represents objectives in a programme of technology development.

4.6   UCAV

Examples of these, shown in Figure 4.26, are the Northrop-Grumman X-47B in development under the auspices of DARPA and the US Navy, and the BAE Systems Taranis being developed with the support of the UK MoD.

UCAV systems will be deployed in advanced strike missions with the aim of destroying enemy air-defence systems in advance of attacks by manned aircraft. The aircraft, therefore, must achieve a compromise between performance and stealth.

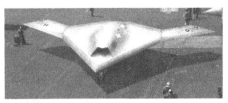

Northrop-Grumman
X-47B

Wing Span	18.92m
AUM	≈21,000kg
Thrust	106kN
Service Ceiling	12,000m
Combat Radius	2,800km
Speed	High Subsonic
Payload	2,050kg
	(EO/IR/SAR/GMTI/ESM)

BAE Systems Taranis

Wing Span	≈10m
AUM	≈8,000kg
Thrust	≈30kN
Service Ceiling	10,000m?
Combat Radius	not known
Speed	M 0.8?

Figure 4.26   Unmanned combat air vehicles (Reproduced by permission of BAE Systems)

The airframe will be of high wing loading and high thrust-to-weight ratio to achieve high penetrating speed without excess power demand and of dart-shaped, low aspect ratio flying wing with internal weapon carriage to minimise its radar signature on the approach to the target.

4.7   Novel Hybrid Aircraft Configurations

As remarked in the previous chapter, for operational reasons, the ideal aircraft is one which can take off and land vertically, yet fly at high speeds. This benefits the whole system in that less infrastructure is required compared with systems in which the aircraft is launched from a runway or ramp.

Helicopter types of rotary wing aircraft are the most efficient in hover flight but, as already explained, have limits to their forward flight ability.

For many years, attempts have been made to produce aircraft which can perform well in both flight regimes. Inevitably the results are compromises where the aircraft are less efficient in both regimes compared with the 'specialist' hover (helicopter) or cruise flight (high wing-loaded fixed-wing) aircraft. Hence the emergence of tilt-rotor and tilt-wing aircraft types. The search for the ideal aircraft continues and is made easier to achieve if no provision has to be made for aircrew to be accommodated or to function. Three different approaches which are aimed at achieving this 'El Dorado' are shown in Figures 4.27 and 4.28.

The Sky Tote

This is essentially a tilt-wing-body aircraft. A configuration similar to this was built, in prototype form, for VTOL fighter aircraft in the 1960s by Convair and Lockheed of the USA. However, both projects were abandoned when it was found how difficult it was for a pilot to land the aircraft whilst lying on his back with his feet in the air.

AeroVironment Inc. of the USA has a prototype Sky Tote UAV under development. Unlike the Convair and Lockheed prototypes which used a delta wing of low aspect ratio, it uses a main wing of relatively high aspect ratio and tail surfaces. At the time of writing the aircraft is undergoing hover tests and very little specification data have yet been released. It will be of interest to see how it fares during the transition

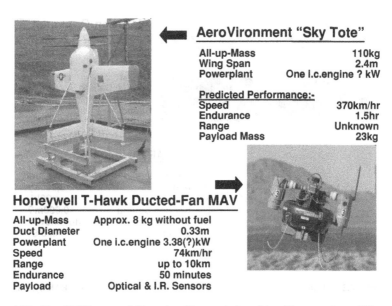

AeroVironment "Sky Tote"

All-up-Mass	110kg
Wing Span	2.4m
Powerplant	One i.c.engine ? kW

Predicted Performance:-

Speed	370km/hr
Endurance	1.5hr
Range	Unknown
Payload Mass	23kg

Honeywell T-Hawk Ducted-Fan MAV

All-up-Mass	Approx. 8 kg without fuel
Duct Diameter	0.33m
Powerplant	One i.c.engine 3.38(?)kW
Speed	74km/hr
Range	up to 10km
Endurance	50 minutes
Payload	Optical & I.R. Sensors

Figure 4.27   Novel UAV systems 1 (Reproduced by permission of AeroVironment Inc and Honeywell)

mode. One problem which may be presented by the configuration could be interference by the rotors of any forward-looking imaging sensors.

Honeywell Ducted-fan MAV

Although designated a micro-AV, one would have thought that it more appropriately comes within the mini-UAV category by nature of the aircraft mass which is likely to be increased before it is qualified

Selex Damselfly

This aircraft is in early development
and little data on it have yet been
released other than the intake duct
is about 1 metre in diameter.
An internal fan, powered by either an
electric motor or an internal
combustion engine, provides an
air flow to four nozzles. These can
swivel through 90 degrees to provide
vertical or horizontal thrust.
It is hoped that the aircraft will
Achieve a forward speed of 150 knots.

Figure 4.28   Novel UAV systems 2. *Source*: Selex

for service. It is creditable that the back-packed system was exposed to a military environment in Iraq so early in its development.

Lessons have been learned already to indicate that it was underpowered, the flight endurance was inadequate and that vibration was causing a sensor problem. These short-comings are not unusual and are to be expected in a relatively novel design, and their discovery is better addressed earlier than later in the programme.

The new engine in development is stated to produce 3.38 kW. A simple estimate indicates that the power required to hover out of ground effect in SL ISA conditions must be of order 2.6 kW. The margin of 0.78 kW may be barely adequate to allow for engine power reduction with increases in ambient temperature and altitude and engine wear with usage in addition to the extra power required in manoeuvres. Those unused to VTOL aircraft operation, for example, seem often to be unaware of the margin of power to be instantly applied in the pull-up from a vertical descent. This is even more critical for small aircraft and ducted systems in particular since they gain little benefit from ground effect.

The designers/developers task in this brave programme will inevitably be centred on mass reduction. Fortunately technological development is on their side.

Jet-lift Aircraft

A model of the Selex S&AS Company's Damselfly is shown in Figure 4.28. This uses the jet-lift principle, in this case employing four separately directable nozzles which, presumably, achieve both lift and control functions. The Company's claims for the product of having 'the hover capability of a helicopter' and 'outstanding wind-gust resistance' remain nonvalidated at the time of publication, with little supporting evidence available in the public domain.

The ultra high jet velocity of the configuration and long ducting must surely result in a large demand for power in a size regime where engines are not known for their frugality. Flight at low speeds must be very expensive in fuel consumption and, with a large wing, vulnerable to gusts, making it a challenge to perform effectively at low speeds in the urban canyons.

The author looks forward to being proven wrong, in which event Selex will have achieved a winner. However, the claim of 150 kt cruise speed hardly justifies the expense of the ambitious new development since a hover-efficient helicopter, let alone a tilt-rotor machine, can achieve that speed.

The model is an 8.75% dynamically scaled version of the Boeing Company's proposed airliner of a radically new configuration

Figure 4.29   Blended wing-body model

4.8   Research UAV

As previously noted, another increasing use of UAV is for research purposes. Using dynamically scaled models of proposed full-size aircraft, the flight characteristics of the new aircraft can be assessed more cheaply, quickly and with less risk and waiting until a full-size prototype is built. Should modifications of a new configuration be found necessary, then those modifications can be made far more quickly and cheaply than if made to a full-size prototype. An example of this use is shown in Figure 4.29 with the proposed blended wing-body airliner of the Boeing Company of the USA. The fully-instrumented scaled model was designed and built by Cranfield Aerospace Ltd, UK, for flight testing at NASA Langley, USA.

The principle of dynamic scaling is explained in Chapter 6.

References

4.1. *Unmanned Vehicles Handbook*, The Shephard Press Ltd, UK.
4.2. W. Barnes. *Aerodynamics, Aeronautics and Flight Mechanics*. McCormick–Wiley, 1979.
4.3. D. Potts. 'The Observer Concept', *Proceedings of the 14th Bristol International UAV Systems Conference*, University of Bristol, UK, 1999.

5

Design Standards and Regulatory Aspects

5.1 Introduction

The introduction of aircraft without aircrew to global skies has understandably, and rightly, raised concern amongst the aviation community and the general public as to the probability of the aircraft becoming out of control and causing injury to persons or property on the ground or in the air. Therefore regulatory authorities have been set up in most of the major countries to monitor the manufacture and operation of UAS.

The regulation is divided into two activities – that of military and that of civilian UAS. This is usually accomplished by different organisations.

Military

The regulation of military UAS is somewhat easier than that of civil UAS.

Military UAS are specified by the military customers and their design and manufacture is contracted to approved teams previously experienced in manned military aircraft and their systems. The design for airworthiness, manufacture, testing and operation will be carried out by methods specified in military documents.

Systems' testing is carried out in the dedicated airspace of military test sites or ranges under the supervision of military controllers. The operation of military systems is, apart from exercises, carried out in theatres of conflict.

Operations are conducted to minimise the risk of civilian injury particularly with respect to the prevention of 'collateral damage victims' in wartime.

This situation, along with the demand for superior weapons for defence, has enabled the numbers of military UAS to multiply, especially in the recent decade.

Civilian

The situation for civilian application is rather different. By definition, civilian or commercial operations, although potentially of wider application than the military, are likely to take place where failure of the UAS could cause death or injury to otherwise uninvolved persons. Therefore their airworthiness and operational control may be required to be, and to be seen to be, of even higher standards than that of the

Unmanned Aircraft Systems – UAVS Design, Development and Deployment   Reg Austin
© 2010 John Wiley & Sons, Ltd

military. Unfortunately there has not been the same degree of funding available nor sufficient commercial imperative to set up appropriate monitoring organisations and to develop the technologies required to ensure the safe operation of UAS within civilian airspace.

In theory, the certification and safe operation of civilian unmanned air vehicle systems is regulated by such authorities, for example, as the European Aviation Safety Agency (EASA) in Europe, the Federal Aviation Administration (FAA) in the USA and the Civil Aviation Safety Authority (CASA) of Australia. Other countries will have their equivalent organisations.

Collaborative meetings regularly take place between these Authorities with the aim of agreeing common international standards for UAV regulation. The several Agencies are agreed that:

> UAV operations must be as safe as manned aircraft insofar as they must not present or create a hazard to persons or property in the air or on the ground greater than that attributable to the operations of manned aircraft of equivalent class or category.

The function of the agencies/authorities may be considered in two complementary parts, i.e.

a) That part which issues approval of aeronautical design organisations, defines and ensures the main-tenance of design standards for airworthiness and for test standards, and issues design certificates for systems or components which comply with those standards.
b) That part which overseas control of airspace to ensure safe operation of airborne systems.

Included in one or other of these parts, probably the latter, may be a requirement for the operator(s) to be suitably qualified.

The author strongly advises any person or organisation proposing to develop or operate a UAS to contact their relevant airworthiness authority in advance of doing so. Indeed this advice also applies to aeromodellers as, since the increase in activities in the UAS field, the authorities have become aware of the dangers that model aircraft may present to the public if not operated in an appropriate and considerate manner. Aeromodellers are advised to join a relevant aircraft modellers' association where they can obtain guidance.

As a generality, the heavier and/or faster that the aircraft is, then the requirements imposed will be the more strict approximately in proportion to the ability of the aircraft to cause damage if it collides with persons or buildings, etc.

In the following sections the author has attempted to interpret and summarise, as examples, the regulations applying to UAS and model aircraft within the United Kingdom, Europe, and the USA.

Rightly, the several countries involved have meetings (for example the UASSG Unmanned Aircraft Study Group of ICAO and JARUS which consists of the Civil Aviation Authorities of Austria, Belgium, Czech Republic, France, Germany, Italy, the Netherlands, Spain and the United Kingdom) where their individual views are represented and so there is much overlap between the respective regulations.

Note that the regulations are constantly being reviewed and so may be out of date when this book is published. The author cannot therefore take responsibility for the accuracy in detail of his summary and a would-be manufacturer or operator is urged to obtain the documents listed as references below and to make early contact with the relevant department of the appropriate authority within his country to obtain their advice.

5.2   United Kingdom

In the United Kingdom, there are two regulatory regimes – military and civil UAS must comply with the regulations of one or the other as applicable.

A document prepared by a Working Group set up in 1983 under the auspices of the Society of British Aerospace Companies was intended to provide a basis for the certification of the design and operation of the several UAS then under construction. The working group included members of the UAS industry and representation from the CAA and Air Traffic Control.

In November 1991, the SBAC issued the document entitled 'A Guide to the Procurement, Design and Operation of UAV Systems' (Reference 5.1). The document, still relevant today, was used to assist in the development of UK UAV systems in that era and was used by the UK Ministy of Defence as the basis to prepare and issue an Aviation Publication, now Defence Standard 00-970 part 9 (Reference 5.2) to cover those aspects for the military procurement of UAV Systems. Whilst the SBAC document included proposed design requirements for rotorcraft, the Defence Standard, however, currently gives scant attention to those types.

5.2.1 Military Systems

It is required that the design and development of military systems for the UK MOD be carried out by an Approved Organisation – that is an organisation assessed by the UK MOD to have sufficient staff with the necessary knowledge and experience to carry out the work to a satisfactory level.

This will require that the organisation has an adequate level of expertise in its management and appropriate design disciplines, e.g. aerostructures, aerodynamics, electronics, mechanisms and to have, or to have access to, the necessary approved manufacturing and test facilities. The organisation must also have the required systems in place to ensure that the required documentation is prepared, recorded and stored and, where classified work is entailed, the security clearance of all involved staff and the necessary security control.

The design staff will have access to and conform to the design requirements of the appropriate Defence Standard documents as nominated in the MOD contractual documents which include the System Specification. The System Specification is a document prepared by the customer (in this case the MOD) which details the customer's requirements for the performance and reliability of the system, any limitations to be imposed on its size, mass, number of operators, support, etc. and the climatic and other conditions in which it will operate. The System Specification will override, if necessary, any of the general requirements called up in the Defence Standard documents.

For the design and certification of military UAV systems in the United Kingdom, the UK MOD has issued a document, Defence Standard 00-970 part 9, titled 'Design and Airworthiness Requirements for Unmanned Air Vehicles'. This document also advises that reference should be made to other parts of Defence Standard 00-970 and to Defence Standard 07-85 (for manned aircraft and guided weapons respectively) which can also be applicable to UAV systems.

Defence Standard 00-970 covers in width and detail the attention to be paid in the design of all elements of the total system under the following three main headings.

System Characteristics

Operational Colouring and Marking, General Operational Environmental Conditions, Reliability and Maintenance, Safety Management, Software, Electromagnetic Compatibility, Resistance to Electronic Counter Measures (ECM), Electronics, Test-Range Interfacing, Design and Assembly of Components, Health Monitoring, Guidance and Control Systems.

Support Facilities

General Requirements, Specific Climatic Conditions, Launch and Recovery Systems, Control Station, Maintenance.

Unmanned Air Vehicle

General Requirements, Climatic Conditions, Flight Performance, Structural Strength Requirements, Airframe, Powerplant, Avionics, Flight Termination, Payload.

Defence Standard 00-970 - Part 9 currently contains 218 pages tabling:

a) the requirements to be met for a specific system, sub-system or component,
b) the method or means of demonstrating the compliance with the requirement(s),
c) guidance, indicating means of achieving that compliance.

An example of one of those pages is shown in Figure 5.1.

Particular documents which the contractor must prepare and keep updated as necessary are:

a) The 'flight envelope', showing the scope and limits of the manoeuvres that the aircraft will be able to perform. This will be required to derive the loads imposed upon the structure and components. An example is shown in Figure 5.2 for a HTOL aircraft.
b) The 'Build-Standard' document which lists all the subsystems/assemblies from which the aircraft and control station are constructed. This will refer to a documents/drawings numbering system, employed to relate each component to its sub-system and system, etc. The system hierarchy diagram shown in Figure 5.3 shows a synthesis of the system down only to the assembly level which is included in the build standard document and is used also for the reliability model. For full definition of the system design, however, the method is extended down to component level. Each component, assembly, subsystem, etc. physically bears its reference number so that its position in the total system is readily determined. This aids the calling up of parts for manufacture, supply, assembly, testing and replacement.
c) The 'Reliability Model', discussed further in Chapter 16 on reliability, like the above documents, is kept up-dated throughout the development and operational phases of the System's life. It begins with the prediction, in its design phase, of the reliability of the UAV system with the expected failure rate of each assembly shown contributing to the failure rate of the total system. These values are updated following results of testing and operational use.
d) The 'Type Record'. This registers the 'pedigree' of the UAV system and includes a summary of all calculations made to determine its performance and structural integrity. It contains a summary of, and reference to, the documents listed above.
e) The 'Operating and Maintenance Manuals' will begin life in the design phase and will be extended during the Development Phase (see Part 2).

Provided that all the requirements and procedures listed above are satisfactorily carried out and supported by successful test results in the development phase, a Design Certificate will be awarded to the system by the appropriate authority. This certifies that the system meets the performance quoted and is a safe system to operate.

The North Atlantic Treaty Organisation (NATO) aims to agree common standards, within the nation members, for the design and operation of UAS and to issue 'Standardisation Agreements' (STANAG). Documents defining those standards are produced in English and French. A listing of a selection of the standards relevant to UAS is provided in the References section at the end of this chapter.

5.2.2   Civil Systems

The EASA was formed in 2003 to take over responsibility for the airworthiness, certification and regulation of all aeronautical products manufactured, maintained or used by persons within the European

REQUIREMENT	COMPLIANCE	GUIDANCE
		.......which should contain a summary document such as an "Operating Limits and Loads" report.
f. The UAV shall employ flight control devices of sufficient effectiveness to achieve the required control authority demanded during each operational phase defined in 4.3.1.a.	(1) In determining the required control effectiveness, the UAV Design Authority shall take into account the flight envelope for all build standards of UAV, and the most adverse environmental effects that can reasonably be expected in operation such as turbulence and airframe icing. (2) The effectiveness of the flight control devices shall be fully predicted by the most appropriate means, then validated by simulation and/or test throughout the flight envelope for each build standard of UAV.	The UAV flight control devices, whether operating aerodynamically or by other means, should be of sufficient power to produce the required control response (such as a rate of rotation about a reference axis) for all conditions of mass, CG, load factor and airspeed in the defined flight envelope. It is implicit in the term "required control authority" that means should similarly be embodied to limit such control devices from exerting excessive force on the UAV in other parts of the envelope, such as at high speed, where extreme structural loads or adverse aerodynamic characteristics could arise. In consideration of control effectiveness, full account should be taken of installation aspects, such as structural flexibility, local aerodynamic effects and flutter.
g. The design shall take into consideration the operating characteristics of the power plant in order to ensure that the performance and controllability of the UAV are not compromised	All power plant related effects shall be considered in the UAV performance and control/handling prediction and validation process, for each applicable flight phase.	The UAV power plant may exert variable or asymmetric effects on the airframe, such as propeller slipstream, rotor downwash, jet/rocket efflux and gyroscopic precession. These should be fully considered in the structural design, aerodynamic design and control system design of the UAV.
4.3.3 FLIGHT ENVELOPES		
a. The permitted operating envelope for all given build standards and role configurations of the UAV shall be declared.	The basic flight envelope shall be defined graphically, without ambiguity, in a recognised format.	(1) The Basic Flight Envelope, conventionally of the form shown in Figure 1 of Clause 4.4. should be expressed in........

Figure 5.1 Compliance documentation

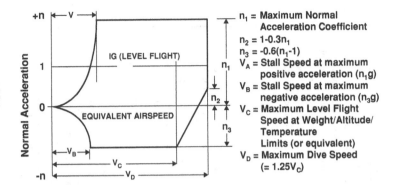

n_1 = Maximum Normal
    Acceleration Coefficient

$n_2 = 1 - 0.3 n_1$

$n_3 = -0.6(n_1 - 1)$

V_A = Stall Speed at maximum
    positive acceleration (n_1g)

V_B = Stall Speed at maximum
    negative acceleration (n_3g)

V_C = Maximum Level Flight
    Speed at Weight/Altitude/
    Temperature
    Limits (or equivalent)

V_D = Maximum Dive Speed
    (= $1.25 V_C$)

NOTE The basic flight envelope is the design envelope for the UAV. It shall be based on the design / maximum All-Up Mass (AUM) of the UAV and should encompass actual flight performance under all conditions. In the case of differing design, AUMs arising from variations in build standard, an envelope should be produced for each.

Figure 5.2 Basic flight envelope

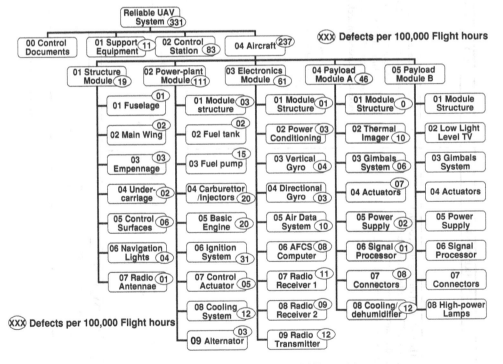

Figure 5.3 System hierarchy and reliability model

Union Member States. Whilst building up its resources, the Agency relies on national aviation authorities who have historically filled this rôle and concludes contractual arrangements to that effect.

In the United Kingdom this is the Civil Aviation Authority (CAA) with documents covering different types of aircraft and bearing the title of the British Civil Airworthiness Requirements.

Within Europe, an EASA Regulation (EC Regulation 216/2008) makes provision for implementing rules dealing with aircraft certification and continuing airworthiness. *These rules, however, do not apply to aircraft engaged in military, customs, police or 'similar' services. These aircraft are known as 'State Aircraft'. EU member states must, however, ensure that such services have due regard as far as is practicable to the objectives of the EASA Regulation.* (The words and phrases shown in italics are extracted directly from Reference 5.3.)

Certain categories of civilian aircraft are also exempt from complying with the EASA Regulation. Those relevant to UAS are:

- *Aircraft designed or modified for research, experimental or scientific purposes and likely to be produced in very limited numbers*
- *Aircraft whose initial design was intended for military purposes only*
- *Unmanned aircraft with an operating mass of less than 150 kg*

Therefore any UAV of mass greater than 150 kg which is not experimental or a 'State Aircraft' will be required to have an EASA airworthiness certificate. Those aircraft (including UAV) which are exempt from EASA regulation must, instead, comply with their national regulations for airworthiness certification and continuing airworthiness. Their equipments, personnel licensing operation within aerodromes and compliance with air traffic, must comply with national regulations.

A process of certification, similar to that for the military, therefore may be required for UAV systems for civilian use (but not for recreational model aircraft) and overseen in the United Kingdom by the CAA with documents covering different categories.

Virtually all UAV systems in use today in the United Kingdom are, however, military which operate on test ranges, or are very light aircraft operating within the restrictions placed on the use of model aircraft. (No airworthiness certification is required for model aircraft except in certain circumstances – see below.) Perhaps for this reason, although detailed design requirements and certification specifications exist for military aircraft in the UK, no such formal requirements currently exist for civil UAV systems.

The CAA, whilst accepting the worth of the available SBAC document, has not adopted it as 'it did not conform to the format' of CAA documents. The CAA has not, itself, prepared an appropriate document since it has, as yet, no funding to do so. There are consequently no defined detailed requirements covering the design and construction of civil UAS within the UK. This situation may change with the future issue of Chapter A8-22 of BCAR section A.

Instead, a document which is advisory only (Civil Aviation Publication 722 – 'Unmanned Aerial Vehicle Operations in UK Airspace – Guidance') has been issued by the CAA and is listed as Reference 5.3.

CAP 722 recommends the use, where applicable, of the appropriate manned aircraft design requirements. It has no legal backing and any contravention of its guidance can only be pursued in the courts if the infringement may be seen to contravene any article in the Air Navigation Order (ANO). The ANO is reproduced as CAA CAP 393.

However, the major hurdle which prevented deployment of UAV systems for civilian uses, even within segregated airspace, in the 1990s and since, and which has still to be resolved, is that of airspace usage. This aspect is also referred to within CAP 722.

The European Organization for Civil Aviation Equipment, EuroCAE, responsible to EASA, has established a Working Group WG73 UAV to start addressing the standards required for civilian UAVs to fly in nonsegregated airspace. The group is currently focusing principally on UAV having an AUM greater than 150 kg, even though many, if not the majority of, UAS seen as serving future civilian operations may have an AUM far less than that.

Airspace may be considered to be divided under four classifications:

a) that which may be used by model aircraft which meet certain required rules and criteria,
b) segregated airspace,
c) uncontrolled airspace,
d) controlled airspace.

Model aircraft and UAV are divided into four categories by the CAA for regulation. The categories are:

1. Model aircraft of less than 7 kg mass when unfuelled.
2. Model aircraft of between 7 and 20 kg mass when unfuelled (alternatively referred to as 'small aircraft'.
3. Model aircraft and 'light UAV' of between 20 and 150 kg mass when ready for take-off.
4. Aircraft of mass greater than 150 kg. These are referred to as 'UAV' and are not regarded as model aircraft.

The CAA see the military broad equivalent of category 1 as being Micro Air Vehicles of up to 5 kg mass and category 2 as Mini Air Vehicles of up to 30 kg mass.

Operation Restricted to Airspace of Category (a)

Aircraft category 1. The regulatory provisions for models (small aircraft) of less than 7 kg are less proscriptive than for those of greater mass but are currently the subject of a further review by the UK CAA.

Small aircraft of less than 7 kg unfuelled mass, if used only for 'sport flying', are currently not required to be built to any standards of airworthiness, but must comply with the following minimum operational constraints:

- clear of controlled airspace, unless with ATC permission,
- clear of any aerodrome traffic zone, unless with ATC permission,
- at less than 400 ft above the point of launch, except with permission as above,
- within 500 m of the operator at all times,
- not within 150 m of any congested area of a city, town or settlement,
- at least 50 m clear of persons, vessels, vehicles or structures. This can be reduced to 30 m for take-off or landing. Other model operators and any assistants or officials may be within this distance; as may vessels, vehicles or structures under their control.

However, if the small aircraft are to be used for commercial operations, i.e. 'for hire and reward' the would-be operators are required to request permission from the CAA to operate. The point of contact is given within Reference 5.3. It is possible that the CAA will then impose a system build standard and operator qualification requirement in addition to the same operational restraints, but this was still under review at the time of publication of this book.

Aircraft category 2. For small aircraft with mass 7–20 kg when unfuelled, further operational constraints, in addition to those listed for small aircraft of category 1, are imposed to ensure adequate safety. As listed in Reference 5.4, these include:

- a serviceable 'fail-safe' mechanism shall be incorporated to terminate the flight following loss of signal or detection of an interfering signal,
- it must be ensured that any load carried on the model is secure,
- flights must comply with any conditions such as byelaws,
- CAA permission is required for any commercial flights, i.e. using the model for 'hire and reward'.

Aircraft category 3. In the United Kingdom, 'light UAV systems', categorised by the CAA as those employing an aircraft with a mass at take-off of no more than 150 kg, may be approved and operated under model aircraft rules, i.e. not be required to meet requirements for formal airworthiness certification, but would be required to be operated under the same restrictions as applied category 2 small aircraft.

For many years the UK CAA has issued exemptions from most provisions of UK national regulation for larger model aircraft that fall outside the definition of a small aircraft. Whilst these exemptions have included relief from the need to comply with airworthiness certification requirements, the granting of the exemption has been conditional upon the CAA being satisfied that the model is designed and built to a satisfactory standard and that operational controls are in place, including the choice of a suitable rural site to preserve adequate separation between the aircraft and third parties, and flight is undertaken under the 'control' of a recognised model association.

For existing large models, the CAA has been satisfied regarding the design and build standards on the basis of recommendations provided by the Large Model Association (LMA), an organisation recognised by the CAA as providing expertise in this field. The process for recommending acceptance starts with the LMA inspecting an aircraft during construction and satisfying itself that it has suitable integrity of design and construction as represented by accepted good practice.

Then, following the granting by the CAA of an 'exemption for flight testing', the LMA will oversee a programme of 'function and reliability' flight trials (at least six flights totalling at least 1 hr for each named pilot). When satisfied that the aircraft has completed this testing without modification or mishap the LMA will normally provide a positive recommendation to the CAA that a renewable exemption should be issued.

For light UAV systems the CAA has concluded that UAV systems that are 'equivalent' to existing large model aircraft and have no greater capability, may be allowed to operate without obtaining airworthiness certification, subject to the UAV system complying with similar limitations and conditions to those applied to large model aircraft.

However, the LMA has advised the CAA that it exists for the benefit of recreational model aircraft enthusiasts only, and will not support commercial UAV systems. Consequently, it will be necessary to establish a body of similar competence to the LMA that will be able to provide similar recommendations on the basis of procedures acceptable to the CAA. In the absence of such a recognised body, the CAA has in the past accepted assurances given by a learned body with aeronautical engineering expertise, such as a University Aeronautical Department, in lieu of a recommendation.

Until such time as an accredited body is established, the CAA may, on an *ad hoc* basis, assess each application, taking into account the experience and knowledge of the applicant/operator in lieu of a recommendation, provided it can be assured that the standards applied are at least as demanding as those applied by the LMA. This absence of such an accredited body should present no problem to a company with the appropriate aeronautical expertise in-house.

To provide a measure of 'equivalence' to a large model aircraft, the regulatory concept developed by the CAA uses impact kinetic energy as a basic criterion. Impact kinetic energy is directly linked to the ability of a UAV to cause damage and injury. It provides both an absolute measure for the showing of compliance and a relative standard for identifying 'equivalence' with model aircraft. Kinetic energy is also an all-encompassing criterion applicable to all aircraft types, is easy to determine and can be readily estimated during the design process from the product $KE = \frac{1}{2}mv^2$, where m and v are in kg and m/s respectively (see the discussion on setting kinetic energy limits below).

To maintain comparability with models in other respects, the UAV system will be subject to design, construction, and flight testing scrutiny at least as demanding as that carried out for large models by the LMA.

A key feature of the involvement of the LMA in the safety of large models is their requirement for 'function and reliability' flight testing of significant duration. It is considered that it is this activity that provides, in the absence of formal requirements, the necessary safeguards against the presence of poor stability, control and performance characteristics. Under these provisions, a light UAV may be granted the necessary permission and exemption to operate commercially in the vicinity of persons and property, (subject to defined minimum separation distances).

To protect these persons and property, it is appropriate to set a safety level somewhat higher than that associated with recreational flying. Also, it is recognised that any flight assessment may be essentially qualitative, and therefore it is considered prudent to supplement the assurance gained through the function and reliability testing with some additional quantitative constraints to address the possible consequences of poor handling qualities.

The proposed additional constraints are as follows:

(a) A light UAV will not be granted an exemption, regardless of its mass, if the maximum speed it can sustain in level flight exceeds 70 kt. This addresses the issue of higher pilot workload inherent in operating at higher speeds, (a problem that increases as aircraft size reduces), and also supports the maintenance of separation from third parties. In common with model aircraft exemptions, those issued for light UAV systems will specify minimum acceptable separation distances from third parties and their property. Imposing the 70 kt limitation should provide a reasonable time period for the pilot to take recovery action in the event of his UAV unexpectedly heading towards a third party. (Note: it is envisaged that compliance with this speed limitation will be demonstrated during flight testing by flying the UAV straight and level at full power and measuring the steady velocity achieved.)

(b) To reduce the possibility of a high kinetic energy impact following loss of control, aerobatics will be prohibited.

(c) The conditions of the exemptions will prohibit tasks that involve aerial inspection of, or flight close to, any object or installation that would present a risk to safety in the event of damage due to any impact by the UAV (e.g. chemical/gas storage areas).

(d) The conditions of the exemptions will prohibit participation in any public flying display (except with the written permission of the CAA).

To set kinetic energy limits for UAV on the basis of equivalence with current large model aircraft, the CAA looked at their previous experience of models operating under exemptions.

Some 94% of large model aircraft carrying out aerial work within line-of-sight restriction, were seen to be within the mass range 20–75 kg. Combining the upper value of 75 kg with 1.4×70 kt maximum speed gives a kinetic energy value of 95 kJ. This value is stipulated to be the maximum acceptable in a crash situation for whatever form the crash takes, i.e.: (i) a free-fall from 400 ft; or (ii) a high-speed impact.

The former would include the results of a complete structural failure, a separation of the rotor from a rotorcraft or the rupture or complete separation of the gas envelope of a lighter-than-air aircraft. On the basis of negligible aerodynamic drag, an object of 80 kg dropped from 400 ft is calculated to have a kinetic energy on impact of 95 kJ.

If the UAV is shown to have a significant aerodynamic drag in free-fall and a lower maximum speed in flight, then a proportionally greater maximum AUM (than 75 kg) would be allowed. Examples of mass and speed combinations which fall within these criteria are given in an appendix to Reference 5.3.

A light UAV system complying with these criteria and operated within the constraints listed in (a)–(d) above, could be eligible for exemption from formal airworthiness certification. UAV systems that do not comply with the criteria or are intended to operate outside the constraints imposed, including beyond 500 m from the UAV pilot, will not be eligible for exemption and will therefore be subject to formal airworthiness certification in accordance with the UK CAA previously published policy, even though no formal civil UAS airworthiness requirements yet exist.

Aircraft category 4. As previously noted, this category of aircraft is required to have an EASA Airworthiness Certificate.

Operation in Segregated Airspace of Category (b)

This is a designated volume of space, beneath a nominated altitude, in which no aircraft is allowed to operate without the express authority of that airspace controller. In the United Kingdom, with the exception of the facility at ParcAberporth in Mid-Wales founded with the support of the Welsh Development Agency (see Appendix A-5, UAV Systems Organisations – Test Site Facilities), the only available segregated airspaces are those owned by the UK MOD. They are all essentially military ranges for testing military equipments for the Navy, Army and Air Force.

UAV systems developers, on application, may be provided with facilities to test their systems at these sites, especially if the systems are under development for the MOD. The systems will be expected to be designed to achieve full military certification and to comply with the range safety requirements.

The same facilities may be made available to assist in the development of civil systems which would be expected to comply with CAA standards, when known, for full civil certification. However, there is currently no provision, in the United Kingdom, to nominate segregated airspace to allow the regular operation of UAV ystems in civilian roles such as, for example, agriculture or power-line inspection.

The situation may be less restricted in other countries. For example, there are more than 1000 Yamaha R Max VTOL UAV carrying out crop-seeding and spraying in Japan on a regular basis.

Operation in Uncontrolled and Controlled Airspace of Categories (c) and (d)

For UAV systems to be operated in either controlled or uncontrolled airspace, full certification, either military or civil, is required. For either operation, it has been ruled by EASA and the FAA that the ability of an aircraft to sense the presence of another aircraft in the vicinity, and to manoeuvre to avoid a collision, is required.

This ability in a UAV must be at least as good as that currently expected in a manned aircraft. The problem may be less acute in controlled airspace as, with the aircraft mounting a transponder, it will be under the watchful eye and intervention of an air traffic controller.

It is probable that, although not indicated in the announcement by Northrop Grumman (see Chapter 16), the Global Hawk aircraft is restricted to controlled airspace.

It is also required that the system maintains control of the aircraft following a total power failure on the aircraft.

Sense and Avoid

This is now usually referred to as detect, sense and avoid (DSA).

Some UAV, such as the Global Hawk, already have a forward-looking video camera dedicated to an operator looking ahead for other aircraft. In uncontrolled airspace it is very difficult for light aircraft pilots to spot another small aircraft approaching if it is on, or nearly on, a collision course. The 'head-on' image that each pilot has for only a few seconds (with a closing speed of about 500 km/hr) is quite small. The view that a pilot may have of a UAV 'head-on' is likely to be even smaller. It may be expected, however, that a sensing system for UAS, using multi-spectral imaging with movement detection, could detect obstacles more effectively than can a human pilot. The system would then have to be programmed to enter an avoid mode which would take into account the UAV manoeuvrability and the possibility of the presence of other obstacles.

A study by the European Defence Agency (EDA) has concluded that a sense and avoid system for long endurance UAV is feasible and that certification of a system is expected by 2015. The study determined that 'no new technologies were needed but that existing technologies had to be designed into UAV-compatible equipment able to fit a UAV's internal volume and power constraints'.

When a sense and avoid system has been developed, and if it can be available at an acceptable cost, might it then be 'spun-off' for use in manned aircraft? The issue of the provision of DSA systems is further discussed in Chapter 27.

Maintain Control after Power Failure

This requirement implies that in the event of engine power, electrical power must remain available to operate the flight control system, communications links and an imaging system in the aircraft in order for the aircraft to be diverted and glided safely to land. All but essential electrical power demand should be cut off.

Electrical power must therefore be supplied for sufficient time for the aircraft to descend and land. This time will depend upon its operating altitude and safe descent rate. For a fixed-wing aircraft, electrical power may be supplied from a battery source or, for example, a ram-air turbine all of which will result in extra mass and cost. In the case of a helicopter, the electrical supply is normally generated by rotor gearbox-driven alternators. These would continue to be driven by the rotor(s) in autorotation, but may be aided by residual power from a battery.

Air Traffic Integration Roadmap

The EDA announced on 8 January 2008 that it had awarded a contract to a consortium of European companies to 'develop a detailed roadmap for integrating UAVs into European airspace so that they can fly routinely with other air traffic by the end of 2015'. No doubt they will be addressing the above issues together with the widespread concern that UAS are consuming an increasingly large amount of communication bandwidth and the need for a dedicated communications band for UAV.

Regulation of Operators

The following is extracted from CAA Document CAP 722, Version 2

> There are currently no regulations governing the qualifications required to operate a civil registered UAV in UK airspace.

This aspect appears not to be further addressed in the current Document (Version 3) other than to say that the operator 'should be suitably qualified'.

For the operation of military UAVs, specific requirements are set out in the Military Aviation Regulatory Document Set (MARDS) and, for contractors, in AvP67. Qualifications for UK Military operators are laid down in JSP 550 Regulation 320.

Individual UAV manufacturers have their own company requirements.

In anticipation of wider operations of UAVs CAA DAP is considering using the word 'crew' to mean flight crew, that is, the UAV Commander and the UAV-p, each of whom is a crew member.

UAV Commander

Every flight of a UAV must be under the command of a UAV Commander. The UAV Commander is a qualified person who is in overall charge of, and responsible for, a particular UAV flight or flights. The UAV Commander can:

- be in direct control of the vehicle by remote controls; or
- co-located with the UAV-p; or
- monitoring the state and progress of the vehicle at the flight deck location in the GCS.

The *UAV-p* is a qualified person who is actively exercising remote control of a nonautonomous UAV flight, or monitoring an autonomous UAV flight. The UAV-p may or may not be the UAV Commander. The UAV-p must meet the training, qualifications, proficiency and currency requirements stated in the approved Flight Operations Manual of the operating organisation.

The UAV Commander is tasked with overall responsibility for the operation and safety of the vehicle in flight and must be fully trained and qualified to assume these responsibilities. The UAV Commander therefore assumes the same operational and safety responsibilities as those of the captain or pilot-in-command of a piloted aircraft performing a similar mission in similar airspace. A UAV Commander may simultaneously assume the prescribed responsibilities for more than one UAV when this can be accomplished safely by directing activities of one or more UAV-p.

The UAV Commander must be licensed and appropriately rated according to airspace classification and meteorological conditions/flight rules. This may mean an instrument rating appropriately endorsed 'UAV'.

The Flight Operations Manual of the UAV Operating Organisation must specify the required qualifications and levels of training and proficiency for flight crew members, that is, for the UAV Commander and UAV-p. The following aspects shall be addressed:

- aeronautical knowledge
- knowledge of flight-critical systems of the relevant UAV
- manned aircraft pilot qualifications
- communications procedures
- UAV flight training levels
- flight proficiency and currency with the relevant UAV
- meteorology

Range Air Controller

For operation on UK Military ranges, a Range Air Controller (RAC) is required (a qualified Air Traffic Control Officer, ATCO, responsible for airspace management within the range). The RAC is responsible for applying the appropriate separation between all aircraft.

Target Air Controller

A RAC or an Assistant Range Air Controller (ARAC), working under the RAC, is responsible for the issuing of instructions to the UAVC team in order to achieve the profile required for the flight."

Other regulatory policy documents of note are listed under references 5.4 to 5.9.

Future Considerations

As UAS operations expand in the future, and civilian operations become accepted, it is to be expected that the Air Accidents Investigation Board (AAIB) will become formally involved in determining the cause(s) of any accidents and advising as to the means of their reduction. This may introduce legislation requiring UAV to carry data recorders (black boxes) as is currently required for manned aircraft, and possibly voice recorders in the control station.

Some UAV currently transmit housekeeping data to the control station where it is recorded. The data is obtained from sensors in the aircraft, and may include aircraft position data, communications status, airspeed, control information, fuel state, electrical power status, temperatures, etc. This information is used to alert the operators to any deterioration in the aircraft condition so that appropriate action can be taken to prevent or minimise the risk of failure. Where the aircraft is operating for periods of autonomy in radio silence and during that period suffers a fatal failure, the data may not have been received. In those circumstances, on-board recording would be needed for later investigation (black boxes).

The current, obviously unsatisfactory, situation where no formal and legal airworthiness and operating rules are available in the UK for civil UAS prompted the setting up in 1997 of the Unmanned Aerial Vehicle Systems Association, a UK trade association for the UAS industry (see Appendix A).

The aim of the Association was for the UAS Industry, collectively, to bring pressure to bear on the UK government to fund the CAA to prepare a document to cover in detail the airworthiness and operating requirements for UAS and to advise and assist the CAA in that work. The author believes that some 40 companies are members of the Association.

5.3   Europe

As previously advised, the regulatory authority for the majority of European countries is EASA. EASA, itself, only legislates for civil UAS with an AUM of over 150 kg. Smaller UAS are regulated individually by the EU's member countries.

A policy statement outlining the airworthiness certification of civil UAS is EASA E.Y013-01 which is listed as Reference 5.11. This statement is an interim solution to providing a means of UAS certification for Type Certificates (TC) and appropriately adapts the principles for certification of manned aircraft as laid down in Part 21, sub-part B of (EC) Regulation No. 1702/2003.

5.4   United States of America

In the USA, the monitoring of civil UAS is through the jurisdiction of the Federal Aviation Administration. Until recently, USA regulations were less prescriptive than those in Europe and are covered in a memorandum, 'AFS-400 UAS Policy 05-01,' listed as Reference 5.12. Before operating an UAS, the would-be operator would apply for, and obtain, a Certificate of Waiver or Authorization (COA) or FAA Form 7711-2.

To obtain this, the operator must show to the satisfaction of the FAA that the UAV has an acceptable level of airworthiness, that the Pilot-in-Command (PIC) has adequate aeronautical knowledge and that the proposed operations 'can be conducted at an acceptable level of safety'. With a COA, UAS operators are not held to rigorous see and avoid requirements.

However, with the proliferation of requests for COAs, this has changed and the COA route is now available only to the government Departments of Defence and Homeland Security. Another route for the operation of small civil UAS in the USA is as a model aircraft. This is more limiting and is covered in an FAA advisory circular AC 91-57 'Model Aircraft Operating Standards'. Reference should also be made to the US Academy of Model Aviation.

For the more general use of civil UAS the FAA are conducting studies with the intent of releasing a revised policy document 'AFS-400 UAS Policy 05-02' in a 2013 to 2020 time-frame and is looking towards Europe and especially the UK for inspiration.

However, the FAA is under rising industry pressure to accelerate the regulatory authorisation of civil commercial UAS. This is especially so for small UAVs which are seen as being the most likely to provide a commercially viable service. This will only become possible when some alternative means can be found to enable the UAV to operate out of the direct sight of the operator. The availability of a light and affordable DSA system is seen as unlikely to materialise for several decades.

As the author has previously advised, any would-be-operator should consult early with the appropriate authority – in this case the FAA. A useful website covering many of the above aspects is given in Reference 5.13.

5.5 Conclusion

As stated above this is an evolving but important area. The author has only been able to provide a snapshot of the current situation – indeed over the time taken to write this book this chapter has been changed dramatically in an attempt to capture the current status of this topic. The drivers are safety and integration with an unwritten need not to compromise manned aviation. Integration is likely to remain a slow process, of necessity, to prevent compromise in safety. It is anticipated however, that the use of unmanned systems will continue to expand beyond the military arena to provide new capabilities or cost savings in civil sectors. The pace of this expansion is likely to be dictated by commercial pressure, limited by the rate of evolution of standards and regulations.

References

5.1. 'A Guide to the Procurement, Design and Operation of UAV Systems'. Society of British Aerospace Companies, November 1991.
5.2. UK MOD Defence Standard 00970 – Part 9.
5.3. CAA Publication CAP 722 'Unmanned Aerial Vehicle Operations in UK Airspace – Guidance', Version 3, April 2009.
5.4. CAA Publication CAP 658 'Model Aircraft – A Guide to Safe Flying'.
5.5. CAA Policy Document 'UK – CAA Policy for Light UAV Systems', Haddon and Whittaker, June 2004.
5.6. Regulation (EC) No.1592/2002 of the European Parliament and of the Council of the 15 July 2002 on common rules in the field of civil aviation and establishing a European Aviation Safety Agency.
5.7. The Joint JAA/Eurocontrol initiative on UAVs. UAV Task-Force Final Report (available to download from the JAA website: www.jaa.nl).
5.8. CAA Paper 'Aircraft Airworthiness Certification Standards for Civil UAVs' D.R.Haddon/C.J.Whittaker; August 2002.
5.9. CAA Airworthiness Information Leaflet AIL/0165 'Applications for the Approval of Aircraft and Modifications to Aircraft'.
5.10. Relevant STANAG Documents:

- STANAG 3680 AAP-6 NATO Glossary of Terms and Definitions
- STANAG 4586 Standard Interface of the Unmanned Control System (UCS) for NATO UAV interoperability (Message formats and Protocols)
- STANAG 4609 (Edition 1, 23 March 2005): NATO Digital Motion Imagery Standard
- STANAG 4626: Modular and Open Avionics Architectures - Part I - Architecture
- STANAG 4660 Interoperable Command and Control Links)
- STANAG 4670 Training UAS Operators
- STANAG 4671 UAS Airworthiness
- STANAG 7085 CDL Communication System.

- STANAG 5066 The adoption of a Profile for HF Data Communications, supporting Selective Repeat ARQ error control, HF E-Mail and IP-over-HF operation
- STANAG 7023 (Edition 3, 16 September 2004): NATO Primary Image Format (NPIF)

5.11. EASA E.Y013-01 'Policy Statement Airworthiness Certification of Unmanned Aircraft Systems'.

5.12. AFS-400 UAS POLICY 05-01 'Unmanned Aircraft Systems Operations in the U. S. National Airspace System – Interim Operational Approval Guidance'.

5.13. Barnard Microsystems Ltd, www.barnardmicrosystems.com./home.htm.

6

Aspects of Airframe Design

The design of an unmanned air vehicle system, and particularly of the airborne element, has similar aims to that of a manned aircraft system, i.e. to achieve the required performance with the necessary integrity and reliability with minimum life-cycle costs (i.e. cost of initial procurement plus the costs of operating it). Therefore the procedures adopted to achieve these aims are similar and, in the main, use similar technology. However, there are differences and these stem mainly from the following aspects:

a) An electronic flight control system which is capable of receiving remote manoeuvre commands, using a pre-planned flight profile and/or acting autonomously to operate the flight controls is lighter and requires less volume than an aircrew and their habitation.
b) The majority of UAV are required to carry only sensor payloads rather than heavy payloads of armament, passengers or freight. More recently, some military UAV have been armed, but with relatively light weapons.
c) There are strong natural scale effects which operate to the advantage of smaller structures and mechanisms and, if used to advantage in design, result in an even lighter aircraft.

6.1   Scale Effects

To date, unmanned aircraft in general have been of lower mass than manned aircraft as indicated in Figure 6.1.

In terms of their all-up mass (AUM), manned aeroplanes range in size from the smallest single-seater such as the Titan Tornado of about 340 kg through the 590 000 kg of the Airbus A380 and the 640 000kg of the Antonov An 225.

UAV systems aeroplanes are on a lower scale, from about 6 kg for the Raphael Skylight, for example, up to the 12 000 kg of the Northrop-Grumman Global Hawk (previously discussed in Chapter 4). Hence the smallest fixed-wing UAV are two orders of magnitude smaller, in terms of mass, than their smallest manned counterparts.

Similarly for rotorcraft – whilst manned versions range in mass from the 623 kg of the Robinson R22 to the 97 000kg of the Mil 12 helicopter, the unmanned versions range only generally from the 20 kg of the EADS Scorpio 6 to the 200 kg of the Schiebel Camcopter.

There are heavier unmanned helicopters as indicated in Figure 6.1 where the original unmanned helicopter, the Gyrodyne DASH naval UAV is shown for reference at 1160 kg, and the more recent Northrop-Grumman Firescout (not shown), at 1430 kg. These latter two, however, are 'unmanned'

Unmanned Aircraft Systems – UAVS Design, Development and Deployment   Reg Austin
© 2010 John Wiley & Sons, Ltd

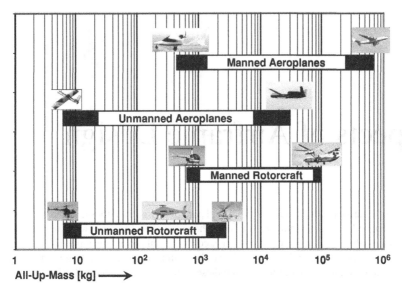

Figure 6.1   Mass domains of manned and unmanned aircraft

versions of aircraft originally designed to carry passengers and so carry, to some extent, a penalty of their heritage. In addition they were (are) used to carry light armament.

Also not included in the figure is the Bell Aerosystems Eagle Eye at 1364 kg AUM (see Chapter 4, Section 4.2.2 and Figure 4.14). This is a tilt-rotor configuration which, in terms of mass, carries a penalty of having elements required for both vertical and horizontal flight.

As can be seen, the lightest helicopter UAV design is, as the fixed-wing UAV, two orders of magnitude lighter than its manned counterpart. However, unlike the fixed-wing UAV, the heaviest specifically designed helicopter UAV does not yet enter at all into the domain of the manned machines.

There are, of course, even lighter UAV (MAV) in development, down to 0.5 kg or even lighter in both fixed-wing and rotorcraft categories. However, care must be taken when extending the scaling laws to such low masses since they can form a discontinuity in the laws as they may use quite different technologies or configurations from mainstream aircraft. For example, some MAV use flexible membrane wings.

In general it is possible to determine the effect that these differences of scale can have in the design of UAV aircraft. This can be done through the use of the principles of "Froude Scaling".

Froude Scaling

An observer may compare, for example, a mouse with an elephant or a wren with a swan. Within each species, both have the same basic structural and locomotion elements, etc., but the frequency of limb or wing movement is vastly higher in the smaller creatures. The bone structure is denser in the larger creatures than in the smaller. These trends may be explained by the following logic using dimensional analysis.

Definitions. Linear dimension ratio is L_a/L_m, $= n$ (scale factor) where subscript a indicates 'actual', m indicates 'model'. For a model system where $n = 10$, an aircraft model represents a full-scale aircraft having linear dimensions 10 times that of the model and areas 100 times those of the model. But the actual size system operates in the same density of air ρ, and gravitation field strength g as the model.

Since the dimensions of acceleration are Lt^2; then for a constant value of g, t will vary as $L^{1/2}$. From this the following examples of relationships result:

Function	Scale factor	For $n = 10$, model represents system having:
Area $\sim L^2$	n^2	Area $\times$ 100
Volume $\sim L^3$	n^3	Volume $\times$ 1000
Mass $\sim \rho L^3$	n^3	Mass $\times$ 1000
Velocity $\sim L/t \sim L^{1/2}$	$n^{1/2}$	Velocity $\times$ 3.16
Dynamic pressure $\sim \rho V^2 \sim L$	n	Dynamic Pressure $\times$ 10
Angular inertia $\sim L^4$	n^4	Angular inertia $\times$ 10 000
Frequency $\sim 1/t \sim 1/L^{1/2}$	$n^{1/2}$	Frequency $\times$ 0.316
etc.		

Thus a small aircraft, built observing these criteria, of the same configuration as one of ten times the size will exhibit similar characteristics when flying at, say, 31.6 m/s to its larger brother flying at 100 m/s.

As noted in Chapter 4, this principle has been used successfully to determine the flight characteristics of full-sized aircraft using fully instrumented small-scale models, usually unmanned, flying at lower speeds. It enables flight experiments, with any subsequent modifications, to be conducted at much lower costs and without the risks sometimes encountered in the initial flight testing of new configurations of aircraft. It is an expanding use of UAV technology (an example is shown in Chapter 4, Figure 4.29).

6.2 Packaging Density

The size and weight of the UAV can be significantly reduced compared with a manned aircraft designed for the same role by taking advantage of the ability to achieve a high density of packaging (aircraft mass/aircraft volume) and the structural and aerodynamic benefits which result.

The specific gravity (SG) of people is a little less than unity, but the effect of providing them with room for access and to operate, reduces the effective SG of the occupied cockpit for most light manned aircraft to an overall value of order 0.1 (i.e. about 100 kg/m^3) or less. In contrast, the electronics and optics for a UAV have densities greater than unity and can be tightly packaged, still allowing room for cooling. The TV camera system or other electro-optic sensor (eyes), AFCS (brain), radio and power supplies, (communication, etc.) and support structure of a UAV will typically have a SG of about 0.7 (700 kg/m^3).

Engines, transmissions, actuators and electrical generators, where applicable, though usually of different scale, are common to both manned and unmanned aircraft and have SG of about 5–6 (5000–6000 kg/m^3), although still requiring some room for access, cooling, etc. Landing gear varies considerably depending upon type, fixed or retractable, wheeled or skids. Gear for VTOL aircraft is usually much lighter than for HTOL aircraft. Therefore it is not possible to generalise for these components.

Structures such as wings, tail booms and empennage tend to have low values of SG. For example, the packaging density of a light aircraft wing, which typically accounts for about 10% of an aircraft mass, may be as low as about 25 kg/m^3 and this increases only slowly as the aircraft size increases. In contrast, the density of a helicopter rotor system, also accounting for about 10% of the aircraft weight, ranges from about 1200 kg/m^3 for a small helicopter UAV to possibly 4000 kg/m^3 for a large manned helicopter.

Fuel is more readily packaged into suitably shaped tanks and a fuel system will have packaging densities, when full, of about 900–1000 kg/m^3. When installed, this will, of course, increase the effective packaging density of the containing wing or fuselage.

The actual overall packaging density of an aircraft depends on configuration and on size. As an example, a light manned surveillance aircraft, such as a two-seat Cessna 152, having an AUM of over 700 kg will have an overall packaging density of about 120 kg/m^3. A fixed-wing UAV, such as the Observer with an AUM of only 36 kg, used to carry out a similar surveillance mission with TV camera equipment, has a packaging density of about 200 kg/m^3. A small coaxial rotor helicopter, such as the Sprite, of 36 kg mass can achieve a packaging density of 600 kg/m^3.

6.3   Aerodynamics

An indicative measure of the response to air turbulence of an aircraft, and to some degree its relative aerodynamic efficiency, may be given by the ratio Λ of its surface area to its mass. The larger the surface area, the more it may be disturbed by aerodynamic forces. The greater its mass, so greater will be its inertia (resistance) to the imposed forces.

Using the scaling laws it may be seen that the area/mass ratio Λ will vary as n/ρ_D, the linear dimension ratio n divided by the packaging density ρ_D. That is to say that Λ will increase as the aircraft becomes smaller, but will be reduced as the packaging density is increased. Thus UAV, being generally smaller than manned aircraft, will tend to suffer more aerodynamic disturbance in turbulent air than will larger aircraft but this can be offset by the ability of the designer to achieve increased packaging density in the UAV.

The achievement of aerodynamic efficiency is obtained by reducing to an acceptable minimum the profile and friction drag of the wings and body. (This is often quantified by the drag-to-weight ratio of the aircraft – low for efficiency – and usually referred to the drag at 30 m/s airspeed.) Again, this tends to be a higher value for aircraft of low packaged densities.

The smaller aircraft suffers a higher friction and profile drag of both wings and body for the same shape as a larger aircraft as it operates at a lower Reynolds number I_R. The Reynolds number is a nondimensional parameter which is the ratio between the inertia forces and viscous forces of a fluid flow, and consequently quantifies the relative importance of these two types of forces within a given flow. The inertia forces are greater downstream over longer surfaces at greater flow speeds and therefore are the dominant forces at higher N_R.

The value of N_R is given by the expression

$$N_R = vl/v$$

where v is the flow (air) speed; l is the characteristic length (e.g. wing chord) and v is the kinematic viscosity of the fluid and has a value of 1.47×10^{-5} m^2/s for air under standard conditions. (For more information see aerodynamic text books such as References 6.1 and 6.2.)

The aerodynamic drag of surfaces operating at lower values of N_R is greater than for those at the higher values. UAV, being generally smaller and generally operating at lower speeds than manned aircraft, operate at lower N_R so therefore generally suffer higher levels of drag.

This is shown graphically in Figure 6.2, where the drag coefficient, based on surface area, of a streamlined aerofoil of 15% thickness to chord ratio is shown for a range of values of N_R. The data is extracted from Reference 6.2 as an indicative example of the trend, but applies in principle to all exposed surfaces.

A number of UAV, ranging from the largest and fastest (Global Hawk) to the smallest UAV (Wasp III) believed to be in current production, are aligned to their respective values of N_R based upon their wing chord and approximate minimum-power speed. It can be seen that the drag coefficients appropriate to the smallest, having a mass of only 0.43 kg, are three times that of the largest – the Global Hawk of about 12 000 kg mass.

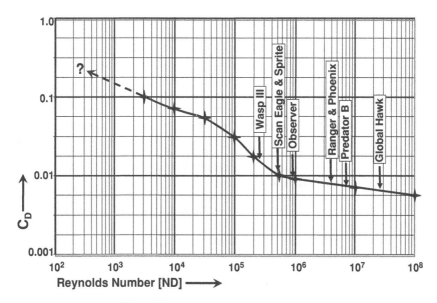

Figure 6.2   Trend of drag coefficient with N_R

For the small MAV, these high values of drag coefficient apply to the surface area of a wing which is large relative to the aircraft mass. (i.e. a very low wing loading of 5 kg/m² necessary to achieve the minimum flight-speed required for hand-launching this aircraft). The other bad news for very small MAV aerodynamics is that the maximum lift coefficients obtainable at low N_R. are considerably lower than are obtainable at the more usual values of N_R.

Designers of MAV therefore have a challenge in making them aerodynamically viable and must alleviate the situation by extracting the maximum benefit through achieving:

a) the highest packaging density possible, and
b) the best aerodynamic shapes of the body.

Proposals for even smaller MAV of masses as low as 100 g will require much research into new approaches to aerodynamics in order to overcome the potential difficulties of operating at even lower Reynolds numbers.

6.4   Structures and Mechanisms

Although the achievement of an efficient aerodynamic solution for the smaller UAV is a challenge to be met by well-considered design, the good news is that structures and mechanisms usually benefit at smaller scale. Loads in larger aircraft are much higher than their smaller counterparts and so require materials of greater specific strength and stiffness to carry those loads over greater distances without failure through direct loading, bending or buckling.

When the structural design for light manned aircraft moved from fabric-covered tubular framework to monocoque construction in light alloys it was found that the direct loads in the shells (or skins) required the use of only very thin gauge material. To prevent local buckling of the skins between frames (or formers), stringers were added. Even this was insufficient as handling of the bodies and wings caused dents in the surfaces. It became necessary to accept the use of thicker skins to prevent this, but at a weight penalty.

The advent of composite materials of lower density helped to alleviate this problem as the skins could be made thicker without an increase in weight. Although, initially, the materials such as glass-fibre epoxy resin matrices had a lower specific stiffness (Young's modulus), the extra thickness of the cross-section more than made up for this and the buckling and handling problem was solved.

This solution is directly applicable to UAV, and composite materials have become *de rigueur* for UAV construction except in areas where the higher loads must be carried within limited space. Examples of these areas can include undercarriages when reversion to light alloys or steels becomes necessary. However, with the development of materials such as carbon-fibre matrices in more specialised resins, cured in autoclaves (heated and pressure controlled chambers), and plastic–aluminium alloy composites, even these components may undergo material changes.

Most UAV benefit through smaller scale in that loads are lighter and are required to be carried over shorter distances. This reduces the probability of buckling though attention must still be paid to ensuring that the structure is robust enough for man-handling.

The reduction of the weight of aircraft components allow for a greater mass of payload and/or fuel to be carried in the same AUM of aircraft. Alternatively, it allows for the task to be carried out with a lighter and smaller aircraft.

An indication of the relative mass of the several components within an aircraft is by the use of the ratio.

Mass of Component/AUM of Aircraft

This is known as the 'weight fraction' of the component. The following expressions for weight fraction, derived from the scaling laws, indicate the benefit of small size and higher packaging density available to UAV.

$$\text{Body structure (for given package density)} = \text{Constant } K$$

$$\text{Body structure (change in package density)} = K\left[1/\sigma^2 + 1/\sigma\right]$$

$$\text{Wing (change in aircraft AUW)} = \left[\frac{K_T AR^{3/2}}{t/c \cdot w^{1/2}} \times \frac{\rho_m}{f_c}\right] W^{1/2}$$

$$\text{Rotorcraft gears and shafts} = K\left[\frac{0.43}{V_T} + \frac{0.0085}{p^{1/2}}\right] W^{1/2}$$

$$\text{Rotor blades and hub} = \left[\frac{22}{\beta_0 V_T^2 p^{1/2}}\right] W^{1/2}$$

where

σ is the package density ratio (dense/less dense);

W is the gross weight of the aircraft (N);

K_T is a constant depending upon the wing geometry, e.g. Taper;

AR is the wing aspect ratio;

ρ_m is the density of the wing material (kg/m^3);

f_c is the allowable direct stress of the wing material (N/m^2);

t/c is the wing thickness to chord ratio;

w is the wing loading (N/m^2);

V_T is the tip speed of the rotor (m/s);

p is the disc loading of the rotor (N/m^2);

β_0 is the coning angle of the rotor blades (rad).

These examples give an indication of the benefits that can be gained in structural and mechanical weight fraction for smaller and lighter aircraft and how achieving a high density of packaging, possible in the UAV, reaps rewards. This does depend, of course, on the appropriate selection of materials and good design.

Structure Design

Although the usual structural design methods, detailed in many textbooks, apply equally to UAV as to manned aircraft, their application may be a little different. The design of any UAV structure must take several requirements into account. Not the least of these is ease of initial manufacture, cost, longevity, reliability, accessibility and maintenance.

For manned aircraft, some of the ancillary equipment that must be accessed may be reached from on-board. The remainder has to be accessed through detachable panels in the aircraft external structure though attempts are made to keep these to a minimum for reasons of structural strength and stiffness and for aerodynamic cleanliness. In UAV, especially the smaller ones, access from on-board is not feasible. Access through external removable covers is limited in scope due to the small size of the structure. For access by human hand, the panels may have to be large in proportion to the surrounding structure and would weaken the structure so that heavy reinforcement would be required. The solution adopted for some manned aircraft is the use of removable stressed-skin panels. The close fitting of these for reliable stress transfer is difficult to achieve at small scale. Therefore, for all but the largest UAV, a more efficient solution is to construct the airframe from detachable modules, using composite materials wherever appropriate.

An example of a commonly used construction method for UAV fuselage modules of small or medium sized UAV is shown in Figure 6.3. Glass fibre is used with a suitable resin as the main material, with stiffening elements constructed as shown with a stiff plastic foam core wrapped with carbon fibre tape.

Manufacturing the skins from carbon would be very expensive and carbon alone has little inherent damping, thus is prone to shattering if subjected to a sudden blow. Matching it with glass provides the necessary element of damping and has been found to provide a practical solution. Whilst carbon cloth is expensive, carbon tape is reasonably cheap. The whole provides a lightweight, stiff, durable and acceptably cheap structure suitable for mass production.

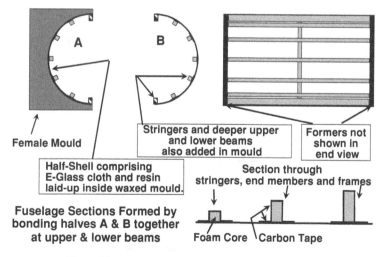

Figure 6.3   Structural techniques using composite materials

An alternative material which has about half the density yet offers similar stiffness to carbon/glass combination is a glass-fibre or carbon-fibre reinforced polycarbonate. It can be thermo-moulded and its forming into structure is less labour intensive than is the case with laid-up materials. Similar construction methods may be applied to the lifting surfaces.

Inter-modular structural connections will usually require loads being diffused into connecting lugs or spigots, etc. which may have to be of light-alloy or even steel material depending upon the stress level at the connection.

Much of this can apply to the larger HALE and MALE aircraft, but because of the larger area of structures and greater load densities present in larger aircraft, a greater use is made of stiffer, carbon composites for outer shells of wings and fuselages. To achieve even greater stiffness and robustness a sandwich construction is another option with a nylon honeycomb layer sandwiched between two layers of carbon-fibre or carbon-fibre and glass-fibre mix.

The continuing development of new materials offers several advantages for UAV design and manufacture and suitable material selection should be a prerequisite of any new UAV design. Advice is available from specialist publications and material manufacturers (see References 6.5 and 6.6).

Mechanical Design

Although the effect of small scale is generally advantageous in terms of stiffness requirements, a downside is that the holding of accurate tolerances in machining operations is more difficult. The approach to fatigue therefore has to be more cautious as stress-raising joints and bearing mountings can be more critical because of the scaled-down radii in corners. It is advisable to use larger reserve factors on joints and larger radii than pure dimensional scaling would indicate.

For example, as shown in the upper diagram in Figure 6.4, the edge radii on a standard small roller bearing mounting would require a 0.5mm internal radius on the shaft. This would result in a greater increase in local stress than the equivalent at a larger scale. A proven solution, shown in the lower diagram, is to use a larger radius on the shaft with an abutment washer between the shaft collar and the bearing. The alternative of using a specially designed and produced bearing with larger edge radii would result in a heavier and more expensive assembly.

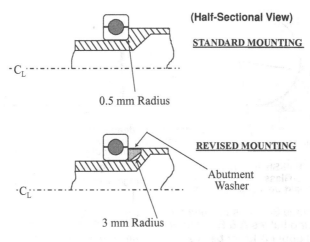

Figure 6.4 Revision of bearing mount design

Wear and Fatigue

Whether the mechanical system in design is a structure or a mechanism, the prerequisite is that it not only adequately and efficiently performs its required function, but that it continues to do so, reliably, for a specified time. Such time values, stemming from the operational requirements, will be specified at the concept of the design.

Components in design will be subjected to calculations aimed at ensuring that the specified lives will be achieved. The lifetime before the replacement of a component becomes necessary will be measured in operating hours for both wear and fatigue.

The wear rate on bearing surfaces is usually estimated from load–velocity charts. Subsequently these components will be tested in the development phase to confirm their operational lives. The number of oscillatory cycles to be applied to a component, before it fails in fatigue, will be calculated and an appropriate reserve factor added to estimate the component life (see References 6.1 and 6.2).

One effect of scale, as listed above, is that the natural frequencies of the UAV and components increase as size reduces. Therefore fatigue life as determined by the number of cycles before failure will be reached earlier in terms of hours of life in a smaller UAV. This trend may, however, be reversed by the same scale effect if the loads and stresses in the materials are lower.

The careful selection of the type of material to be used can be an important factor in reducing the occurrence of fatigue failure. Composite materials tend to have high resistance to fatigue and so their use instead of light alloys can be of advantage in extending the fatigue lives of airframes.

Areas where a complete change from metallic components may be delayed for some time are in engines, rotorcraft hubs and transmission systems where, apart from high load densities occurring, other factors such as high temperatures and wear apply.

With the greater number of stress cycles per hour occurring in the smaller aircraft, if it is required to achieve a fatigue life comparable with a larger aircraft, it may be worthwhile considering the use of steels rather than light alloys in some applications. This may be especially true for rotorcraft. Most steels have a stress level at or below which the component will have an infinite fatigue life. This is not true of light alloys and to obtain a long (though still not infinite) life with a light alloy may require that it be operated at such a low stress level that the component is heavier than if designed in steel.

As with components subjected to wear, those subject to fatigue will undergo rigorous testing in the development phases to establish their operational lives.

Undercarriage Design

The functions of the undercarriage are broadly the same for both manned and unmanned aircraft, the main purpose of the undercarriage being to absorb and dissipate the energy of impact on landing and to provide a stable base for the aircraft on the ground. For HTOL aircraft it will provide support whilst accelerating to flight speed on launch and decelerating after landing.

Whilst the design techniques are the same for manned and unmanned aircraft, the structural scale effect alone would favour the weight fraction of the UAV undercarriage. However, other effects more than negate that potential advantage.

Impact velocities tend to be greater for UAV due to the following causes.

a) Smaller HTOL UAV tend to have lower wing loading than manned aircraft and so they can be affected more by air turbulence on landing,

b) Unless the UAV landing is automatically controlled by accurate sensors, the judgement of a remote human pilot in assessing the UAV height above ground can be poor compared with an on-board pilot who has more positional and acceleration cues available.

Hence a wise UAV designer will factor this in when considering the design of the undercarriage. The author recommends designing for a vertical impact velocity of no less than 2 m/s and an undercarriage deflection yielding a maximum of $4g$ deceleration.

If the undercarriage has wheels, they are likely to be smaller than the wheels of a manned aircraft. Imperfections of the taxi-way surface, such as drainage channels, present relatively large discontinuities compared with the wheel diameter and entry into them can impose very large drag loads onto the undercarriage and its supporting structure.

Undercarriage configurations range from the very simple to the relatively complex. The simplest will be those employed in land-based VTOL aircraft. These may be a fixed tubular skid type with flexible plates to absorb the initial energy of impact, and use friction between metal plates or the hysteresis in composites to provide a limited amount of damping.

Ship-based VTOL aircraft may require a greater length of undercarriage deflection to cater for higher impact velocities and almost certainly a greater amount of damping to prevent bouncing on deck. This latter may be achieved in various different ways. A composite material of very high natural damping may be employed or discrete pneumatic or hydraulic dampers may be incorporated. An alternative is to design the undercarriage to 'splay' its feet or wheels on impact to provide damping through friction with the deck.

HTOL aircraft will require a wheeled undercarriage which may have similar forms of energy absorption and damping to those outlined above. The wheels will have to be capable of carrying the aircraft to high speeds on take-off and landing and absorb greater shocks than the smaller wheels which may be used on VTOL aircraft merely to assist in manoeuvring the aircraft on the ground or deck.

Both HTOL and VTOL aircraft may require the undercarriage to be retracted into the aircraft wings, body or nacelles, as appropriate, in order to reduce its aerodynamic drag and/or destabilising effect in high-speed flight.

Consideration will be given as to the direction of retraction. It is usually preferable for the undercarriage to be retracted forwards or sideways rather than aft. A forward retraction in a HTOL aircraft will move the aircraft centre of mass forwards and so improve its stability in flight. Also the airflow will aid its subsequent extension, particularly pertinent if any power failure has occurred during the flight.

Another aspect to consider with aircraft operating from land is the degree of firmness of the ground on which it will have to be moved. This is usually specified as the Californian bearing ratio (CBR) and the reader is referred to books on undercarriage design where it is evaluated. Suffice it to say that operation on or from soft ground will require larger wheels and lower pressure tyres than for operations from firm ground.

The requirement that an undercarriage provide a stable base for the aircraft when on the ground or ship deck can have a large impact upon the undercarriage design. The first aspect that must be determined is the likely degree of destabilising force. This may arise from wind forces imposed upon the airframe or rotor system or unevenness or movement of the ground or ship's deck.

Naval operations from off-board impose the most adverse conditions for both HTOL and VTOL aircraft due to high wind turbulence, tilt of the ship's deck and movement of the deck. The calculation and addition of these effects will derive the effective displacement of the aircraft centre of mass (CoM) which must be contained by the undercarriage to prevent the aircraft toppling.

The stability cone of the undercarriage is defined as the cone whose vertex is at the aircraft CoM and its circular base is the circle contained within the undercarriage footprint. The larger the vertex angle, so is the stability increased. A large vertex angle is obtained by having a low aircraft CoM, i.e. the CoM being as close to the ground as possible, and a large base to the stability cone.

It may be noted that, as shown in Figure 6.12, a four-legged undercarriage will provide a larger stability cone for a similar undercarriage spread as does a three-legged undercarriage. For this reason, several naval helicopters have four-legged undercarriages.

The design of an undercarriage for rotary wing aircraft must take into account the interaction between the dynamics of the rotor, if it is articulated, and the aircraft on its undercarriage. This phenomenon is known as ground resonance and its prevention requires the provision of a degree of damping in both the undercarriage and the rotor in-plane motion (see Reference 6.4).

Hence the initial statement that undercarriage configurations may range from the simple to the very complex. This will be reflected in the mass of the undercarriage as a proportion of the aircraft gross mass and also the extent of the testing required during its development. The mass fraction may range from about 2% for the simple to 5% for the complex (see Reference 6.7).

Helicopter Rotor Design

The discussion above has indicated areas where the design of UAV components may differ in detail from those of manned aircraft, and especially effects of scale when the UAV is smaller than the manned counterpart. The difference in scale can bring about significant differences in the design parameters of the aircraft.

Regarding the effects of scale on the parameters of a helicopter rotor can provide a good example. The weight fraction equations, shown above, indicate that the weight fraction of elements in a helicopter rotor and transmission system vary as the square root of the aircraft design gross mass (DGM). This knowledge is used in the optimisation of the complete aircraft where the mass of one system can be traded against another.

The mass (or weight) fraction of a helicopter rotor increases severely with its diameter, principally because much of the mechanical elements are designed to withstand the torque applied to them. The speed at the rotor tip is limited by aerodynamic factors and so tends to be similar for all sizes of rotor. Hence the torque per unit power applied increases with rotor size and so therefore does the rotor mass fraction.

As the aircraft DGM is increased, the designer may elect to limit the increase in the rotor diameter in order to save weight. However that will increase the disc loading p of the rotor and require more power and thus a larger and heavier engine. The weight increase with size of a rotor is at a generally greater rate than an engine weight increases with power. Therefore there is a trade-off of power-plant weight with rotor weight.

The result is that the designer will choose a disc loading which produces the lowest combined rotor and power-plant weight for the endurance required of the aircraft. As in Figure 6.13, disc loading will be seen to tend upwards with aircraft DGM although aircraft designed for long endurance will be below that trend.

6.5   Selection of power-plants

As the old song goes – 'Life is nothing without music'- and one might add 'an aircraft is nothing without a power-plant'.

The siren voices promising the virtues of a projected high-technology, but nonexisting, engine have lured more new aircraft programmes onto the rocks of delay, over-budget and cancellation than any other technical reason. Aircraft designer beware! The design and development of a new engine and its integration into a viable power-plant often takes longer than the development of an airframe.

The resulting attempted development of a new aircraft and new engine in parallel can result in the engine being late in arrival – it may not meet its promised performance and reliability or, in order for the engine developers to get it to work, may have been changed in form and/or have different output characteristics from those for which the aircraft was designed. The new engine may have become incompatible with installation in the aircraft.

Although it is very tempting to have the latest technology engine in a new aircraft, the aircraft designer must exercise judgement as to the probability of a new engine arriving in time and to specification. It is

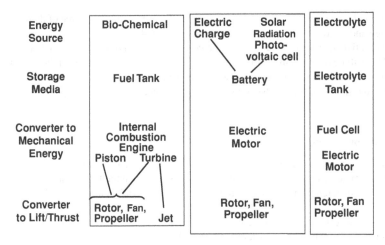

Figure 6.5   Power-generation systems

probably unwise to consider any new power-plant until it has been run satisfactorily on a test-rig for at least 100 hours.

The power system for a UAV, as for any aircraft, includes an energy source, a means of converting that energy into mechanical energy and a means of converting that into a lift or thrust force, as shown in Figure 6.5. A power-plant will include means of engine speed and/or power output control, engine temperature control and, usually for fixed-wing aircraft, a means of electrical power generation. (Rotorcraft will have electrical generators driven instead from the rotor system, thus ensuring that electrical power remains available, even in the event of engine failure.)

The great majority of UAV in operation are powered by internal combustion engines and most of those have piston engines.

6.5.1   Piston Engines

These may be considered to exist in four main types although there are sub-types of each.

a) Two-stroke (two-cycle) engines
b) Four-stroke engines
c) Stepped piston engines
d) Rotary engines.

Two-stroke and Four-stroke Engines

There is probably a greater range of sub-types of two-stroke engines than there is for four-stroke units. For simplicity, for example, some two-stroke units achieve lubrication by mixing lubricating oil with the fuel. Others have a separate oil system, as do four-stroke units.

Some two-stroke units use valves for controlling the airflow, others do not.

Both types can be designed to use petroleum fuels or, with higher compression, to use diesel or other 'heavy' fuels.

Both types can be equipped, if necessary, with turbo-charging. Both types may be air-cooled or water-cooled.

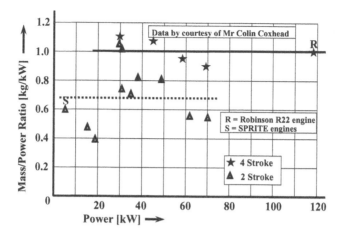

Figure 6.6   Air-cooled piston engines: mass/power ratios

The only basic difference between the two types is that the two-stroke engine has a power-stroke on each revolution of the crank-shaft whereas the four-stroke has a power-stroke every other revolution. Thus the two-stroke, in theory at least, tends to produce twice the power in unit time at the same rotational speed compared with the four-stroke unit.

The two-stroke tends to have greater specific fuel consumption than the four-stroke since the intake gases can be contaminated by the outgoing exhaust gases. Typically a four-stroke engine will consume between 0.3 and 0.4 kg of fuel per kW hr (specific fuel consumption). The two-stroke engine typically will have a sfc between 0.4 and 0.6 kg/kW hr.

However, much depends upon the detailed design of the engine and its carburettor or fuel injection system and also how it is operated. Running a piston engine at full power and speed will tend to increase the specific consumption compared with lower powers and speeds.

As Figure 6.6 shows, at least for those units for which mass/power data is more readily available, there is considerable scatter of data, the more so for the two-stroke engines. This also applies to any attempt to show any trend in fuel consumption and therefore only generalised comments may be made.

The reasons for this scatter of results is that not only are there many variants of the configuration, e.g. one, two, or more cylinders, lubrication from a sump or oil mixed with the fuel especially for the two-stroke units and carburetion etc., but the background ancestry of the units is different.

The smaller units have been mostly the products of companies selling to the model market and have little or no aviation heritage. They are not designed to meet any defined manufacturing or reliability standards and, often, the manufacturers have not measured the performance data of their products.

With the more recent increase in funding for UAV systems, there are signs of new manufacturers entering the market place with better products. However, it behoves the UAV designer to exercise caution when selecting an engine. Engines for the larger aircraft may be available from approved sources of light aircraft engines, well designed to aeronautical standards and present fewer problems.

The two-stroke unit tends to run hotter than the four-stroke and may require more cooling facilities than the four-stroke. The four-stroke unit tends to be heavier than the two-stroke (see Figure 6.6).

Both types will pay for higher performance and higher fuel-efficiency with greater complexity, weight and cost. It is up to the designer to decide the priority. It may be that a two-stroke installation is more suited to the smaller, shorter-range aircraft whilst the four-stroke is more appropriate to the larger, longer-range aircraft.

There is an important advantage that the two-stroke unit offers compared with a four-stroke engine. Neither type produces power with smooth torque (as does a turbine engine), but the torque of both varies during each revolution. However, the torque peaks of the two-stroke unit are much smaller than those of the comparable four-stroke unit. This is of particular concern for a rotorcraft transmission and rotor system which must be the more robust if driven by a four-stroke engine. To a lesser extent this will affect the design of propeller or fan propulsors.

Linear vibration has also to be considered and will be largely a function of the number of cylinders of the engine. Although the oscillating mass of a single cylinder may be balanced to some extent by a mass on the other side of the crank-shaft, it is far from perfect. On the other hand, two horizontally opposed twin cylinders will give almost complete balance.

Stepped Piston, Two-stroke, Engines

These are a relatively new development where each cylinder has two bores of different diameter, the larger at the base. Each cylinder has a piston with an upper and lower section and rings to fit the two bores. Each cylinder has a double purpose where the lower section of each cylinder supplies compressed air to the upper section of an adjacent cylinder where the air is further compressed and burnt.

It is claimed that this arrangement achieves the better features of two-stroke and four-stroke engines, i.e. with the lighter mass/power ratio and torque smoothness of the two-stroke and the better fuel efficiency of the four-stroke. Although prototype units have been built and tested, no such unit has yet been fitted to a UAV. The type offers great promise, but is not in production for aircraft use and so is covered further in Chapter 27.

Rotary Engines

Rotary engines have suffered a chequered history but are now becoming more reliable and accepted. Initially, and for a long period, there were problems with seals between the rotor(s) and the casing breaking down. 'Chattering' of the rotor lobes in the casing caused early wear, and engine life was short. Although lateral vibration emanating from the engine was low, torsional vibration was high. However, most of these problems seem now to have been overcome.

In the 1980s, a Canadair UAV used a rotary engine, but it was soon replaced with a turbo-shaft engine. Today the Israeli Hermes 180 and Hermes 450 UAV use rotary engines, in 28 and 38 kW form respectively, and they are reported to offer a long life and low specific fuel consumption (0.35 kg/kW hr).

Although the basic engines are of low mass/power ratio, because the engines operate at a high rotational speed, a reduction gearbox is usually necessary this, together with high levels of cooling equipment required, increases the mass towards that of a conventional four-stroke engine.

Other reports speak of high fuel consumption, noise and high cost of operation of other engines of the rotary type. It is necessary, therefore, for the designer who is considering such installation to be aware of the possible advantages and down-side of the generic type and to assess the situation with care. Currently the type seems to be available only within a very limited power range. There appear to be no rotary engines below a power rating of about 28 kW or above 60 kW available for aircraft.

6.5.2 Gas-turbine Engines

Gas-turbine engines are fundamentally quieter than piston engines and produce smooth power at low mass/power ratios. They may be considered as two generic types:

(i) *turbo-jet units* which are designed to produce thrust (kN) from a high-velocity jet for direct propulsion;

(ii) *Turbo-shaft units* which produce power (kW) in an output shaft which may drive a propeller or helicopter rotor to provide thrust.

In their simplest form they employ a compressor set and turbine set on a single output shaft. Their disadvantage is that any increased load on the output which will slow the turbine also slows the compressor set, thus reducing the power available to accelerate the engine back to operating speed until an increase in fuel injection can take effect. The result is a lag in response which is bad for a propeller-driven aircraft, but can be disastrous for a helicopter.

Most turbo-shaft engines of today therefore are of the free-power-turbine (FPT) configuration. Here the output shaft is a second, separate, shaft from that mounting the power-generating compressor/turbine sets. Thus when the output demand is increased, the compressor is not slowed and an increase in injected fuel accelerates the 'compressor spool' more rapidly, giving a speedy response to extra power demand.

A *turbo-fan unit*, possibly to be regarded as a third type, is in effect a mixture of the turbo-jet and turbo-shaft engines in so far as some of the combustion energy is extracted as a jet whilst some energy is converted to mechanical power to drive a fan which produces a slower-flowing, but larger volume, jet of air.

A thrust-producing jet is at its most efficient when the minimum amount of jet velocity is left in the ambient air mass after the aircraft has passed. Therefore the turbo-jet engine is most appropriate for the higher speed aircraft, the turbo-fan engine for intermediate speed aircraft and the turbo-shaft engine, driving a propeller with its much lower efflux velocity, for the slower aircraft.

Hence the Global Hawk HALE UAV uses a turbo-fan engine, and the Predator B MALE UAV uses a turbo-shaft engine driving a propeller.

The choice of a turbo-fan and turbo-prop engine for the HALE and MALE aircraft respectively is appropriate since both aircraft, being at the upper range of size for UAV, require more powerful engines than do the smaller, medium-range aircraft. Gas turbine units are available in the higher power ranges, but due to scale effects are not economic for lower power requirements.

Smaller, medium- and close-range aircraft are invariably powered by piston engines, but they would benefit from the low mass/power ratio of the turbine engine. Unfortunately, there are no small turbo-shaft engines available below the approximately 500 kW power of the Predator or Firescout engines. Of the current medium- and short-range aircraft, a few would require installed power levels of about 120 kW, several in the 30–40 kW range and a number as low as 5–10 kW. A turbine engine is at its most fuel-efficient when operating near maximum power, and its specific fuel consumption deteriorates sharply if operated at part-power. Therefore attempting to use an over-size engine for the smaller aircraft would impose not only a mass and bulk penalty, but an unacceptable level of fuel consumption.

It is perhaps ironic that a number of adaptable turbo-shaft engines were available during the 1950s–1970s in powers down to 60 kW, having been developed as auxiliary power units (APU) for manned aircraft. These fell out of use when the increase in size of aircraft required more powerful APUs or their engine-starting and other on-board services were obtained by other means.

Two examples of those engines were the 60 kw (single-shaft) Rover Neptune and the 120 kW (FPT) Rover Aurora. Both of them used centrifugal compressors and centripetal turbine stages, and it may be expected that any new small turbo-shaft units would adopt that technology since a scaled-down axial-flow compressor/turbine system would suffer very high fuel consumption through unsolvable tip-clearance and manufacturing difficulties as well as other scale-effect problems.

Modern materials and fuel monitoring together with new manufacturing techniques and the upsurge in demand for UAV systems may yet make the development of small turbine engines viable. Indeed, there are encouraging signs of activity such as the 'small uninhabited air vehicle engine' (SUAVE) program in the USA which aims to develop gas turbine engines with power generation as low as 8 kW, using ceramic materials, and Microjet UAV Ltd in the UK with similar objectives.

6.5.3 Electric Motors

Electric motors, of course, convert electrical energy into mechanical energy to drive a propeller, fan or rotor. The electrical power may be supplied by battery, a solar-powered photovoltaic cell or a fuel cell.

They have the particular advantage of being the quietest of all the engines and with the smallest thermal signature.

Battery Power

Currently only micro- and mini-UAV are powered by batteries and electric motors. Typical examples being the Desert Hawk and Skylight shown in Chapter 4, Figure 4.23.

Although considerable improvements are continuously being made in battery design and production, the demand on the battery is made not only by the motor, but also by the payload and communication system. Therefore the flight endurance and speed of such UAV systems and the capability of their payload and communication systems are limited. The systems are small and light enough to be back-packed so they have a place in very short range operations under relatively benign conditions. Back-up batteries must be carried and regularly charged to ensure an electrical supply.

Other means of obtaining a continuous electrical supply are being sought in order to extend the range and capability of electrically powered systems and to this end research is underway to develop solar-powered photovoltaic cells and fuel cells compatible with UAV systems requirements. Both systems have been flown in a UAV, but it is early days for the technology and they are better discussed in Chapter 27.

Thus there are many factors for the UAV designer to consider when selecting an engine and there is no comprehensive reference book to consult. It is necessary to liaise with the several engine manufacturers in order to assess the suitability of their products.

6.5.4 Power-plant Integration

The engine is only a part of a power-plant. It must be suitably mounted, possibly on flexible tuned mounts; provided with means of starting; fed with fuel; controlled; cooled; and provision made for fire detection and suppression, an exhaust system and silencing system if required. The characteristics of the engine will inevitably determine the complexity or otherwise of these sub-systems and this will influence the choice of engine.

In addition the UAV services such as electrical power alternators or air blowers may be engine-driven and mounted within the powerplant.

Larger engines, for larger UAV, and qualified for aircraft use will most probably come with much of that equipment, but smaller engines, for the smaller UAV, may not. In the latter case that equipment must be separately sourced. Unfortunately starter motors, alternators, carburettors, etc. are not generally available in the small sizes required by small UAV. In the 1980s M L Aviation were forced to design and manufacture their own alternators and cooling fans for their UAV, but the situation has improved and specialist companies are now appearing to provide appropriate items.

6.6 Modular Construction

In addition to ready access to components, as already discussed, the modular construction concept allows for the separate manufacture and bench proving of the several modules. This saves factory space and readily allows for the manufacture and testing, to agreed standards, of complete modules by different suppliers, and in different countries. Final assembly will then be carried out at the system's main contractor. Here the total UAV system will be integrated, ground and flight tested before delivery to the customer.

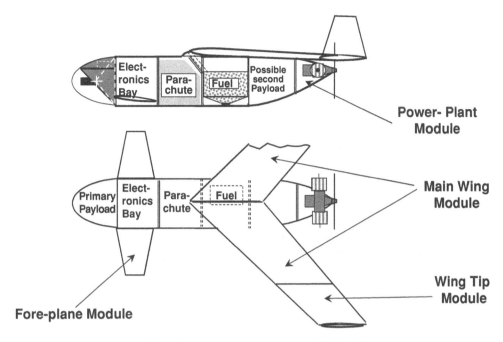

Figure 6.7 Modular construction of HTOL aircraft

The attributes and details of the contents of the different modules are covered in other chapters, but they must all be mounted and held together by the structure.

Figure 6.7 shows a possible modular configuration of a medium- or close-range fixed-wing (HTOL) UAV. The advantages of employing modular construction have already been listed and an indication of how this might be realised in such an aircraft is illustrated.

Viewing the layout of the aircraft, access to the main payload in the aircraft nose might be via a removable cover. Replacement of it in its entirety in favour of a different type of payload, or replacement if faulty, would be by removing it at fixing(s) at its rear. The structural connection may involve, say, quick-acting pip-pin(s), and electrical connection by suitable connectors. Built-in test equipment would register on the housekeeping display in the control station whether or not a satisfactory mechanical and electrical connection had been effected. Each exchangeable payload would have to be balanced to be of the same mass moment about the aircraft centre of mass. Thus a change of payload can be effected in minutes.

Removing or hinging forward the payload gives access to the electronics module which could then be slid out in its entirety for bench testing if required. Similarly, the power-plant module could be removed, if required by separating suitable structural, fuel supply and control connections. The lifting-surface modules would also be removable, by disconnecting suitable structural and control connections. It may be considered advisable to have separate outer-wing modules as the wing-tips are often vulnerable to damage on landing or even take-off (launch). Electric cabling, as required, would be part of each module with suitable inter-module connections.

In the case of a larger, for example, MALE type UAV, which is required to carry far more equipment, modular construction is less easy to achieve. The internal view of a typical MALE UAV is shown in Figure 6.8 with numbers being allocated to the several components. A listing of these components is carried in Figure 6.9.

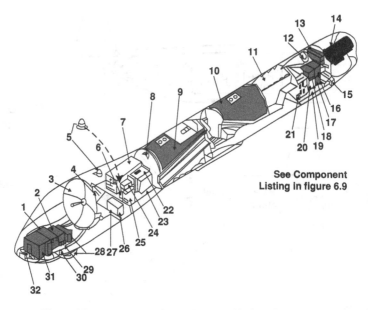

See Component Listing in figure 6.9

Figure 6.8   Internal view of MALE UAV and listing of components

1. Synthetic Aperture Radar (SAR) Antenna
2. Inertial Navigation System/GPS
3. Ku-Band Satellite Communications Antenna
4. Video Cassette Recorder
5. GPS Antennas (Left and Right)
6. Identification Friend or Foe Transponder
7. Ku-Band Satellite Communications Sensor Processor Modem Assembly
8. C-Band Upper Omni-directional Antenna Bracket
9. Forward Fuel Cell Assembly
10. Aft Fuel Cell Assembly
11. Ancillaries Bay
12. Ancillaries Cooling Fan
13. Oil Cooler/Radiator
14. Engine
15. Tail Servo (Left and Right)
16. Battery Assembly
17. Power Supply
18. Battery Assembly
19. Aft Equipment Bay Tray
20. Secondary Control Module
21. Synthetic Aperture Radar Processor / Electronics Assembly
22. Primary Control Module
23. Front Bay Avionics Tray
24. Receiver/Transmitter
25. Flight Sensor Unit
26. Video Encoder
27. De-ice Controller
28. Electro-Optical/Infrared Sensor Turret / Electronics Assembly
29. Front Bay Payload Tray
30. Ice Detector
31. Synthetic Aperture Radar (SAR) Receiver/Transmitter
32. Nose Camera Assembly

Figure 6.9   Internal view of MALE UAV and listing of components

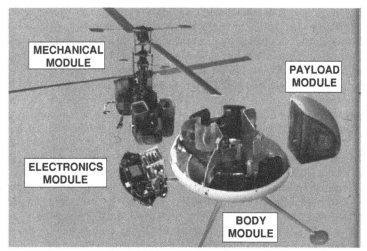

Sprite VTOL UAV Main Modules
(M L Aviation Photograph)

Figure 6.10   Modular construction of VTOL aircraft (M L Aviation)

The considerable mass of fuel required for long-range operation must be carried near the aircraft centre of mass, i.e. near the wing centre of pressure. The forward-looking cameras must be mounted in the nose and the engine and accessories in the rear. Other electronic equipment will be positioned with regard to longitudinal balance and the closeness or otherwise to other components to reduce cable lengths and possible EMC problems.

The installation and removal of these several modules poses an access problem. Too many removable panels or hatches will require considerable local reinforcement of the monocoque airframe structure, and recourse may have to be made to stress-carrying removable panels in order to limit the weight penalty incurred by the maintenance needs of ready access.

Whereas the illustration of the possible modular construction of the HTOL aircraft shown in Figure 6.7 was hypothetical, the modular arrangement of the VTOL aircraft shown in Figure 6.10 is that of an actual Sprite aircraft, a type which had extensive operating experience and which proved the value of its modular construction. The aircraft is of coaxial helicopter configuration which, of all aircraft types, is probably the easiest to construct and deploy in modular form.

As the illustration shows, the aircraft comprises four main modules:

a) the body module,
b) the mechanical module,
c) the electronics module, and
d) a range of alternative payload modules.

(These were known, colloquially, as the 'body', 'brawn', 'brains' and the 'business' end respectively).

The body is constructed entirely of composite material, being a matrix of resin-reinforced glass and carbon fibres. It is by far the cheapest of the four modules. It is essentially an endoskeletal structure with a central box having a vertical pillar at each corner. The mechanical module attaches to the upper end of each pillar, whilst one leg of the four-legged undercarriage plugs into the lower end of each tubular pillar.

The volume of the circular body is essentially divided into four segments by four vertical diaphragms which act as 'brackets' to carry the electronics and payload modules, and to provide insulation from the other two segments which are occupied by each power-plant. There is no 'front' of the aircraft as it is symmetric in plan view and can fly equally readily in any direction. However, to simplify the description, we will refer to the segment which carries the detachable payload as the 'front'.

Opposite the payload module, and balancing it, is the electronics module, whilst the mechanical module gear-box straddles the central box section. The rotor drive shafts rise centrally from the gear-box and the two power-plants occupy each of the lateral segments. The body is completed by four removable covers which form the upper surface of the body and provide access to the modules for attention *in situ* or removal.

This arrangement enables the mechanical module to carry the weight of the body and its contents from the top of the box-pillars in flight, whilst the weight of the whole aircraft in landing is transmitted directly down through the pillars to the undercarriage. The design concept provides a lightweight, cheap structure and a very compact aircraft with the desirably high density of packaging. These are the features which give the helicopter UAV its small size, for a given payload and performance, compared with the equivalent small aeroplane UAV. It is a trend which is likely to continue into the future as electronic components become ever smaller (for further details of the aircraft see Figure 4.19 in Chapter 4).

Another advantage of the coaxial rotor configuration of helicopter is that the complete power and lift provision can be invested in one module independently of the remainder of the components of the aircraft. This enables a range of sub-configurations to be provided and is exampled in Figure 6.11.

On the right-hand side of Figure 6.11 is shown a two-view drawing of the version of the Sprite UAV discussed above. The configuration offers the ability readily to fly in any direction and with great stealth. However, that compromises its ability to fly at higher speed because of its higher-drag body shape.

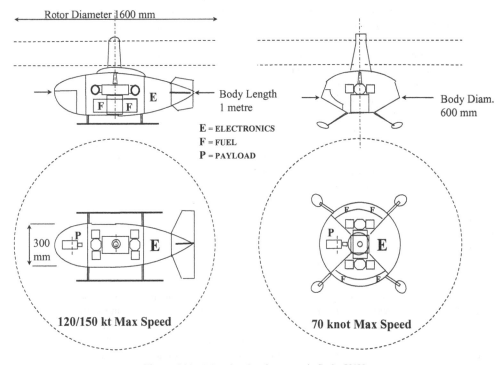

Figure 6.11   Directional and symmetric Sprite UAV

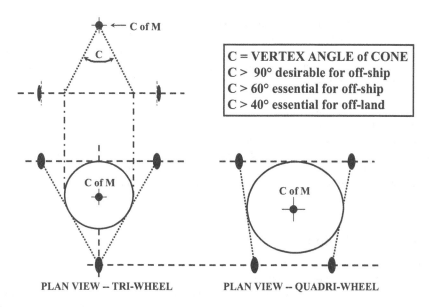

C = VERTEX ANGLE of CONE
C > 90° desirable for off-ship
C > 60° essential for off-ship
C > 40° essential for off-land

PLAN VIEW -- TRI-WHEEL   PLAN VIEW -- QUADRI-WHEEL

Figure 6.12   Undercarriage stability

Making the change to a more streamlined body, whilst retaining the same mechanical, electronic and payload modules, confers higher speed and range to the aircraft, but at reduced stealth and sideways flight ability. Some reprogramming of the AFCS would be necessary.

Another configuration considered was the installation of the modules in a tilt-wing airframe somewhat similar to the Aerovironment Sky Tote of Figure 4.27, but with a delta-wing. The advantage of this ability of the aircraft to emerge in different configurations is that an operator can operate more than one type and choose the variant which provides the best capability for a particular role. The airframe body is by far the cheapest of the modules and the others can literally be moved into whichever airframe is to be operated.

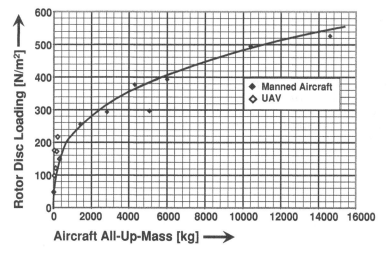

Figure 6.13   Rotor scale effect

This flexibility did, at first, introduce some complexity in operating the Sprite UAV, a 'problem' which could apply to other modular UAV. To meet airworthiness regulations it is necessary to define the 'build standard' of each aircraft before flight, and for a UAV system, probably of the whole system. Usually in an aircraft the airframe is the primary element and is allocated a series number which is 'attached' to the airframe. Components, such as power-plants, are replaced as required and logged as sub-units.

In the case of Sprite, apart from the payload modules being changed, often between missions to best suit the requirements of the mission, an electronics or mechanical module might be replaced at any time to allow it to be checked or serviced. Thus, it was questioned, which was the primary module and how did one define the build standard of an aircraft at any one time?

The solution which was adopted was to give each module a letter and serial number and these were grouped on the build standard document as Exxx-Myyy-Bzzz and the payload A, C, etc., depending upon type, followed by its own serial number nnnn. The airframe still carries its serial number (Bzzz) externally.

6.7   Ancillary Equipment

UAV need a similar range of ancillary equipment as manned aircraft except for those items relevant to aircrew accommodation and functioning. For the largest UAV this is seldom a problem as they are in a similar mass domain as the lighter manned aircraft. For the smaller UAV problems may arise in sourcing appropriately sized electrical alternators, actuators, air-data systems, attitude and altitude sensors, batteries, fasteners, external lighting, antennae, etc. Existing aircraft-approved and certified equipments, available for even the lightest manned aircraft, are too large and heavy to appropriately meet the requirements of the smaller UAV. Some of these equipments may be available to a limited extent from model aircraft suppliers, but they seldom meet the airworthiness standards required for UAV.

This has certainly been a problem in the past and UAV manufacturers have often had to resort to developing the equipment, and achieving its certification, themselves. Fortunately, following the demand from a burgeoning UAV industry, specialist equipment suppliers have appeared on the scene and, even if they do not have suitable equipment available off-the-shelf for a specific requirement, they are usually able to develop suitable equipment under a sub-contract arrangement. It is not appropriate to list these suppliers in this book but readers are referred again to Reference 6.3.

References

6.1. J. H. Faupel and F. E. Fisher, *Engineering Design*. Wiley-Interscience, 1981.

6.2. J. Bannantine, J. Comer and J. Hanrock, *Fundamentals of Metal Fatigue Analysis*. Prentice Hall, 1990.

6.3. *Unmanned Vehicles Handbook*. Published annually by the Shephard Press.

6.4. A. R. S. Bramwell, Edward Arnold, *Helicopter Aerodynamics*, 1976.

6.5. B. Hoskin, *Composite Materials for Aircraft Structures*. AIAA Education Series.

6.6. M.C Y. Niu, *Airframe Structural Design*. Conmilit Press Ltd, ISBN 962-7128-04-X, 1988.

6.7. C. J. Burgh and J. L. Pritchard., *Component Design: Handbook of Aeronautics No. 2*. 1954. Sir Isaac Pitman and Sons Ltd.

7

Design for Stealth

There are three main reasons why it is desirable that a UAV system remains undetected in operation. They apply principally to the air vehicle although other components of the system may be involved.

(a) It is desirable that the air vehicle remain undetected whilst on a reconnaissance/surveillance mission in order not to alert the enemy (military) or criminals (policing) to the forthcoming operation.
(b) Principally in military use, it is necessary to protect the air vehicle from loss due to enemy counter-measures.
(c) Mostly applicable to civilian operations, low-detectable signatures will result in minimising environmental disturbance.

However, the other side of the coin is the need for the UAV to be readily visible when operating in civil airspace. In some operations, therefore, it may be necessary to have the UAV offering 'stealth' which can be changed to overtness, at perhaps 'the flick of a switch'! This will be taken further in Part 3, Deployment.

The principal means of detecting an air vehicle are through its 'signatures', i.e. its acoustic or electromagnetic emissions at the following wavelengths:

a) noise (acoustic) [16 m–2 cm, or 20–16000 Hz]
b) optical (visible) [0.4–0.7 μm]
c) infrared (thermal) [0.75 μm–1 mm]
d) radar (radio) [3 mm–3 cm]

Hence to reduce the air vehicle detectability to an acceptable level, it is necessary to reduce the received emission or reflection of the above frequencies below a threshold value which, itself, is often a function of the operation – principally the operating height of the air vehicle.

An unmanned air vehicle has advantages, compared with a manned equivalent, in its inherent ability to achieve low signatures. These virtues should be pursued to best effect by the competent UAV system designer. They are:

(a) The removal of aircrew and their support equipment enables the designer to achieve a more densely packaged and thus dimensionally smaller machine. For a similar mission, the mass of the air vehicle, therefore, will normally also be less than that of the manned equivalent.

Unmanned Aircraft Systems – UAVS Design, Development and Deployment   Reg Austin
© 2010 John Wiley & Sons, Ltd

(b) Without the need to accommodate aircrew and their need for access and external vision, the shape of the air vehicle may be more specifically designed for aerodynamic efficiency and signature reduction, and the resulting shape will often be a compromise between these two.

7.1   Acoustic Signature

The noise emanating from an aircraft may be the first warning of its presence, although not usually so directionally locatable as to immediately lead to its detection. As a means of alerting other means of detection it merits suppression.

Aerodynamic noise from aircraft emanates predominantly from vortices, principally at the tips of wings, rotors or propellers. This increases with wing or blade span loading and speed, so that low values of both enhance acoustic stealth.

Usually, however, noise from the power-plant(s) is of greater concern, and results from noise of combustion in piston engines and compressor noise or wake noise in turbo-jets and to a lesser degree in turbo-shaft and turbo-fans. Electric motors, of course, develop virtually no noise, but their use, with few exceptions, as discussed in Section 6.3 of Chapter 6, is largely limited to short-endurance MAV.

Noise generally increases with power-plant power usage level, so that keeping the mass and aerodynamic drag of the aircraft as low as possible is a good first step to achieving low noise generation.

The human ear is most sensitive to frequencies of around 3500 Hz and can hear sound down to a practical threshold of about 10 dB. For a given sound pressure level, attenuation of sound with distance in air (and in insulating media) varies as the square of the sound frequency. Hence low-frequency sound usually presents the greater problem.

Turbo-jet and turbo-fan engines tend to be used for MALE or HALE systems and the noise from these high-frequency generators is usually well attenuated by the time it has reached earth.

For the larger tactical UAV operating at lower altitudes, a mitigating solution is to mount the turbo engines above the wing to shield the compressor and efflux noise from the ground. The greater noise problem is posed by smaller aircraft using piston engines where no practical turbine engines currently exist.

Water-cooled engines are quieter than air-cooled engines since the coolant serves as an absorptive blanket. However, they are significantly heavier than air-cooled engines and are therefore seldom used in aircraft.

The choice of piston engines is thereby reduced to conventional two-cycle or four-cycle units, rotary (Wankel-type) units being, as yet, inadequately reliable and difficult to cool. Although four-cycle engines offer slightly lower fuel consumption than do two-cycle units – possibly of order 75% – they have a far worse power-to-weight ratio (see Chapter 6, Section 6.5.1).

Sound is emitted principally from the internal combustion and from the exhaust, although intake noise can be an added factor.

The combustion frequency of a small two-cycle engine operating at, typically, 5000 rpm will produce noise predominantly at about 100 Hz. The equivalent four-cycle engine will generate noise at about 50 Hz. Their noise is less attenuated than the two-cycle by a factor of four, since attenuation of sound in media is proportional to the square of the sound frequency. The two-cycle engine is therefore usually favoured but requires silencing if stealth is needed.

Combustion noise can be attenuated by blanketing appropriate areas of the engine in sound-absorptive materials, though this comes at the price of extra weight. The possibility of surrounding the emitting areas by the fuel tanks, ideally of absorptive materials, should be considered. Whilst there is fuel in the tanks, this fuel will attenuate the sound, depending of course, on the thickness of the fuel 'wall'.

The exhaust noise can be reduced by silencers (or mufflers) and, if possible, both air intake and exhaust should be directed skyward. However, the level of sound that can be detected also depends upon the level and character of the background noise, i.e. sound contrast. The background noise on a battlefield, for example, may readily drown out the noise emanating from a 'quiet' UAV.

7.2   Visual Signature

The most common means of visibly detecting the aircraft is initially by the unaided eye. The human eye operates within a small range of the electromagnetic spectrum, i.e. between about 0.4 and 0.7 μm, peaking at maximum efficiency at about 0.55 μm wavelength.

Criteria which determine the ability of the eye to see an airborne object against an open sky or cloud background include:

a) the size and shape of the object,
b) its contrast against the background and the sharpness of the edges of that contrast,
c) the effect of the atmosphere,
d) any movement of the object,
e) exposure time,
f) the stability and diligence of the observer,
g) glint.

This is a complex subject, on which many volumes have been written, and research continues to refine the current understanding. Even so, the physics of ground-to-air observation are simpler and somewhat better understood than that of air-to-ground. An introduction to the physics is given in Reference 7.1.

(a) The Size and Shape of the Object

These combine to determine the threshold of detectability of the airborne object.

(b) Contrast

Contrast C is defined as the ratio of the difference in luminance between an object and its background to the luminance of its background,

$$C = (B - B_1)/B^1 = \Delta B/B^1$$

where B is the luminance of the object and B^1 is the luminance of the background in units of cd/m^2.

For example, if the luminance of object and background is the same, $\Delta B = 0$ and $C = 0$. Thus the object would be undetectable.

In ground-to-air observation, the luminance of the object is generally less than that of the background and ΔB is negative.

The threshold of contrast C, where the average observer has a 50% chance of detecting the object is given the symbol ε. However, in real operations, predicting the background luminance and that of the object is problematic. The background luminance will depend upon the atmospheric conditions at the time, and the position of the object relative to the position and height of the sun.

The luminance of the object, although dependent upon its shape and surface texture, is also dependent upon atmosphere-reflected illumination which is unpredictable.

Although absolute prediction of visual detection of an object is not possible, some approximation is possible and relative detectability can be determined for various aircraft from the size, shape and angular velocity of the object as presented to a ground observer. The larger the object, other things being equal, so the object becomes easier to see – that is, the value of ε reduces.

Laboratory tests have shown that for a black, sharp-edged, circular disc against a well-lit background, and within a limited field, there is a 50% probability of its detection by the average unaided human eye if it subtends an angle of about $\alpha = 0.15$ mrad. For such circular objects, which subtend a small angle α (of the order of 1 mrad or less), the simple law $\varepsilon \alpha^2 = $ constant applies.

Such a condition is unlikely to be realised in practice and, for realistic contrast values, a large field of view and atmospheric attenuation, it may be assumed that there is little probability of detection for a stationary object in a large field if α is less than 0.5 mrad.

(c) Atmospheric Effects

The atmospheric water content or pollution, apart from affecting object visibility to a ground observer through attenuation, may also increase the level of luminance of the object through reflecting the sun's rays upwards. Although subject to variation, a general conclusion from observation is that an object at 2500 m altitude will have a contrast typically reduced to 40% of its contrast at 200 m due to increased luminance.

(d and e) Movement of the Object and Its Exposure Time

Field experiments have shown that movement of the object at angular speeds of about 40 mrad/s offer the greatest stimulus to the eye to aid detection. The stimulus reduces rapidly at speeds lower than that, and speeds above that steadily reduce the exposure time, thus reducing the probability of detection. 40 mrad/s equates to about 80 kt at 1000 m height, and the object would take about 80 s to traverse the whole bowl of the sky from horizon to horizon.

(f) Stability, Awareness and Diligence of the Observer

Under field conditions, the observer may not be aware of the impending presence of an aircraft. Even if he is, if no aircraft appears, his attention may lessen. If he is mounted on a vehicle, for example, he may not have a stable platform from which to observe.

All these factors can reduce, significantly, the probability of detection unless the observer is alerted by another indicator, in particular, noise.

(g) Glint

Glint results from the sun being reflected from glossy surfaces, especially from glazed canopies. UAV have an advantage in not requiring canopies for aircrew vision. However, attention should be paid to ensuring that camera windows are as small as possible and that the aircraft exterior has a matt finish, preferably in light grey.

7.3   Thermal Signature

Infrared (IR) radiation is emitted from a heat source and propagates in much the same way as light. The detectability of the radiant body is similarly determined by its contrast with the background and the radiating area. However IR radiation is more readily absorbed by the atmosphere than is light, particularly by water and CO_2 molecules. The major windows for IR radiation to pass through the atmosphere are in the 3–4 and 8–12 μm wavelength bands. Thus IR detectors are designed to receive within one or the other of these bands.

Heat in an aircraft is generated principally as waste heat from the power-plant and to a lesser degree from electronic components. Heat can be generated at the stagnation points at the leading edge of wings, propellers and rotors. However, this is negligible until high-subsonic speeds and above are reached, and will always be of lower order than heat energy emitted from power-plants, and so will be ignored in this volume.

Heat is radiated or conducted from the engine carcass and from the exhaust. Dealing with the former first, it is necessary to prevent this being radiated from the aircraft to ground by containing it within the aircraft and emitting it skywards – away from ground-based detectors. Materials of low emissivity such as silver or aluminium may be used to prevent radiation in adverse directions.

The exhausts pose the greater problem as they will be particularly hot and will inevitably be made of high-emissivity materials such as steel or Nimonic alloys. They should be screened as much as possible by other airframe components.

As a background for IR, the sky is very cold, offering a base for an excellent contrast. However, although more advanced detectors are under development, long-range detection of low-emission targets is only achievable with a small (5°) field of view. So that, unless other indicators show the direction of the target, the detection of an object at high altitude is unlikely. The detection range is a function of temperature (K) to the fourth power. Reducing the target emission temperature is very effective in evading detection by IR.

7.4   Radio/Radar Signature

The radio signature relates to radio frequency emissions from the aircraft (and also from the control station) and these must be minimised to prevent detection. This is discussed in Chapter 9, Communications.

The radar signature, like the visible spectrum signature is a reflected frequency, in this case from radio frequency pulses generated by an enemy emitter which is scanning the sky, from the ground, looking for return (reflected) pulses from a body entering its sector. Therefore the emitted radiation will be approaching the aircraft from the lower hemisphere. Except for when the aircraft is close to overhead the transmitter, the radiation will be arriving at the aircraft from a small angle beneath the horizontal.

The UAV designer's aim is to prevent the pulses being reflected back to a detector. As with the visible spectrum, the UAV usually offers the advantage, compared with a manned aircraft, of being smaller (less reflective area) and not having its shape constrained by the necessity of accommodating aircrew.

There are basically three methods of minimising the reflection of pulses back to a receptor:

a) To manufacture appropriate areas of the UAV from radar-translucent material such as Kevlar or glass composite as used in radomes which house radar scanners. This method can seldom be effectively employed except in the case of very small MAV where the wing 'skins' are very thin since these materials are translucent, and not transparent as is sometimes mistakenly thought. The proportion of the received pulse which is reflected increases with thickness until at a skin thickness of typically 15 mm, depending upon the material matrix composition, a sufficient proportion of the received energy is reflected to be detected by the searching radar. Also, of course, the material cannot be used to house metallic or other components as these will return, and even amplify the pulses back through the translucent cover.

b) To cover the external surfaces of the aircraft with RAM (radar absorptive material). This material absorbs radio energy in much the same way as sound energy is absorbed in anechoic chambers, turning the energy into heat. The material is usually comprised of foam sandwich, is rather bulky, somewhat fragile and can add significant weight if used extensively. Another problem is that it is usually designed to absorb a limited range of frequency. Any frequency outside that range is not absorbed but reflected. It is best used in small amounts in critical areas.

c) To shape the aircraft externally to reflect radar pulses in a direction away from the transmitter. This is probably the most effective method of the three and is particularly suited to UAV where the external shape is not conditioned by aircrew needs but is reduced to obtaining the best compromise between aerodynamic and radar signature requirements.

It is desirable that little or no surface area of the aircraft will present at, or approaching, a right-angles to the radiation. As the radiation will usually be received from a small angle beneath the horizontal, vertical surfaces such as fins are the most critical. It is also desirable to avoid having surfaces meeting at 90° as these will act as a radar 'corner reflector' to give strong returns.

If shaping is to be used successfully, it clearly must be addressed at a very early stage of design. Having determined the proposed shape, models will be constructed in a reflective material, tested in a wind-tunnel and in a radar test facility to check that the theoretically determined shape does meet both the required aerodynamic and radar signature criteria. Shaping for radar 'diversion' is a specialist, and usually classified, technology not appropriately covered in this book. However, an approximate idea of the effectiveness of the shaping can be obtained by mounting a polished model in a dark chamber and directing a beam of light onto it. Any light which is reflected back to source will indicate the area of the model which may return radar and will need to be modified.

7.5   Examples in Practice

With so many unknown variables the prediction of aircraft detectability is not an exact science! It may be noted, also, that the priority in signature reduction may depend upon the operational circumstance of the system. As an oversimplification, in terms of their needs and means of achieving stealth, there might be considered to be three basic operational types:

a) HALE and MALE higher-altitude reconnaissance and surveillance systems,
b) Lower-altitude, close- and mid-range reconnaissance and surveillance systems,
c) UCAV high-speed attack air vehicle systems.

7.5.1   HALE AND MALE Systems

The former, by nature of carrying sophisticated sensor suites, to survey from high altitudes, and great mass of fuel, are large aircraft but, due to the altitudes at which they fly, they are not greatly at risk of visual detection.

Figure 7.1 shows the basic theoretical trend of the threshold area for visual detection increasing with altitude. The threshold increases still further at altitude due to the reflective effect of the atmosphere in increasing the luminance, and hence reducing the contrast, of the under-surfaces of the aircraft. The figure is based upon data for circular discs of the same total area and could be modified by the higher aspect ratio of wings and bodies. There is some evidence that the visibility is reduced by up to 10% as the aspect ratio of a body is increased. However, there is other evidence that straight lines attract the attention of the human eye more readily than do rounded shapes.

For the aircraft shown in the figure it is possible that these two effects may mutually cancel out. Certainly they would all appear to be beyond the threshold of visual detection at their operating altitudes.

Radar probably offers the most likely means of detection of these high-flyers but, once detected, there is still the problem of interdiction against them. This problem is to be exacerbated by the Polecat Research UAV, currently under development by Lockheed-Martin and shown in Figure 7.2, which is eventually expected to be capable of operating at altitudes in excess of 30 000 m and looks to have a low radar signature with its engines mounted above the wing. However, a much more sophisticated sensor suite will have to be developed to carry out surveillance from those altitudes.

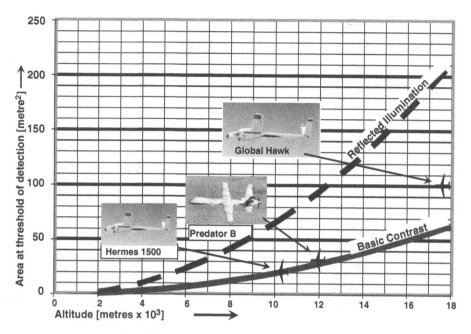

Figure 7.1   Threshold of visual detection (high altitudes)

7.5.2   *Tactical or Battlefield-launched UAV*

These are required to operate at low altitudes, often below cloud-level, and are open to detection by all four signatures unless those signatures can be reduced below threshold levels. Observer and Sprite UAV (each were presented in detail in Chapter 3) are compared in Figure 7.3 as examples of contending configurations for this role. They have similar mass and performance. The needs and means of achieving stealth for each of them are considered. From the author's experience, however, it is far more difficult to achieve low signatures in all frequency ranges for a small fixed-wing aircraft than for a coaxial rotor helicopter.

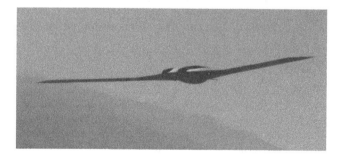

Wing-span - 28 metres;   AUM - 4090 Kg
Powerplant – Two, 13.4 KN thrust Williams FJ44 Turbo-fans
Predicted Operational Altitude - 30,000 metres (~100,000 ft)

Figure 7.2   Polecat Research UAV

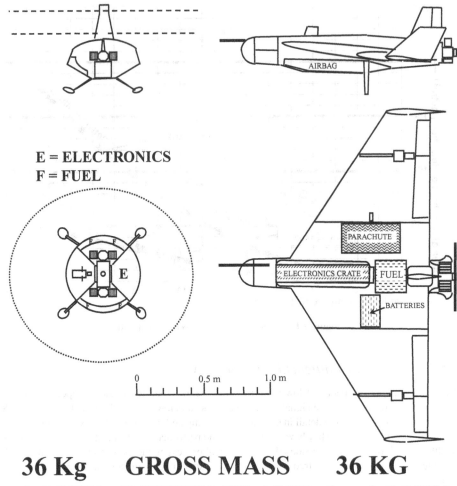

E = ELECTRONICS
F = FUEL

0   0.5 m   1.0 m

36 Kg   GROSS MASS   36 KG
2 x 4.6 KW   POWER PLANT 1 x 4.6 KW

Figure 7.3   Sprite and Observer compared (Reproduced by permission of Cranfield Aerospace Ltd)

(a) Visual Signature

The greatest obvious difference between the two configurations probably lies in their visual signatures.

Referring to Figure 7.4, it can be seen that virtually all of the lower surface of the Observer wing (about 1.65 m²) will be in shadow, thus presenting a high contrast with the sky background. On the other hand, the Sprite has a maximum body diameter of 0.6 m, giving a projected area of only 0.28 m². However, only a central area of about 0.025 m² of this is horizontal and in full shadow, the remainder benefiting from a lightening from reflected luminance. The resulting difference in visual signature from below is shown in Figure 7.4.

The image of Sprite can be reduced even further if the option of under-body illumination by light-emitting diode (LED) is adopted, as shown in Figure 7.5. Photoelectric cells are used to measure the

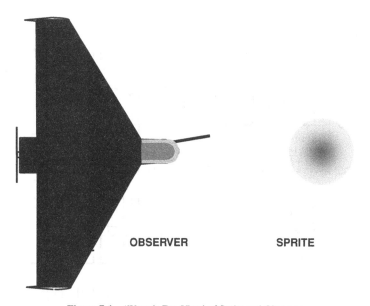

Figure 7.4   'Worm's Eye View' of Sprite and Observer

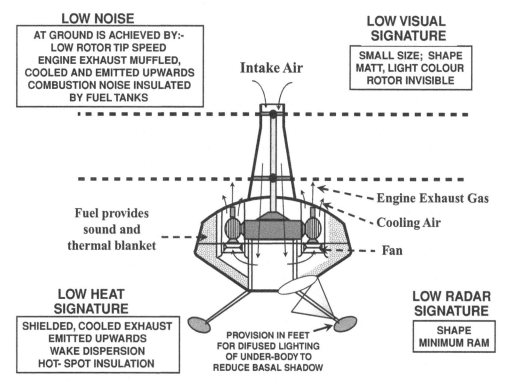

LOW NOISE

AT GROUND IS ACHIEVED BY:-
LOW ROTOR TIP SPEED
ENGINE EXHAUST MUFFLED,
COOLED AND EMITTED UPWARDS
COMBUSTION NOISE INSULATED
BY FUEL TANKS

Intake Air

**LOW VISUAL
SIGNATURE**

SMALL SIZE; SHAPE
MATT, LIGHT COLOUR
ROTOR INVISIBLE

Engine Exhaust Gas

Cooling Air

**Fuel provides
sound and
thermal blanket**

Fan

**LOW HEAT
SIGNATURE**

SHIELDED, COOLED EXHAUST
EMITTED UPWARDS
WAKE DISPERSION
HOT- SPOT INSULATION

PROVISION IN FEET
FOR DIFUSED LIGHTING
OF UNDER-BODY TO
REDUCE BASAL SHADOW

**LOW RADAR
SIGNATURE**

SHAPE
MINIMUM RAM

Figure 7.5   Reduction of detectable signatures in Sprite UAV

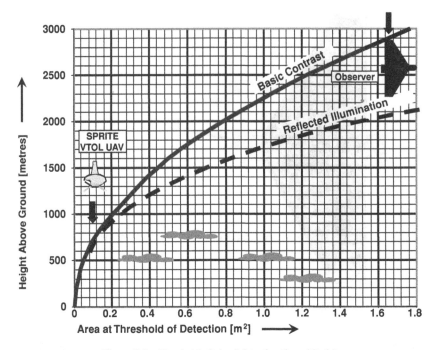

Figure 7.6   Threshold of visual detection (low altitude)

luminance of the background sky and the aircraft under-body and adjust the lighting to achieve equality. To use the same technique beneath the larger area of a wing is not as simple.

A graph plotting the threshold area for detection for lower altitudes is shown in Figure 7.6. This indicates that with its wing area of 1.65 m², Observer would have to operate at at least 2000 m altitude to escape visual detection, whilst Sprite, which has an effective basal shadow much smaller than its 0.28 m² plan area can operate at 300 m or lower without visual detection. Also, in actual operation it is confirmed that straight lines in a silhouette attract the eye more than do rounded, amorphous shapes.

As recorded earlier, an ability to move slowly in situations requiring covertness is of great benefit in aiding the avoidance of detection.

(b) Radar Signature

The means of reducing the signatures of the two aircraft is explored in Figures 7.5 and 7.7 for Sprite and Observer respectively.

Radar, as for visual, signature results from a reflected medium, as discussed in Section 7.4. Unless the radar detection system is mounted in an attacking aircraft, the radar will generally be beamed from a hemisphere beneath the aircraft. In the case of Sprite, the aircraft body is totally covered with a thin coating of silver to form a Faraday cage.

The only situation in which any significant reflection will be returned to the emitting radar system from the aircraft body is if it were emitting directly upwards and Sprite was directly above it, presenting the small, 170 mm diamter, flat under-surface normal to it. This is a most unlikely, if not impossible, situation since radar dishes do not normally have the ability to transmit vertically upwards.

The maximum likely reflection is if the beam momentarily impinges at right-angles to the slope of the conical surface. This would produce an extremely small 'blip', less than that received from a small bird, and unlikely to be repeated on any subsequent sweep of the beam. The orbit of the rotor blades will not present a normal surface unless immediately above the radar. As with the body, such a situation is impossible in practice. This has been proven in several trials when Sprites have remained undetected by radars.

Looking now at Observer, if the surfaces of the wing were to be silvered, in the manner of Sprite, with the aircraft in level flight the radar pulse would be reflected away from the radar receiver, unless, of course, the aircraft were flying in a banked turn and so presenting the wing area normal to the axis of radar emission.

The fins, propeller and engine would present the greater problem which might possibly be addressed by covering the fins in RAM and enclosing the engine within a duct, as suggested in Figure 7.7. This would still leave a problem with the propeller, and, of course, carrying and balancing the extra weight.

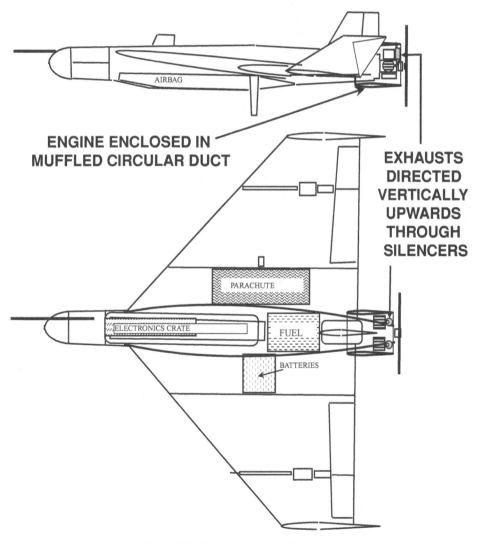

Figure 7.7   Observer reduction of signatures

In both aircraft configurations, attention must be paid to the payload. The sensor apertures must be made as small as possible and probably covered in a thin film of a material which will reflect radio frequencies, but allow the sensor frequencies to pass through.

(c) Thermal and Acoustic Signatures

These are discussed together as they are both emitted from the aircraft and the means of reducing them are similar.

First, looking again at Sprite in Figure 7.5, the majority of the excess heat generated is from the two engines, with some from the electronic equipment. The engine exhaust stacks and silencers are surrounded by cooling air supplied by engine-driven fans with the gas mix being exhausted vertically upwards to minimise levels of both noise and heat radiated earthwards.

The cooling gases mix with the engine exhaust gases in the ratio of about 10/1 and on leaving the aircraft are further mixed into the rotor downwash with a further dilution of up to 1000/1. The fan-induced cooling air is also used to extract the heat from the electronic heat sinks.

Fuel tanks surround each engine and absorb the radiated heat. The muffled and cooled engine exhaust emitted upwards and the blanketing fuel tanks similarly reduce the aircraft noise.

Noise from rotor blades is proportionate to the rotor thrust and to the sixth power of the tip speed. The typical manned helicopter blade tip speed is about 220 m/s and a small helicopter rotor will generate typically about 20 kN thrust.

By comparison the Sprite UAV has a blade tip speed of only 134 m/s and generates only 0.35 kN thrust in steady flight. Hence the rotor generates no measurable noise. At ground level the aircraft records 85 dB at 15 m laterally, principally power-plant-generated noise, but this reduces at a measured rate of 0.33 dB/m as the aircraft climbs. Hence the aircraft becomes inaudible from the ground long before it reaches 300 m height. The author currently has no comparable figures for the Observer, but the aircraft does not appear to have had any attempt made to screen either heat or noise.

These signatures could be reduced somewhat if the engine was shrouded, as suggested by the author in Figure 7.7, but at the expense of aircraft performance, mass and mass balance. An alternative approach might be to bury the engine within the fuselage and use a drive-shaft aft to the propeller. It might then be necessary to accommodate the fuel within the wings in shallow tanks which would present other problems, and engine accessibility would be compromised.

7.5.3   Medium-range UAV

In between the HALE and close-range UAV systems are a plethora of medium-range systems such as the IAI Malat Searcher, the AAI Corporation Shadow series, and SAGEM Sperwer to name but a few. To remain visually undetected, they must operate at altitudes of 3000–6000 m, depending upon their size, but to deploy their electro-optical payloads may force them to descend below cloud level when they become vulnerable to shoulder-launched missiles.

7.5.4   MAV and NAV Systems

MAV aircraft are essentially small and expected to operate at low level in urban warfare or within buildings carrying lightweight, low-performance camera systems.

Detection by radar is not appropriate to their role and, as most of them are electrically powered, detection by noise, or heat is probably not applicable. There remains the problem of visual detection and, though they are by definition of order 15 cm wing-span, it is difficult to see how they will remain unseen in the low-level action for which they are intended.

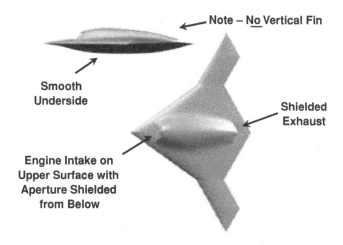

Note – No Vertical Fin

Smooth Underside

Shielded Exhaust

Engine Intake on Upper Surface with Aperture Shielded from Below

Figure 7.8   UCAV shape

However, it is also interesting to ponder what type of weapon might be deployed against them in a military operation. Suffice it to say that these systems are still of research status.

NAV systems are looking even more to the future and are envisaged as carrying optical payloads as small as 30 g within an air vehicle of MTOW of 100 g and with dimensions not exceeding 40 mm. If this becomes possible, their detection will require even more futuristic means.

7.5.5   UCAV

The same rules for signature suppression apply as for the previous aircraft types. As with the others, to suppress a radar return, it is necessary to ensure that there are no apertures on the lower surface of the airframe. To this end, as shown in Figure 7.8, the turbine engine air intake and exhaust is set into the upper surface of the wing.

This arrangement also benefits the reduction in thermal and acoustic signatures since radiated heat and sound from the intake and exhaust can only be upwards. The exhaust gases are cooled as far as is possible and any heat transfer to the wing will be limited to the upper surface.

7.5.6   Conclusions

(a) Military Operations

For military operations it may be concluded that HALE UAV can currently escape from physical countermeasures due to their extreme operating altitudes. However, historically, all military offensive equipment has been rendered obsolete/vulnerable by the development of countermeasures which have spurred the development of new offensive equipment and so on. Whilst research is in hand to extend the HALE UAV operating altitude, so it can be expected that new countermeasures will inevitably follow.

MALE systems and medium-range systems are militarily more vulnerable to armed interdictor aircraft, either manned or unmanned.

Close-range UAV, operating at relatively low levels over the battlefield, can be most vulnerable to shoulder-launched weapons or even small arms fire unless they can remain covert.

It would appear that certain configurations of rotorcraft offer the best chance of achieving an effective level of 'stealth' in this role.

MAV and NAV may be more at risk from natural elements such as wind and precipitation than from human intervention.

UCAV in their upper altitude approach mode may remain undetected and invulnerable to attack. It is in their descent to lower altitudes in their attack mode where they may become more at risk and may have to rely upon their high speed and manoeuvrability to evade attack by missile or manned or unmanned interdiction aircraft.

(b) Civilian Operations

For civilian operations, issues other than 'stealth' may be paramount, including survival in the natural elements, but also operating in airspace populated by manned aircraft (see Chapter 18).

However, there are many diverse civilian roles for UAV (Chapter 24) many of which require the UAV to operate at low level for long periods. Here, the low visual and acoustic signatures of the close-range aircraft are most necessary to minimise disturbance to the human and animal populations.

(c) Night-time Operation

The above discussion has been limited to day-time operation. However, the arguments generally hold true in darkness with respect to all signatures except the visual ones. Here the ambient light may result from over-flying a well-lit rural area or from moonlight. In this case, aircraft contrast can be reduced through using a darker matt surface finish instead of matt light grey.

Reference

7.1. *Vision and Acquisition*. Overington, Crane, Russak & Company Inc., 1976

8

Payload Types

First, it is necessary to define what is meant by 'payload'. In times past, the term probably arose from the load carried by an aircraft the transport of which was paid for by the customer. In that context, the author prefers to consider the 'payload' as only that part of an aircraft which is specifically carried to achieve the mission, i.e. that part often known as 'rôle-fit'. The aircraft should be capable of flight with the payload removed.

For civilian operations, for crop-spraying for example, the payload would comprise the spray fluid, its tank and pumping and distribution beams and nozzles.

In the earlier days of UAV systems, and perhaps even sometimes today, some manufacturers declare a 'payload mass' which includes not only the mission payload (rôle-fit) but the AFCS and its sensors and even the radio communication systems. This, of course, can mislead by vastly inflating the perceived payload-carrying ability of the aircraft, especially in mini-UAV, and the potential customer is advised to have the payload claim clarified.

There is often a trade-off between the mass of payload carried and the mass of fuel carried. In this case the manufacturer should specify the 'payload–range' graph of the aircraft.

Accepting the definition of 'payload' as only that part of the UAV required for a specific mission, it is interesting to ask the question – 'which comes first, the mission, the aircraft or the payload?'

In the development of an ideal system the customer's mission(s) requirement will determine the optimum combination of payload sensor and platform (aircraft) to perform a specific rôle. A new total system would then be designed to include a new control station and a new aircraft designed to carry and support the payload at the required speed, endurance and atmospheric conditions etc. This would happen in a perfect world.

In reality, a UAV system usually will be required to be capable of a range of different roles, requiring different payloads and different capability in the aircraft. Thus compromises will be necessary.

Payloads may be required to house a number of different sensors or, in the interests of saving weight and reducing the level of cost which is exposed to loss (the cost of some payloads may exceed the cost of the remainder of the equipped aircraft) single-sensor payloads may be designed to be quickly interchangeable between missions.

In the author's experience, once an aircraft and its payload have been developed for a specific mission, other potential operators may then ask 'with that aircraft and payload, could we do that?' Others will ask if a payload that is required for another specific task could be developed, integrated into and successfully operated in an available aircraft.

Hence several processes occur. New missions demand new payloads. New payloads can often be developed more rapidly than new platforms to carry them. The development and certification of a new aircraft with its control station is not cheap and may take several years to reach fruition.

Therefore, with a number of platforms (aircraft) and control stations now already in existence, it has become more the norm to develop a payload for a mission and concurrently to select the most appropriate platform, to which it will subsequently be fitted, to form a system. This may require considerable thought in the structuring of the payload and also the possibility of accepting a less than optimum systems solution, but with the greater 'malleability' of electronics it will generally be a quicker, cheaper and, initially, more reliable solution than the development of an entirely new, dedicated system.

Hence, especially with the rapid advances in electronics technology, new developments in payloads may be expected more frequently than new developments in aircraft. Only occasionally may a new requirement be so different in its mission as to need a radically new total system.

Payloads will be considered in two basic types:

a) sensors, cameras, etc. which remain with the aircraft, and
b) dispensable loads such as armament for the military and crop-spray fluid or firefighting materials for civilian use. There may also be a future requirement for UAV to deliver mail or materials into difficult terrain.

8.1   Nondispensable Payloads

The approach taken here is to describe the various payloads whose design will be driven by several factors which will include:

- the range, endurance and altitude required in the operation of the platform;
- mass and drag constraints imposed by a selected platform performance;
- installation and field-of-view constraints;
- type of target or surveillance modes;
- range and area of surveillance needed;
- resolution of the imagery;
- discrimination of the target from the background;
- need for tracking (track while scan, lock-on, etc.);
- need for map data base correlation;
- need for recording;
- need for superimposed latitude/longitude, date, time data for admissible evidence;
- need for weapon-aiming interface.

8.1.1   Electro-optic Payload Systems

These range from simple monochrome or colour single-frame cameras through colour TV systems, low-light-level television (LLLTV), (also sometimes abbreviated to LLTV), thermal imaging video systems to multi-spectral systems.

Considerable advances in miniaturisation have been made in all of these in the past two decades and their development is continuing so fast and in so many forms that it is pointless to attempt to list them here. Reference should be made, of course, to the several manufacturers offering devices for both civilian and military markets.

'Daylight' Cameras

Optical, or 'visible light' cameras operate in the 0.4–0.7 μm wavelength range. The human eye sees light, generated from the sun, reflected at different frequencies (colours) from different objects, thus providing recognition of patterns.

The light is focused through ground-glass lenses onto receptors (charged-coupled devices, CCD) from which the image is transformed into a format which can be displayed onto a monitor or coded on to a radio carrier frequency for transmission to a receiving station.

Low-light-level Cameras

LLL cameras function in the same manner as standard optical cameras, but are fed an amplified level of light, using fibre optics which accept light from a larger area and concentrate it onto the camera lens. The remaining process is the same as 'ordinary' optical cameras, but LLLTV cameras can provide images at as much as one-tenth of the light necessary for the operation of ordinary cameras.

Thermal Imagers

The infrared, or heat radiation, spectrum covers longer wavelengths – in the range of about 0.7 μm (near infrared) to 1000 μm (far infrared). The boundaries between the optical and infrared spectra at one end and the infrared and, still longer wavelength, radio spectra at the other end are not precise and are open to slightly different interpretations, depending upon the authority. The infrared spectrum is further sub-divided into divisions. Again, depending upon the authority, the number of divisions and their names will differ. The most common division is three, but four and five divisions are to be found.

For thermal imaging cameras, only the two divisions of about 3–5 μm and about 7–15 μm are significant. These are usually known as short-wave IR and long-wave IR respectively. This is because absorption of infrared emission by water molecules in the atmosphere is at a minimum at these wavelengths, which provide a 'window' through which heat can be transmitted most effectively.

Unlike light which is reflected from objects, heat is principally emitted by the object itself, and which is not dependent upon being illuminated. The object(s) continue to emit heat in the dark. The intensity of heat radiated from an object is proportional to the ability of the object to radiate (emissivity) and the temperature of the object. The temperature of the object will be determined by it having been heated by external means (the sun's radiation for example) or from internal generation of heat (internal combustion engines or human bodies for example).

The heat radiation is focused onto special receptors in the camera which convert it into a format which is displayed on a monitor in monochrome which is recognisable by the human eye. The objects emitting the greatest intensity of heat are usually presented as the darkest (black) in the greyscale, i.e. known as 'black-hot'. Many cameras have a function whereby the presentation can be switched from 'black-hot' to 'white-hot' and back again at the operator's wish. This can assist with interpretation of a scene (Figure 8.1).

Different conditions can determine the clarity of the presented picture. During the day the landscape will be heated by the sun so that objects with a greater ability to absorb heat will become progressively hotter than neighbouring objects. As the sky and air temperatures reduce at nightfall, those hotter objects are likely to transmit heat more intensely. Thus greater contrast will usually be found on thermal imagers then than just before dawn.

Probably the greatest 'enemy' of thermal imaging is extended rainfall since that has the effect of cooling all inanimate objects and severely reducing contrast.

Almost all current commercial cameras now operate in the 8–14 μm (long-wave) spectrum rather than the 3–5 μm (short-wave) spectrum as this offers the better penetration of other atmospheric contaminants such as carbon in smoke. It does require, however, that heat 'rays' are focussed onto receptors through lenses manufactured from germanium. Germanium is a heavy and quite rare and therefore expensive material (about £400/kg) and the lens has to be covered with an anti-reflective material, a process which is also expensive.

Hence the better-performing long-wave cameras are far more expensive than the less effective small, short-wave, short-range cameras often used in micro-UAV. However, considerable research is underway to improve the performance of the short-wave systems and in the longer term the situation may change.

Figure 8.1   Thermal image (white-hot)

In the past, in order for the infrared receptors to remain cool to give good image discrimination in spite of absorbing heat, they were cooled by a cryogenic, gas-refrigerant system and required regular replacement of gas cylinders. The cryogenic system is heavy and expensive and replacement in the field of gas cylinders in the field is a logistic burden.

An alternative cooling method is by use of what is often known as a 'reverse Stirling engine'. In this method an electric motor drives an air compressor and the heat is removed from the compressed air with an air-to-air heat exchanger. The air is then allowed to expand, cooling in the process, and applied to the receptors. This method removes the logistic burden and the need for the handling precautions of the cold liquid gas cylinders, but adds complexity to the system. The degree of cooling achieved is likely to be less than with the gas cylinders, producing a less sharp image.

Although still occasionally used, as cooled systems still give the sharper image, the technology behind the lighter, less expensive, uncooled systems has progressed to make them usually the preferred choice.

Resolution

The area of the receptor of the camera and the number of picture elements, pixels, that it contains will determine the acuity or resolution of the image. The larger the receptor area, the more light (or heat) it will accept and so will operate successfully at lower levels of target illumination. The more pixels receiving light from a target area so will the sharpness of the image be improved.

For a given number of pixels on a receptor, a greater number can be deployed per area of target by focusing them onto a small area of target by use of a 'zoom', 'telephoto', i.e. narrow field-of-view (FOV) lens. However, the use of a narrow FOV lens will require a more accurate stabilisation of the sight-line than will a wider FOV lens or the aircraft will have to approach more closely to its target to obtain a sufficient sharpness of image.

Working rules for the determination of the necessary resolution of an image vary somewhat, depending upon the source, but an approximate idea is given by the Johnson criteria which attempt to define the number of pixels needed in two coordinates for each level of discrimination with a 50% probability of success, definition of these levels being:

Detection – something is there,

Recognition – (for example) it is a vehicle,

Identification – it is a Humvee, Landrover, Centurion tank, etc.

Use of the Johnson criteria is somewhat open to interpretation, however. One point of interpretation is that detection of a target is possible if it is covered by two pixel-pairs, recognition by four pixel-pairs and identification by eight pixel-pairs. Another is that the object area be covered by 6, 80 and 200 pixels respectively.

It is generally recognised that such simplistic criteria can lead to very optimistic conclusions since much depends upon the level and direction of illumination of the object, its orientation, its contrast with the background and the type of background, i.e. cluttered or uncluttered.

A more sophisticated model, the targeting task performance (TTP) is proposed in Reference 8.1. The means by which this more complex method may be employed is another issue.

For the purpose of this book it may suffice to say that in a typical reconnaissance operation the detection of a target will first be achieved by viewing the larger area through a wide FOV lens. Zooming in to a narrower FOV or closing the UAV towards the target, to put more pixels over it, should lead to its recognition. Finally identification will be achieved by reducing the FOV still further by further zooming or moving the UAV even more closely to the target.

Identification of a target from a longer range will require a well-stabilised, narrow FOV lens, focusing onto a receptor with a large number of pixels. A higher quality of image will be obtained by a more expensive, larger and heavier system, possibly requiring greater electrical power for its operation.

Thus the heavier, more expensive equipment, whether it be electro-optical, thermal or radar imaging or a combination of these, will require to be carried by a larger, heavier, more expensive UAV which will have to stand off further away from the area of survey for reasons of overtness and survivability than will its smaller UAV cousin. The smaller cousin will have to approach more closely to the target to achieve results with its lighter, less capable equipment.

As a generality, the former extreme will be seen as the longer-range MALE or HALE UAV, with the latter the mini or close-range UAV. Hence compromises have to be made though steady advances in electronic technology will continue to make these less onerous.

Lenses

Lenses for visual light cameras are ground from high-quality optical glass and the majority of cameras employ zoom lenses. These, typically, may offer a field of view range of from a wide-angle of about $45 \times 60°$ up to a telephoto of, say, $1 \times 1.3°$.

As mentioned above, germanium is heavy and expensive, so thermal cameras may compromise by having a general purpose $30 \times 40°$ lens and another, possibly $15 \times 20°$ lens which move into position as commanded.

8.1.2   Electro-optic Systems Integration

Mounting

Electro-optic systems may be mounted in one or more of three ways.

a) Essentially forward-looking from a mounting in the nose of the aircraft for purposes of obstacle avoidance. A nose-mounted camera will preferably be mounted on gimbals with their actuators to provide a field of regard in elevation, to cover elevation angles usually measured from the vertical downwards. The range will be upwards and forwards (positive) to above the horizon and possibly upwards and rearwards (negative). Gimbals may also be added to allow a limited pan range in azimuth for reconnaissance/surveillance purposes, and in roll for purposes of image stabilisation.

b) A rotatable turret is mounted beneath the aircraft to obtain a 360° azimuth field of regard for reconnaissance/surveillance. Elevation and roll gimbals, carrying the sensor(s), are mounted with their actuators within the turret.

c) A system applicable only to specific types of rotorcraft such as the Canadair CL84 and the M L Aviation Sprite which are symmetric in plan view. This type of aircraft is essentially a 'flying turret' and 360° of azimuth pan is obtained by rotating the complete aircraft. Actuated elevation and roll gimbals carrying the sensors are included.

Some aircraft missions may require both (a) and (b) installations. It is less likely that a plan-symmetric helicopter (PSH) will require two installations, but such a configuration would be possible.

Integration

In addition to physical integration of the sensing system, the payload design and integration will need to accommodate the processing functions associated with detection and reporting of 'targets'. Most of these will need an integration with the aircraft automatic flight control system (AFCS) and navigation system. This applies to the ability to lock the sensor sight-line on to a target whilst the aircraft is moving, a correlation with a map data base and the need to superpose position and time data onto the recorded sensor image for use in military missions or to provide admissible evidence for legal purposes.

As an example of (a), a nose-mounted payload, a Controp Payload in a Aeronautics Orbiter is shown in Figure 8.2 with an example of position and time data superposed on its sensor image.

There are many different types of electro-optic payload systems available from several suppliers and it is beyond the scope of this book to describe them in detail. Recourse should be made to those suppliers for

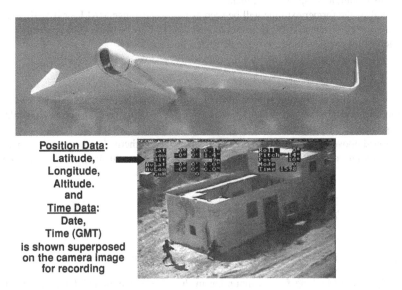

Position Data:
Latitude,
Longitude,
Altitude.
and
Time Data:
Date,
Time (GMT)
is shown superposed
on the camera image
for recording

Figure 8.2   EO payload in Orbiter UAV. (Reproduced by permission of Controp Precision Technologies Inc.)

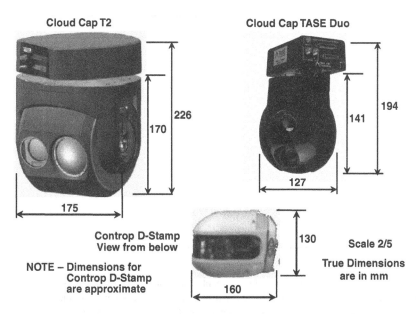

Figure 8.3 EO turrets for close- and medium-range UAV. (Reproduced by permission of Cloud Cap Technologies and Controp Precision Technologies Inc.)

information. However, a small selection of gimballed, turreted payloads is shown, with their dimensions, in Figure 8.3 for installation in the smaller, close-range and medium-range UAV.

Some available data relating to them is shown in the following table:

	Cloud Cap TASE Duo	Cloud Cap T2	Controp D-Stamp
E/O camera	Sony CCD	Sony CCD	CCD
Pixels	380 K	380 K	
Lens	26 × optical	26 × optical	10 × optical
FOV [°]	40 × 55 to 1.7 × 2.3	40 × 55 to 1.7 × 2.3	35 × 47 to 3.9 × 5.2
Thermal imager			
Pixels	324 × 256	640 × 480	384 × 288
Lens	35 or 19 mm		35 or continuous zoom
FOV [°]	20 or 36	14	15, 36°–9°
Laser pointer/illumin	No	Yes	No
Gimbal limits [°]			
Azimuth/roll	Azimuth continuous	Azimuth continuous	Roll +/− 60
Elevation [°]	+23 to −203	+15 to −195	+70 to −30
Electrical power [W]	22 max	100 max	8
Overall mass [kg]	1.05	2.6	0.75; 1.00; 1.200

Controp offer alternatives in single-sensor lightweight payloads. Interchangeable alternatives to D-Stamp are U-Stamp and U-Stamp-Z with thermal imager sensors. All three Stamps can be installed as nose-mount or belly-mount.

Two of the heavier and more capable payload systems appropriate to HALE and MALE UAV are shown in Figure 8.4 with data relating to them shown in the following table.

	Wescam X15	Controp DSP1
E/O camera	CCD	CCD
Resolution	470 TV lines	480 TV lines
Lens	× 23	× 20 and × 3
FOV[°]	46 × 35 to 2 × 1.5	13.6 × 10.2 to 0.23 × 0.17
Thermal imager		
Pixels	640 × 512	320 × 256 or 640 × 512
Lens	4 steps	Continuous Optical Zoom × 36
FOV [°]	26.7 × 20 to 0.36 × 0.27	27 × 20.6 to 0.25 × 0.19
Laser ranger/illumin	Yes	Optional
Gimbal limits		
Azimuth[°]	360	360
Elevation [°]	+90 to –120 from vertical	+70 to –115 from vertical
Electrical power [W]	280–900	110
Overall mass [kg]	45	26

Note that both turrets offer a number of equipment options and the data shown is for example only.

A typical Wescam Turret is seen in Figure 8.7 installed beneath a General Atomics Predator.

As an example (c) of a plan-symmetric helicopter (PSH), an M L Aviation Sprite is shown in Figure 8.5. An internal view of a Sprite payload, with outer cover and window removed, is shown in Figure 8.6.

Stabilisation

Whichever type of camera is used, in nose-mounted or turret-mounted configuration, in order to obtain a useful quality of image, it will require that the sight-line be directed to, and stabilised in, angular space at whichever angle is demanded. This is achieved by the rotation of the gimbals by their actuators in response to commands from the stabilisation system.

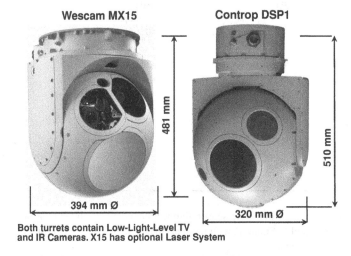

Wescam MX15   **Controp DSP1**

481 mm

510 mm

394 mm Ø

320 mm Ø

Both turrets contain Low-Light-Level TV and IR Cameras. X15 has optional Laser System

Figure 8.4   EO turrets for HALE and MALE UAV. (Reproduced by permission of Controp Precision Technologies Inc.) *Source*: Wescam

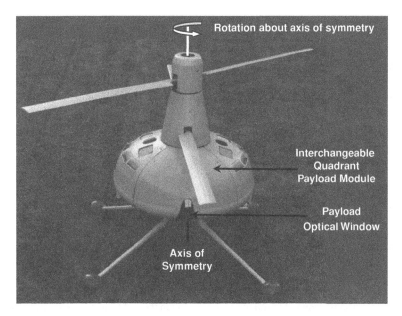

Figure 8.5   Sprite UAV – flying turret

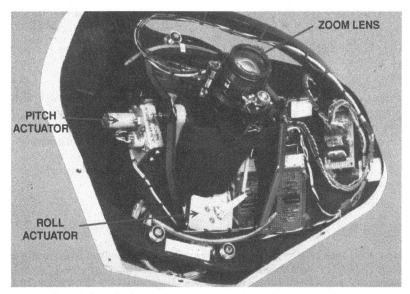

**Contains Colour TV camera with 10/1 zoom lens.
Field of regard 110° in elevation
Stabilised in Pitch and Roll**

Figure 8.6   Sprite interchangeable payload module

The narrower the lens field of view so the more precisely must that sight-line be held. Stabilisation of the sight-line is required to off-set the effects of both aircraft manoeuvres and displacements of the aircraft due to turbulent air. It is most probable that three-axes stability will be required.

There are two basic methods of maintaining a stabilised sight-line.

 (i) The whole camera assembly is mounted on a stabilised platform with attitude and rate sensors;
(ii) A camera assembly is mounted on gimbals remote from the attitude sensors.

In method (i), the camera system payload with its gimbals is mounted upon a platform which is gyro-stabilised to remain horizontal. Quick-acting actuators are tasked to correct any movement away from the horizontal. The camera sight-line is driven by actuators to adopt the required direction of view relative to the horizontal base.

One disadvantage of this system is that the interchange of payloads may require replacement of the stabilised platform with the payload or, at best, a sophisticated control of the mass and balance of individual payloads to ensure the correct functioning of the platform. In the latter option, replacement of the payload will take much longer than for method (ii) and both options will be more expensive than (ii).

In method (ii), the whole system is contained within an exchangeable payload module with the gimbals' base fixed to the structure of that module. If the system is fitted with elevation and azimuth pivots, then these same pivots may be used in driving the sight-line for both direction and stabilisation. A third pivot may also be required in roll in order to maintain the sight-line steady in that axis to provide a 'horizontal' picture. The aircraft attitude data is taken from the aircraft stability and control system and used to drive the gimbal actuators appropriately.

The advantage of method (ii) is that it greatly facilitates quick changes of payload and is less expensive than method (i). However, it is best suited to aircraft with a low response to air turbulence since there may be a lag in payload response where the aircraft is subject to large and sudden deviation from course.

Transmission

As discussed in Chapter 9, images and/or data from the various payloads must be transmitted, usually by radio, to the control station and/or elsewhere in a networked system. The output from the several different types of sensors within the payload will require processing and converting to radio signals. Different sensors will require different amounts of radio bandwidth and at different data rates. As the use of UAS grows, a larger demand is made on the available radio bandwidth and this may be at a premium. Therefore it is necessary that as much processing of the signals as is practical is carried out within the UAV in order to reduce the demands made upon the communication system.

8.1.3   Radar Imaging Payloads

Unlike these camera systems which are 'passive', i.e. they receive sunlight reflected from the target objects, or heat emissions from the objects, radar is an 'active' system which emits pulsed radio beams and interprets the returned reflections from those objects.

Unlike the camera systems, it can see through clouds. This is an operational advantage. Being an active system, however, the emitted energy is detectable and, in certain circumstances, is an operational disadvantage. The image resolution, though being steadily improved by developing technology, is not as good as that of the electro-optical camera systems.

The radar system used for ground target surveillance from UAV is known as synthetic aperture radar (SAR). The same antenna is used for emitting the outgoing signal and receiving the returned signal. It is mounted looking sideways (normal to) the direction of flight, and is scanned downwards from a depression angle of typically 60° below the horizontal to the vertical. Radar wavelengths of typically 5–15 cm are used.

The acronym SLAR will be met which stands for 'side-looking airborne radar'. This operates with an antenna of discrete length which may be as long as 6 m in order to obtain a useful spatial resolution which is proportional to the antenna length. As this is incompatible with an installation in most UAV, an electronic 'trick' is used to increase the effective antenna length by electronic means and allow a much smaller physical antenna to be adopted. Hence the use of the 'synthetic aperture' nomenclature. (Aperture is analogous to field of view).

The operation relies upon the UAV moving forward at a steady speed so that the electronics can accurately integrate the returning signals over the traversed range.

The radar pulses undergo various degrees of backscattering when they reach an object. The amount of returned signal determines the relative lightness of the image.

A smooth surface, such as still water, at low angles to the 'look' direction of the pulse will reflect most of the pulses away from the receiver so that that part of the image will look dark. A rough surface will scatter the pulse beam into many directions, some of which will return to the radar receiver and so the image will appear darker. A surface normal to the look direction will return the greatest number of pulses and so their image will appear darkest.

The overall image therefore produces a monochrome picture which is not easy to interpret, but test flights over known terrain have accumulated data to assist in providing solutions. This is a simplified explanation of the technique and other, more sophisticated, measures have been developed to improve the interpretation of the images.

Recent years have seen considerable advances in the miniaturisation of SAR systems for operation in UAV, with their volume and mass being steadily reduced. As examples of these, the General Atomics (GASI) Lynx I and II, and the Scandia International Laboratories (SNL) Mini-SAR with the Insitu Nano-SAR are shown in Figures 8.7 and 8.8 respectively.

GASI, by miniaturising the electronics and using composite materials for the antenna and its mounting, have reduced the mass of the Lynx I (52 kg) to that of the Lynx II (34 kg) without sacrificing any performance.

Accepting a reduction in range and resolution performance has enabled these two companies to produce SAR systems of only 13 and 0.5 kg respectively. This makes them available now for use in close-range and mini-UAV systems.

Another development has been that of multi-spectral imaging where images received from a range of frequencies are fused together. This technique has the advantage that elements invisible to one frequency are seen by another. The result of the integration is a more inclusive picture.

8.1.4   Other Nondispensable Payloads

With increasing deployment of UAS in both military and civilian operations, the possibility of further uses is seen. This has given rise to a requirement for an extended range of payload types. Some are briefly discussed below.

Laser Target Designation

There is a military need to illuminate ground targets for attack by laser-guided missiles. Although this has been done by the manned aircraft which launches the missiles, it is better accomplished by one aircraft (manned or unmanned) laser-marking the target for others which release the weapons.

Laser designation has also been undertaken by Special Forces on the ground but, apart from being a very hazardous task, ground objects or terrain can obscure the path of the laser illumination. Generated

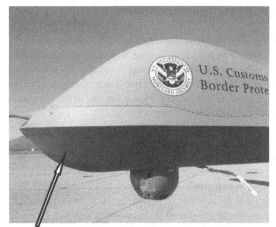

LYNX I installed is installed in Predator Nose

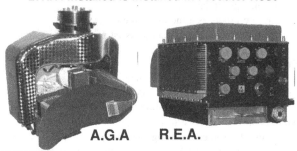

A.G.A   R.E.A.

LYNX II is a lighter development of Lynx I

Figure 8.7 Lynx synthetic aperture radar systems

INSITU / ImSAR
Nano-SAR System

Sandia National Laboratories Inc. Mini-SAR

The Antenna Gimbal Assembly is seen on the left
of the picture with the Radar Electronics Assembly
on the right.
The combined assembly has a mass of about 13 kg.
It offers a resolution of 10 cm at a range of 15 km.

Nano-SAR operates at
X-Band offering resolution
of 35 cm at a range of
one kilometre and a system
mass of less than 0.5 kg.

Figure 8.8 Lightweight SAR developments (Reproduced by permission of Sandia National Laboratories). *Source*: ImSAR LLC

from the ground, it will not be at an optimum angle for the incoming missile or projectile and may be ineffective. It is a task better done by a small UAV and is discussed further in Part 3.

The UAV payload will consist of a laser generator ganged to an appropriate electro-optic TV camera with an appropriate sight-line for the UAV operator. Very accurate control of laser pointing and good stabilisation of the payload is necessary. Such a payload should be achievable within a 6 kg mass limit.

Pollution Monitoring

Armies use devices for detecting pollutants in the form of nuclear, biological or chemical (NBC) contaminants on the ground or in the atmosphere. This is done using hand-held equipment which sucks in air samples and analyses them before displaying the data. This is a dangerous occupation and time consuming. It is readily possible to mount such a device in a UAV to carry out the task more expeditiously and without danger to personnel.

The same type of equipment may be applied by civilian authorities for monitoring contaminants from, for example, industrial waste. The payload would include an appropriate electro-optic TV camera system, and the UAV would transmit the data back to a control station. Such a payload is readily achievable within an overall 5 kg mass.

Public Address System

A loudspeaker system and TV camera can be mounted in an integrated payload which, again could be achieved from 5 kg mass depending upon the sound level required.

Radio Relay System

A UAV positioned at an appropriate altitude, carrying a payload of radio receiver, amplifier and transmitter can significantly increase the range of radio communication for both military and civilian authorities operating in the field.

Electronic Intelligence

A covert operation with a UAV carrying a radio receiver capable of frequency scanning can intercept enemy radio transmissions for intelligence purposes.

Radar Confusion

A radar transmitter in a UAV is able not only to jam a radar receiver, but also confuse it into believing that objects, forces, etc. are present when they are not.

Magnetic Anomaly Detection

Such systems are readily convertible for carriage by UAV and can be used for various purposes including the location of, for example, sunken wrecks at sea.

Alternative or Additional Payloads

The discussion above outlines some of the payloads already possible and many more are likely to arise as the versatility of UAV systems becomes even more apparent. Depending upon requirements and the

**General Atomics Inc MQ9 "Reaper"
armed with Guided bombs and Hellfire missiles**

**Northrop Grumman RQ8 "Fire Scout"
Armed with rockets in side-mounted pods**

Figure 8.9   UAV carrying ordnance

carrying capacity of the UAV, the payloads may be carried as alternative, readily changeable payload modules or as a number of payloads mounted on the same air vehicle.

Obviously there is usually a trade-off between the number and total mass of payload(s) carried and vehicle range or endurance. On a vehicle with multi-payload capability, one or more of the exchangeable payloads could be an extra fuel cell.

Payload Development

As electronic technology advances, lighter, smaller, more capable payloads will be developed, as is already being seen, to enable missions to be carried out using smaller, cheaper, less-vulnerable and easier-to-operate platforms.

8.2   Dispensable Payloads

In the civil field these may include agricultural crop-spraying with pesticides, fungicides or possibly fertilisers and anti-frost measures. Provided that the UAV has an adequate dispensable payload, water or other fire suppressant material may be carried and dropped. Other dispensable payloads may include flares, life-rafts and other materials all of which is discussed further in Part 3.

In all of these, care must be taken in the design to ensure that material cannot be inadvertently dropped and is released cleanly,without upsetting the balance of the aircraft, and without entering vulnerable parts of the aircraft.

Thought has to be given as to the method of loading the material and its storage medium if the material is toxic.

In the military environment, the dispensable loads will usually be armament: bombs, rockets or missiles. These will usually be under-wing-mounted on fixed-wing aircraft and side-mounted on rotary wing aircraft. MA means of weapon-aiming will be necessary and integrated with the electro-optic sensor(s).

Precautions to be taken in providing for the carriage and release of military ordnance will follow the design and manufacturing requirements and procedures for military manned aircraft. These are well documented in the appropriate STANAG and Defence Standard documents. The equipments such as release units need be no different from those currently in use on manned aircraft.

There are different issues, however, on the protocol of releasing weapons to strike targets and these issues are discussed in Part 3.

Reference

8.1. Richard H Vollmerhausen and Eddie Jacobs. 'The Targeting Task Performance Metric'. *Technical Report AMSEL-NV-TR-230*, US Army CERDEC, Fort Belvoir, VA 22060, 2004.

9

Communications

As introduced in previous chapters, the communications between the UAS control station (CS) and the UAV consist primarily of an 'up-link' transmitting command and control from the operators to the UAV (or UAVs in multiple operations) and a 'down-link' which returns data showing UAV, including payload, status and images from the UAV to the CS and to any other 'satellite' receiving stations. The UAV status data is also known as 'housekeeping' data.

The maintenance of the communications is of paramount importance in UAS operations. Without the ability to communicate, the UAS is reduced merely to a drone system and loses the versatility and wide capability of the UAS. Loss of communication during operations may result from:

a) failure of all or part of the system due to lack of reliability,
b) loss of line-of-sight (LOS) due to geographic features blocking the signals,
c) weakening of received power due to the distance from the UAV to the control station becoming too great,
d) intentional or inadvertent jamming of the signals.

These aspects will be addressed later. The specifications for communications performance will include two fundamental parameters:

(a) 'data rate' which is the amount of data transferred per second by a communications channel and is measured in bytes per second (Bps), and
(b) 'bandwidth' which is the difference between the highest and lowest frequencies of a communications channel, i.e. the width of its allocated band of frequencies and is measured in MHz or GHz as appropriate.

9.1 Communication Media

The communication between the GCS and aircraft and between the aircraft and GCS may be achieved by three different media: by radio, by fibre optics or by laser beam. All are required to transmit data at an adequate rate, reliably and securely. All have been attempted.

By Laser

The laser method seems currently to have been abandoned, principally because of atmospheric absorption limiting the range and reducing reliability.

Unmanned Aircraft Systems – UAVS Design, Development and Deployment   Reg Austin
© 2010 John Wiley & Sons, Ltd

By Fibre-optics

Data transmission by fibre-optics remains a possibility for special roles which require flight at low altitude, high data rate transmission and high security from detection and data interception. Such a role might be detection and measurement of nuclear, biological or chemical (NBC) contamination on a battlefield ahead of an infantry attack.

The fibre would be expected to be housed in a spool mounted in the UAV – not in the ground control station (GCS). This is because it must be laid down onto the ground rather than being dragged over it, when it might be caught on obstacles and severed. The method is probably better suited to VTOL UAV operation, and necessarily limited in range to a few kilometres.

Data would be transmitted securely back to the GCS and at the completion of the mission the fibre would be severed from the UAV which would climb and return automatically to the GCS. Such a system was simulated, designed and partly constructed in 1990, under US Army contract, for the Sprite UAV system.

By Radio

Currently, the only system known to be operative is communication by radio between the UAV and its controller, directly or via satellites or other means of radio relay.

9.2   Radio Communication

As discussed in Chapter 5, the regulation of UAS, including radio communication, is effected in the USA by the FAA which is advised by the Radio Technical Commission for Aeronautics (RTCA). In Europe EASA is the overall regulating authority, and it delegates various aspects of regulation in the UK to CAA which again is advised by OFCOM, the authority within the UK for the allocation of radio frequency. This is discussed in more detail below.

Radio Frequencies

Electromagnetic waves generally considered usable as radio carriers lie below the infrared spectrum in the range of 300 GHz down to about 3 Hz (Figures 9.1 and see Figure 9.4)

Frequencies in the range 3 Hz (extremely low frequency, ELF) to 3 GHz (ultra-high frequency, UHF) are generally considered to be the true radio frequencies as they are refracted in the lower atmosphere to curve to some degree around the earth's circumference, increasing the effective earth radius (EER) by up to 4/3.

Frequencies above this range, 3–300 GHz (super-high frequency, SHF and extremely high frequency, EHF) are known as microwave frequencies and, though they may be used to carry radio and radar signals, they are not refracted and therefore operate only line-of-sight.

It is necessary to transmit high rates of data, especially from imaging-sensor payloads, from the aircraft to its control station or other receiving station. Only the higher radio frequencies are capable of doing that and, unfortunately, these depend progressively towards requiring a direct and uninterrupted line-of-sight (LOS) between the transmitting and receiving antennas. There is therefore a compromise to be made when selecting an operating frequency – a lower frequency, offering better and more reliable propagation, but having reduced data-rate ability and the higher frequencies capable of carrying high data rates, but requiring increasingly direct LOS and generally higher power to propagate the signal.

UHF frequencies in the range 1–3 GHz are, in most circumstances, a desirable compromise, but due to increasing demand by domestic services, such as television broadcasting, for the use of frequencies in the VHF, UHF ranges, the frequency allocation agencies are requiring that communication systems use increasingly higher frequencies into the SHF microwave band of 5 GHz or above.

Band Name (Frequency)	Abbr.	ITU Band	Frequency	Wave Length	Typical Uses
Extremely Low	ELF	1	3-30Hz	100,000km-10,000km	Submarine Communications
Super Low	SLF	2	30-300Hz	10000 - 1000km	Submarine Communications
Ultra Low	ULF	3	300-3000Hz	1000 -100km	Comm. in mines
Very Low	VLF	4	3-30kHz	100-10km	Heart Monitors
Low	LF	5	30-300kHz	10km-1km	AM Broadcast
Medium	MF	6	300-3000kHz	1km-100m	AM Broadcast
High	HF	7	3-30MHz	100m -10m	Amateur Radio
Very High	VHF	8	30-300MHz	10m-1m	TV Broadcast
Ultra High	UHF	9	300-3000MHz	1m-100mm	TV, phones, air to air comm. 2-way radios
Super High	SHF	10	3-30GHz *	100-10mm	Radars, LAN *
Extremely High	EHF	11	30-300GHz *	10mm-1mm	Astronomy *

* Note that these are microwave frequencies and are also used in domestic devices

Figure 9.1   Radio frequency spectra

The radio range in terms of effective LOS available between the air vehicle and the GCS can be calculated by simple geometry which is derived in Figure 9.2, and results in the following expression:

$$\text{LOS Range} = \sqrt{(2 \times (\text{EER}) \times H_1) + H_1^2} + \sqrt{(2 \times (\text{EER}) \times H_2) + H_2^2}$$

where H_1 and H_2 represent the heights of the radio antenna and air vehicle respectively.

For the higher, microwave, frequencies the EER is the true earth radius of about 6400 km, while for the lower, radio, frequencies a value EER of 8500 km is appropriate.

The results are shown in Figure 9.3 for a UAV operating at relatively low altitudes (up to 1000 m) and using radio frequencies. It may be seen, for example, that using a ground-vehicle-mounted transmitting

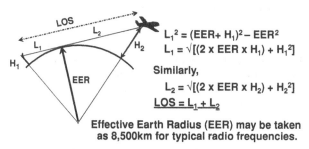

$$L_1^2 = (\text{EER}+ H_1)^2 - \text{EER}^2$$
$$L_1 = \sqrt{[(2 \times \text{EER} \times H_1) + H_1^2]}$$

Similarly,

$$L_2 = \sqrt{[(2 \times \text{EER} \times H_2) + H_2^2]}$$
$$\underline{\text{LOS} = L_1 + L_2}$$

Effective Earth Radius (EER) may be taken as 8,500km for typical radio frequencies.

LOS Range = $\sqrt{[(2 \times \text{EER} \times H_1) + H_1^2]} + \sqrt{[(2 \times \text{EER} \times H_2) + H_2^2]}$
where H1 and H2 represent the heights of the radio antenna and air vehicle respectively.

Figure 9.2   Radio LOS derivation

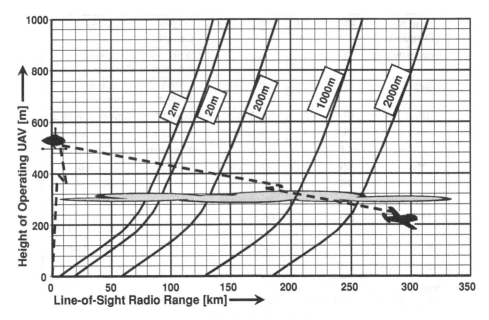

Figure 9.3   Radio line-of-sight

antenna of typical height (say 2–4 m) and even with the UAV at 1000 m, the communication range will be little more than 130 km.

To achieve a greater range (say 600 km), calculation will show that the aircraft would have to fly at a height greater than 20 000 m. For longer ranges, therefore, it is necessary to use an earth satellite or another UAV to act as a relay station.

There may often be clouds at, say, 300 m height and, in order to operate electro-optical cameras the air vehicle must remain below the cloud. In that case, the effective radio range will be little more than 50 km, even over level terrain. So for even short/medium range operations a means of relay may have to be employed.

Radio Frequency Band Designations

There are at least three systems in use to designate frequency bands (Figure 9.1 and see Figure 9.4):

a) The International Telecommunication Union (ITU) designations, shown in Figure 9.1, cover the wide spectrum from extremely low frequencies from 3Hz up to the microwave bands.
b) The Institute of Electrical and Electronics Engineering (IEEE) designations were the original band ranges developed in World War 2, but do not cover the lower radio ranges below HF.
c) The NATO and EU Designations are the more recent series, but do not cover the VHF and HF radio frequencies; (b) and (c) are shown in Figure 9.4.

Reference to bands is made when it is not necessary to refer to a specific frequency or it is inadvisable for security reasons.

Radio/Microwave Frequency Allocation

The international forum for worldwide agreement on the use of the radio spectrum and satellite orbits is the World Radiocommunication Conference (WRC). It is organised every two or three years by the

IEEE		EU, NATO, US ECM.	
BAND	FREQUENCY RANGE	BAND	FREQUENCY RANGE
HF	3 to 30MHz	A	0 to 0.25GHz
VHF	30 to 3MHz	B	0.25 to 0.5GHZ
UHF	0.3 to 1.0GHz	C	0.5 to 1.0GHz
L	1 to 2GHz	D	1 to 2GHz
S	2 to 4GHz	E	2 to 3GHz
C	4 to 8GHz	F	3 to 4GHz
X	8 to 12GHz	G	4 to 6GHz
K_u	12 to 18GHz	H	6 to 8GHz
K	18 to 26GHz	I	8 to 10GHz
K_A	26 to 40GHz	J	10 to 20GHz
V	40 to 75GHz	K	20 to 40GHz
W	75 to 111GHz	L	40 to 60GHz
		M	60 to 100GHz

Figure 9.4   Radio frequency band designation

International Telecommunication Union (ITU) of the United Nations Organization. The conference seeks to make the most efficient use of the radio spectrum and to regulate access to it internationally, taking account of emerging radio communication needs arising from technological, economic, industrial and other developments.

With increasing demand for access to the radio spectrum for commercial, scientific development and other purposes, the conference is attended by telecommunication providers, TV and radio broadcasting and equipment industries. It is equally attended by the military, as defence capabilities are largely dependent on the provision of sufficient frequencies. Well in advance of each conference, consultations in the Frequency Management Sub-committee allow NATO member states to adopt common positions on each agenda item affecting the military, in order to protect Alliance interests in the use of the radio spectrum for military purposes.

NATO contributed to drafting the European Union's Radio Spectrum Policy with a view to maintaining a balance between commercial frequency demands and military spectrum requirements.

Regional coordination of the radio spectrum in Europe is carried out by the European Conference of Postal and Telecommunications Administrations (CEPT).

The coordinating body covering the United States and Canada is the Inter-American Telecommunications Commission.

Frequency managers based at NATO headquarters are actively involved in the work of the European Radiocommunications Committee (ERC) of the CEPT and provide advice on NATO's interests in the military use of radio frequencies.

Since the Communications Act of 2003, the coordinating body within the UK is the Office of Communications (OFCOM).

Although the allocation of the use of frequencies for civilian or military communication purposes is made by the appropriate authority (or authorities) in different countries and attempts are being made to coordinate the allocation worldwide, the allocation for the same specific purpose may yet be different in different countries. This can pose a problem for exporters of UAV systems and it is a wise design initiative to configure the communications system in the aircraft and control station to be modular so that frequency changes for export are facilitated.

The testing of an export system in the field in the manufacturer's homeland, using the export frequencies, may not be possible. Hence early consultation with the local regulator, e.g OFCOM in the UK, is advised.

Radio Range Limited by Power

Having established the radio range, as limited by LOS, and available frequencies for the UAV system, the successful operation of the UAV communication system will depend upon the integration of the various components of the system to supply adequate RF energy to achieve the required range. For this, the system designer will take into account the following factors:

1) Transmitter power output and receiver sensitivity.
   Line losses – a loss of power will result from the escape of energy through imperfect shielding of the coaxial cables and imperfect line-couplers as the RF energy is sent to and from the antennae. Minimising the distance between the antenna and transmitter and receiver is advisable.
2) Antenna gain – antennae can be constructed to focus the RF energy in a specific plane or pattern to produce an effective gain in a particular direction, thus maximising the range obtained with a given power output. Depending upon the application, an omnidirectional or a unidirectional, antenna, such as a Yagi or a narrow beam parabolic dish antenna may be appropriate. Although briefly discussed in Section 9.5, antenna design is a very specialist technology, and antennae are best acquired from specialist companies following detailed discussion of the system requirements and options.
3) Path loss – this is the loss of power that occurs to the signal as it propagates through free space from the transmitter to the receiver. The calculation of the path loss must take into account: the distance that the radio wave travels; the operating frequency since the higher frequencies suffer a greater loss than the lower frequencies; and the height of the transmitting and receiving antennae if either is close to the ground.

For details on the above factors it is necessary to refer to specialist radio publications or to specialists in the field.

Multi-path Propagation

Another problem that may occur is known as 'multi-path propagation' whereby two signals displaced in time by microseconds are received at the image display, causing blurring of the image. This may occur, for example, if the transmission is reflected off nearby obstacles. Either very narrow beam transmission or very sophisticated processing is needed to overcome this problem.

Radio Tracking

One of several means of navigating a UAV is by tracking it by radio. This requires the UAV to be fitted with a transponder which will receive, amplify and return a signal from the control or tracking station or to have the UAV down-link transmit a suitable pulsed signal.

The control station transmit/receive antennae would, in fact, consist of two parallel-mounted off-set directional antennae. A signal processing system then detects whether the signals received by the two antennae are in or out of phase, and command the rotation of the antenna system to bring their signals into phase.

At that point, the antenna system would be pointing directly to the UAV and the UAV azimuth bearing relative to the Control Station (CS) would be known. Depending upon the transmitted beam width in elevation, it may be necessary to have a similar arrangement to ensure continuation of contact in elevation also, though there are other means of maintaining direction in elevation if, for example, the altitude of the UAV is known. The inclined distance of the UAV from the CS is obtained by timing the pulse travelling between the two.

Loss of Communication Link Between Control Station and UAV

The antenna systems of both the CS and the UAV may be capable of scanning in azimuth and/or elevation as appropriate. Thus, following loss of link, and depending upon the transmitted beam-width of each, one would scan for the other, both knowing the last recorded position of the other. In the event that contact was not resumed after a given programmed time, the UAV may be programmed to return to base and, if necessary, recovered using a stand-by short-distance omnidirectional VHF link, especially if the loss was due to failure of the CS primary transmission. This aspect is again a specialist area where appropriate organisations would be involved.

Vulnerability

There are two ways in which a UAV system may be vulnerable. One is that an enemy detection of the signal from either UAV or CS will warn that enemy of the presence of the system. At the least this will eliminate the element of surprise and alert the enemy to the possibility of an impending attack. It may also lead to countermeasures and the destruction of the UAV and/or the CS. The other is that radio transmission between the CS and the UAV may be subject to inadvertent or intentional jamming of the signal.

The risk of the former may be reduced by the use of very narrow beam transmissions and/or the use of automatic or autonomous systems whereby the transmission is only used in occasional short bursts of radio communication.

Signals beamed downwards are at more risk than those beamed upwards unless a sophisticated airborne detection system patrols over the area. This is unlikely unless the confrontation is with a very sophisticated enemy and then the airborne patrol would be extremely vulnerable to countermeasures. Signals beamed down from relay aircraft or from satellites would be more open to detection.

The latter risk may be reduced by three types of anti-jam (AJ) measures:

a) high transmitter power,
b) antenna gain/narrow beam-width,
c) processor gain.

(a) Using high power transmission to out-power a dedicated jammer system in a contest is not very practical, especially for the UAV down-link which will be limited by weight, size and electrical power available.

(b) For higher frequency, LOS links, the available transmitter power can be concentrated into a narrow beam using a suitable antenna as discussed in Section 9.5. This requires the antennae on both CS and UAV to be steerable for the beam to be maintained directed at the receiver. A high gain obtained through use of very narrow beams will require the CS and UAV to know the position of each other very accurately in three dimensions. It will also need the beams to be held in position with great stability. This can be assisted by the receivers of both UAV and control station seeking the maximum RF power to be found on the centre-line of the received beam by appropriately steering the antennae. However, a compromise must usually be accepted with beams having a width within which connectivity can be assured. The beam width from the UAV will usually be wider than that from the CS as the size of antenna which may, in practice, be carried on the UAV will be smaller than that available at the CS.

Lower-frequency, omnidirectional and long-range non-LOS links are at a much greater risk of jamming since there are ample paths for the jammer power to be inserted. Such links must rely on the third type of AJ measures. These measures include frequency hopping of the communication links so that the transmission frequency is randomly changed at short intervals, thus making signals difficult to intercept or to jam.

An alternative is to adopt 'band jumping', where the transmission moves from one band to another, say UHF to S band, at short intervals or when interception or jamming is detected. This, however, requires two parallel radio systems with individual antennae. Such a system was used with great advantage by the Sprite UAV system which enabled the UAV to fly through transmissions such as ship's radars and missile guidance systems without inadvertent interference.

A further technique is the use of a so-called spread-spectrum system, where the signal is spread over a small range of frequencies with 'noise' signals interposed. The receiver is aware of the distribution codes of the noise and is able to extract the genuine signal from it. Not knowing the codes, an intercepting enemy would not be able to decipher the signal.

Antenna and AJ technology is a very complex subject, having the possibility of combining many options – RF frequencies, compounded antennae of different types and sizes and signal processing – outside the scope of this book. The reader is referred to References 9.3 and 9.4 which carry a somewhat more detailed appraisal of the subject, but the design and integration of such systems into a UAS can only realistically be achieved through the cooperation of a specialist organisation in the field.

Multi-agent Communication and Interoperability

So far we have considered only one-to-one communication, i.e. that between one CS and one UAV, which is sometimes known as 'stove-pipe' operation. Whilst this situation may often occur, other agents are likely to be involved, with information being sent to and received by one another to mutual advantage. This arrangement will often be the case for military operation and also may be the situation for some civilian applications. This latter may apply, for example, to policing where a larger area has to be covered than is possible with one UAV and with the information needed at different positions.

Such operations may employ a number of interoperable systems, as illustrated in Figure 9.5, and give rise to the term 'system of systems' (SoS).

Figure 9.5   Interoperable systems

For such SoS and even more so with network-centric systems (see Chapter 22), it is vital that the several diverse systems are interoperable. Previously systems' integrators have relied upon adopting proprietary telemetry and sensor data streams which resulted in the inability of systems to interoperate with each other.

NATO recognised the need to ensure interoperability between the forces of its member nations and recommended that a UAV control station Standardisation Agreement (STANAG) be set up to achieve this. The outcome was NATO STANAG 4586, 'UAS Control System Architecture', which document was developed as an interface control definition (ICD). This defines a number of common data elements for two primary system interfaces.

These are the command and control interface (CCI) between the UAS control station (UCS) and the other systems within the network, and the data-link interface (DLI) between the UCS and the UAV(s). STANAG 4586 defines five levels of interoperability between UAS of different origins within NATO. These vary from 100% interoperability whereby one nation's UCS can fully control another's UAV including its payload, down to being limited merely to the receipt of another's payload data.

STANAG 4586 also refers to other STANAG such as 4545 covering imagery formats, 4575 – data storage, etc. STANAG 4586 is now substantially accepted in the UAS industry and is often called up as a requirement for UAS defence contracts. It is also available for commercial contracts.

9.3   Mid-air Collision (MAC) Avoidance

Another issue which is, in effect, a communications issue is the avoidance of mid-air collisions between UAV and other aircraft in the event that UAV are allowed to operate in unrestricted airspace. Manned aircraft currently operating are required to carry an avionic system known as the Traffic Alert and Collision Avoidance System (TCAS) if the gross mass of the aircraft exceeds 5700 kg or it is authorised to carry more than 19 passengers. The issue is further discussed in Chapter 5.

9.4   Communications Data Rate and Bandwidth Usage

As also noted in Chapters 5 and 8, there is concern that military UAS are currently consuming large amounts of communication bandwidth. If the hopes of introducing more civilian systems into operation are to be realised, then the situation may be exacerbated.

There is a need for the technology, such as bandwidth compression techniques, urgently to be developed to reduce the bandwidth required by UAS communication systems. Much of the work on autonomy for UAV is also driven by the need to reduce the time-critical dependency of communications and the bandwidth needed (Chapter 10).

A high-resolution TV camera or infrared imager will produce a data rate of order 75 megabytes per second. It is believed that with its several sensor systems, including the high-definition imaging sensors required to view potential targets from very high altitudes, a Global Hawk HALE UAS uses up to 500 megabytes per second. The bandwidth required to accommodate this with, for example, anti-jam methods such as spread-spectrum techniques added, will be excessive.

Although shorter-range UAV operating at lower altitudes do not use such a huge amount of bandwidth, there is growing danger that radio interference between systems will limit the number of UAS operable in one theatre. It is therefore desirable that as much data processing as possible is carried out within the UAV and, with bandwidth compression, UAS bandwidth usage can be reduced to an acceptable level. Fortunately, developments in electronic technologies make this possible. For further background to the regulation of radio communication, see Reference 9.1.

To ensure safety from inadvertent interference, there is an urgent need for a dedicated communications band for civilian UAS. Most UAS communications currently operate mostly within the L to C bands along with other users, but UAS air traffic integration working groups such as the EUROCAE WG-73 and the RTCA SC-203 are cooperating in preparing a case for a dedicated bandwidth allocation for civilian UAS. This case is intended for discussion at the ITU world radio congress meeting in 2010–11 in the hope of securing an appropriate bandwidth from that date.

9.5 Antenna Types

Antennae of the same configuration are used both to transmit and to receive RF signals. Unless an omnidirectional antenna is used at the UAV, it will be necessary to mount the antenna(e) in a rotatable turret in order for LOS to be maintained between CS and UAV for all manoeuvres of the UAV. In some cases it may be necessary to install the antenna(e) in more than one position on the UAV.

The most usual types of antennae to be adopted for UAS are:

a) the quarter-wave vertical antenna,
b) the Yagi (or to give it the correct name, Yagi-Uda) antenna,
c) the parabolic dish antenna,
d) and less commonly, the lens antenna and the phased array rectangular microstrip or patch antenna.

These are illustrated in Figure 9.6. However, as previously noted, antenna design is a highly specialised, complex technology and the following must be seen by the reader as an over-simplified elementary introduction. More specialist information on the subject is available in Reference 9.2, but companies that specialise in the design and manufacture of antennas should be involved in their choice at an early stage of any development.

(a) The quarter-wavelength antenna erected vertically is vertically polarised and requires a receiving antenna to be similarly polarised or a significant loss of signal strength will result. This type of antenna is omnidirectional; that is it radiates at equal strength in all directions. Because of this, the received power rapidly reduces with distance. This type of antenna is used in RC model aircraft systems where the aircraft is always within sight of the operator. Their use in UAS will generally be limited to local launch and recovery operations where there is little risk of enemy jamming, and they have the advantage of not requiring the CS and UAV antennae to be rapidly steered to maintain

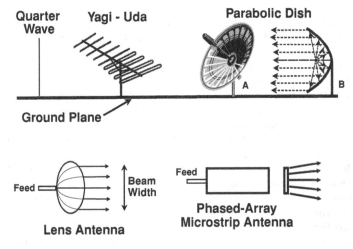

Figure 9.6 Applicable antenna types

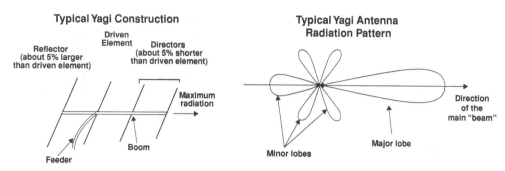

Variation of Parabolic Antenna Beamwidth with Radio Frequency and Antenna Diameter.

Diameter

Frequency	0.3 m	0.6 m	1.2 m	1.8 m	2.4 m	3 m	3.7 m	4.5 m
2 GHz	35	17.5	8.75	5.83	4.38	3.5	2.84	2.33
6 GHz	11.67	5.83	2.92	1.94	1.46	1.17	0.95	0.78
8 GHz	8.75	4.38	2.19	1.46	1	0.88	0.71	0.58
11 GHz	6.36	3.18	1.59	1	0.8	0.64	0.52	0.42
14 GHz	5	2.5	1.25	0.83	0.63	0.5	0.41	0.33
18 GHz	3.89	1.94	0.97	0.65	0.49	0.39	0.32	0.26
23 GHz	3	1.52	0.76	0.51	0.38	0.3	0.25	0.2
38 GHz	1.84	0.92	0.46	0.31	0.23	0.18	0.15	0.12

Beamwidth in Degrees

Figure 9.7   Antenna characteristics

contact in close-proximity manoeuvres. The down-side is that additional equipment, though small and light, must be added to both UAV and CS.

(b) The Yagi-Uda antenna contains only one active dipole element backed up by a number of passive, reflector elements which modify the basic radiation pattern to a predominantly directional beam with, however, small side-lobe radiations. The side-lobes of antennae are the easiest route for jamming RF to enter the system. Therefore, for UAS use, particularly, the antenna designer must apply his knowledge of arranging antenna elements to minimise the size of the side-lobes. The Yagi type of antenna is the type usually seen on rooftops for receiving TV signals as it is operable generally in the frequency range of from about 500 MHz to 2 Ghz. A typical Yagi antenna construction and its radiation pattern are shown in Figure 9.7.

(c) Parabolic dish antennae, as the name implies, are so formed, and as a pure parabola, would reflect power from a point source emitter out as a beam, as shown in diagram B in Figure 9.6. By changing the disc diameter, for a given radio frequency, beams of various widths may be generated as listed in Figure 9.7. This type of antenna is practical only for microwave frequencies in UAS usage. For lower frequencies, the dish diameter becomes unacceptably large, especially for mounting in a UAV turret.

(d) The lens antenna works similarly to an optical lens in focusing RF waves instead of light waves. It uses dielectric material instead of glass and is appropriate for use with microwave frequencies. Beam shaping is achieved by asymmetric forming of the lens. The dielectric material is expensive and/or heavy and developments are continuing to reduce both of these factors for UAV application. Patch antennae use a patch (or patches) which are a little less than a half-wavelength long, mounted over a ground plane with a constant separation of order 1 cm, depending upon the frequency and bandwidth required. The patch is generally formed upon a dielectric substrate using lithographic printing methods similar to that used for printed circuit boards. With these techniques it is easy to create complex arrays of patch antennae producing high gain and customised beams at light weight and low cost. A square patch will produce an antenna with equal beam width in vertical and horizontal directions whilst beams of different width in the two planes will result from rectangular patches.

References

9.1. Andre L Clot, 'Communications Command and Control – The Crowded Spectrum'. *Proceedings of the NATO Development and Operation of UAVs Course*, ISBN 92-837-1033. 1999.

9.2. C. Balanis. *Antenna Theory*, 3rd edn. Wiley, 2005.

9.3. Hamid, R., Saeedipour, Md Azlin, Md. Said and P. Sathyanarayana. School of Aerospace Engineering, University of Science Malaysia. *Data Link Functions and Atrributes of a UAV System*, 2009.

9.4. P.G. Fahlstrom and T.J. Gleason, *Introduction to UAV Systems*. DAR Corporation Aeronautical Engineering Books, ISBN 9780521865746. 2001.

10

Control and Stability

It is not the intention here to present a detailed discussion on the theory and practice of aircraft control and stability. The discipline is rigorously covered in aircraft design textbooks (References 10.1 and 10.2). The intention is simply to indicate how its necessity and means of achievement may affect the design of UAV systems. Nor is it possible to cover all types of aircraft configurations as discussed in Chapter 3. Only the most popular types will be used as exemplars.

For ease of understanding the control examples given are illustrated with reference to heading and vertical gyroscopes which are the traditional ways of implementing attitude control and still used in many systems. In more recent systems there is an increased tendency to use 'strapped down' sensors fixed to the body axes. These directly provide body rates for control, but need mathematical integration to give attitude and heading as undertaken within inertial navigation systems.

The functions of the control and stability of a UAV will depend in nature on the different aircraft configurations and the characteristics required of them. 'Control' may be defined for our purposes as the means of directing the aircraft into the required position, orientation and velocity, whilst 'stability' is the ability of the system to maintain the aircraft in those states. Control and stability are inexorably linked within the system, but it is necessary to understand the difference.

The overall system may be considered for convenience in two parts:

i) The thinking part of the system which accepts the commands from the operator (in short-term or long-term), compares the orientation, etc. of the aircraft with what is commanded, and instructs the other part of the system to make appropriate correction. This is often referred to as the automatic flight control system (AFCS) or FCS logic, and contains the memory to store mission and localised flight programs
ii) The 'muscles' of the system which accept the instructions of (i) and apply input to the engine(s) controls and / or aerodynamic control surfaces.

Another distinction which must be made is whether the aircraft orientation, etc. is to be maintained relative to the air mass in which the aircraft is flying or relative to space coordinates.

10.1 HTOL Aircraft

For a HTOL aircraft the flight variables are basically:

a) direction,
b) horizontal speed,

Unmanned Aircraft Systems – UAVS Design, Development and Deployment Reg Austin
© 2010 John Wiley & Sons, Ltd

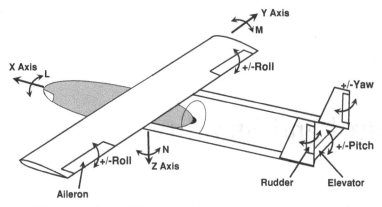

Ailerons produce rolling moment L
Elevators produce pitching moment M
Rudders produce yawing moment N

Figure 10.1   HTOL aircraft aerodynamic control surfaces

c) altitude,
d) rate of climb.

The direction of flight (or heading) will be controlled by a combination of deflection of the rudder(s) and ailerons. The horizontal speed will be controlled by adjustment to the propulsor thrust and elevator deflection, The rate of climb to a given altitude is achieved by the application of a combination of elevator deflection and propulsor thrust.

The arrangement of the aerodynamic control surfaces is shown in Figure 10.1 for a typical, aerodynamically stable, HTOL aircraft configuration. Other HTOL configurations will utilise specific arrangements. For example, a 'flying wing' configuration will use 'elevons' which deflect in the same direction for pitch control and differentially for roll control.

It is somewhat simpler to maintain orientation relative to the air mass, i.e. to configure the aircraft to be 'aerodynamically stable'. This generally requires tailplane and vertical fin areas to provide 'weathercock' stability in both pitch and yaw and requires wing dihedral in fixed-wing aircraft to provide coupling between side-slip and roll motion to give stability in the roll sense. The downside of this is that the aircraft will move with the air mass, i.e. respond to gusts (air turbulence). This movement usually includes linear translations and angular rotations relative to the earth. This will make for greater difficulty in maintaining, for example, a camera sight-line on a ground fixed target.

The alternative is to design the aircraft to be aerodynamically neutrally stable with, in particular, little or no rotation generated by the fixed aerodynamic surfaces in response to gusts. The response now becomes one mainly of translation, so reducing the angular stabilisation requirements for the sensors. The movable control surfaces are used to steer and stabilise the aircraft in the normal manner relative to spatial coordinates.

Thus the latter configuration has the advantage of providing a steadier platform for payload functions. In reality, however, it is virtually impossible to make an aircraft aerodynamically unresponsive to gusts in all modes, but it may be possible to make it unresponsive in some modes and have only little response in others.

Another advantage of the neutrally aero-stable design is that the aerodynamic tailplane and fin surfaces, for example, when replaced by much smaller movable surfaces for control, will save the drag of the larger areas and make the aircraft more efficient in cruise flight. Passenger aircraft manufacturers are moving

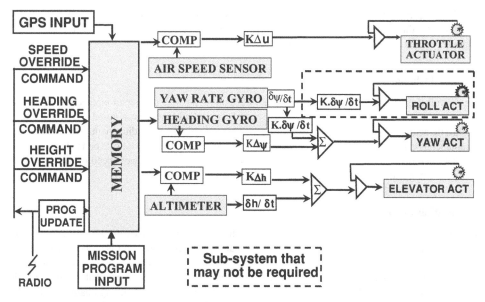

Figure 10.2   HTOL aircraft basic AFCS

this way now to provide more comfort for passengers and to reduce fuel-burn and improve economy of operation.

The downside of this approach is that more sophisticated sensors and computing power is necessary in the 'brain' of the control system in order to determine the orientation of the UAV in flight and apply the correct amount of the appropriate control or combination of controls. This may increase the system first cost compared with the aerodynamically stable system, but should pay dividends in greater operational effectiveness and reduced operating costs.

The control and stability systems will now be addressed according to the chosen coordinate reference (air mass or spatial).

10.1.1   HTOL Aero-stable Configuration

A typical basic flight control system (FCS) is shown in block diagram form in Figure 10.2. Before flight the mission program may be copied into the FCS computer memory. A very basic program may consist of a series of 'way-points' which the aircraft is to over-fly before returning to base, and the transit speeds between those points. It may be more complex in that a flight pattern about those points may be scheduled along with operation of the payload.

If the operators are in radio communication with the aircraft (directly or via a relay) the program commands may be overridden, for example, to carry out a more detailed 'manual' surveillance of a target. Provision also may be made to update the mission program during the aircraft flight.

For take-off and landing the aircraft may be controlled by an initial and terminal part of the program or 'manually' by using the overrides. Currently most systems employ the latter approach since making automatic allowance for the effect of cross-winds in those modes is difficult.

As shown in the figure, the aircraft is maintained on condition usually by use of a nulled-error method. By this means the FCS enables the commands of the controller to be accepted and executed and the aircraft to be stabilised onto that commanded condition of speed, direction and altitude.

A damping term can be added to any or all modes, for example $\delta\psi/\delta t$ in yaw, to ensure that, following a disturbance, the return to condition is rapid and without undue oscillation. Consider the three control 'channels' of Figure 10.2 as described below.

The Speed Channel

The aircraft airspeed command from the memory is compared with the actual airspeed as sensed and any error between the two is obtained. A multiplier K is applied to the error signal which is passed to the throttle actuator system with its feedback loop. This makes a throttle adjustment proportional to the instantaneous error until equilibrium is achieved. Provided that the power unit response is progressive, and that the correction takes place at an airspeed above the minimum power speed of the aircraft, the motion is stable and normally will need no damping term.

The Heading Channel

A similar principle applies. The actual heading of the aircraft can be measured by a magnetometer-monitored attitude gyro and compared with the commanded heading. Any error is processed as before to operate the aircraft rudder via a yaw actuator. In this case, however, damping may be required to prevent the aircraft oscillating in yaw and, in an extreme case, diverging in that mode. The probability of oscillation occurring depends upon the actuation system and aircraft aerodynamic damping characteristics. This phenomenon is covered fully in the specialist textbooks. Should extra damping be required, it may be incorporated by the differentiation respect to time of the gyro position signal or, possibly more readily, through the inclusion of a yaw-rate gyro.

In a turn, if using rudder alone, an aircraft will tend to slip outwards unless its wing dihedral rolls it back into the turn. Most aircraft will be so designed that a coordinated turn occurs naturally. In less conventional configurations, this may not be possible and application of ailerons is required proportional to the rate of turn. A method of achieving this, if required, is also shown in Figure 10.2.

The Height or Altitude Channel

The height of an aircraft is recognised as its vertical distance above ground as measured, for example, by a radio 'altimeter' and is often referred to as 'tape height'. Its altitude, also known as the 'pressure height' is its height above mean sea level and this is obtained by measuring the ambient air pressure outside the aircraft and comparing that with the ambient air pressure at mean sea level. Either can be used, depending upon the mission needs.

Pressure altitude is more appropriate for use when traversing long distances at greater altitudes but is relatively inaccurate for low altitude operation. It cannot respond to the presence of hilly or mountainous terrain.

Operating using tape height measurement is more appropriate for low-altitude, shorter-range operations when the aircraft will follow the contours of the landscape. It gives a far more accurate measure of height than does a pressure altimeter.

Both can be employed in a FCS with the most appropriate sensor being selected for a given phase of the mission.

The same nulled-error method may be used for the height channel with a climb to commanded height being achieved by actuation of an upward deflection of the elevator(s). Entry into a climb will demand more thrust from the propulsor and the aircraft will rapidly lose speed unless the engine throttle is quickly opened. If the response of the engine to the demand of the speed control channel is not adequate then a link from the error signal of the height channel must be taken to the throttle actuator. This will increase the engine power in a timely manner to prevent undue airspeed loss. The reverse, of course, will be ensured when a demand for a descent is made.

In addition to the above, control of the rate of climb will be necessary. The rate of climb (or descent) can be obtained by differentiating the change in measured height with respect to time. A cap must be placed on the allowed rate of climb (and descent) to prevent excessive or unavailable power being demanded from the engine(s) and to prevent the aircraft exceeding its design speed limit in descent.

The cap value that is necessary for protection will vary, depending upon the aircraft weight and speed at the time. For best performance it would be necessary for the cap value to be changed with those parameters. An input of speed to the equation is fairly simple, but determination of the aircraft weight at any time during a mission may be possible, but is not as easy. Therefore a compromise may have to be reached in setting the cap value.

The aircraft speed, rate of climb and engine power needed are inextricably linked. A demand for increased speed will increase the lift on the wing and may initiate a climb. The height channel may react to that and demand a deflection downwards of the elevator to prevent it. However, in similar manner to the advance link to the engine throttle from the height channel, it may be necessary to link the elevator to the error signal from the speed channel to prevent the development of any large height excursion.

Thus, the development of even a relatively simple FCS is no mean task and will require careful study and simulation before commitment to prototype build. The logic within the system will, today, be digital and software based. Until recently, the aircraft developers had to develop their own FCS systems but, with the expansion of the industry, companies specialising in FCS design and development have arisen. These organisations are now available to work with the aircraft developers in the creation of applicable FCS.

The several stability derivatives in the computation will be obtained from calculations and, depending upon the degree of novelty of the aircraft configuration, may also be obtained from testing a model in a wind tunnel. Many UAV are of a size that the model used may be of full scale which has the advantage of avoiding the necessity to correct for scale-effect inaccuracies which may obtain in manned aircraft testing.

10.1.2   HTOL Spatially Stabilised Configuration

For this configuration, the aircraft will be designed to have a minimal response to air gusts. For example, the fin aerodynamic surfaces will be reduced in size so that they merely offset the directional instability of the forward fuselage to provide effectively neutral directional stability overall. Preferably the smaller fins will be fully pivoting (all-flying) to retain adequate yaw control. Horizontal tail surfaces will be similarly treated to provide neutral pitch stability but adequate pitch control.

Wing dihedral will be sensibly zero to prevent a roll response to side-gusts. In many respects, this could move the configuration towards an all-wing or delta wing. However, as described, the aircraft is completely unstable and could, of its own volition, pitch or roll fully over and continue to 'wander' in those modes.

It is necessary to provide a spatial datum in those modes by including such means in the FCS. This is usually done by adding a vertical attitude gyroscope to the pitch and roll channels of the FCS, as shown in Figure 10.3.

10.2   Helicopters

10.2.1   Single-main-rotor Helicopter

The majority of manned helicopters are in this category, principally because, as explained in Chapter 3, there is a greater number of small to medium-sized machines required than large machines. The single-main-rotor (SMR) configuration is best suited to the former whilst tandem-rotor machines are best suited to the latter, larger category. The aerodynamic control arrangement for a SMR is shown diagrammatically in Figure 10.4. and a typical FCS block diagram in Figure 10.5.

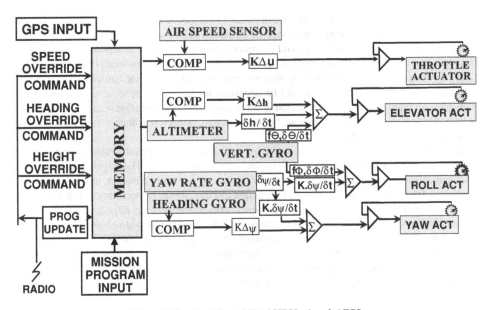

Figure 10.3   Spatially stabilised HTOL aircraft AFCS

More detailed discussion of the characteristics of SMR helicopters is available in References 10.3, 10.4 and 10.5. In them, little mention is included of tandem or coaxial rotor helicopters. These seem to be covered principally in any reliable detail in helicopter manufacturers' unpublished internal reports, for example as Reference 10.6.

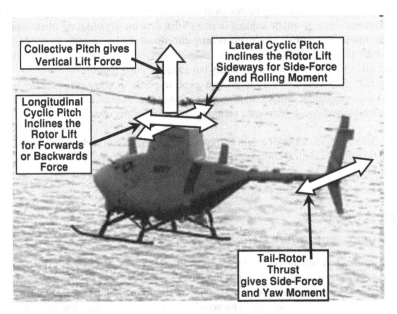

Figure 10.4   SMR helicopter controls

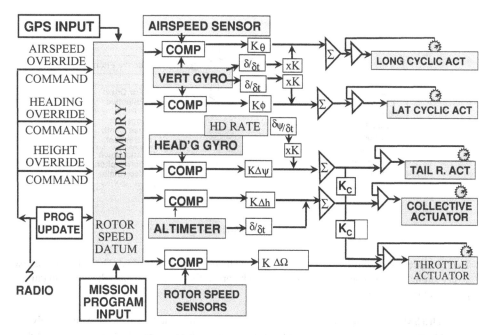

Figure 10.5   AFCS diagram for SMR helicopter

It appears that, at least until recently, most manufacturers of unmanned helicopters have opted for the SMR probably because it is seen as the most understandable technology. In a few cases, existing small passenger-carrying machines have been converted to a UAV by replacement of the crew and their support equipment with an automatic FCS. This latter approach removes much of the development costs and risk of a totally new airframe and systems. The SMR configuration, however, has its shortcomings as a candidate for 'unmanning'. These are principally as follows:

In the smaller sizes, especially, the tail rotor is relatively fragile and vulnerable, particularly during landings on uneven terrain and scrub land.

Means of ensuring the adequate control and stability of the configuration are complicated and caused by its inherent asymmetry compared with the above fixed-wing aircraft which are essentially symmetric. For example:

a) Execution of a climb requires an increase to be made in the collective pitch of the rotor blades which, in turn, requires more engine power to be applied. In its own right, that constitutes no problem. However, more power implies more torque at the rotor which, if uncorrected, will rotate the aircraft rapidly in the direction opposite to that of the main rotor's rotation.

Therefore the thrust of the tail rotor must be increased to counteract this. Unfortunately, this increase in lateral force will move the aircraft sideways and probably also cause it to begin to roll. To prevent this happening, the main rotor must be tilted to oppose the new increment in lateral force.

In a piloted aircraft, the pilot learns to make these corrections, after much training, instinctively. For the UAV FCS, suitable algorithms must be added to achieve accurate and steady flight.

b) In forward flight, the rotor will flap sideways rising on the 'down-wind' side. This will produce a lateral force which must be corrected by application of opposing lateral cyclic pitch. The value of this correction will be different at each level of forward speed and aircraft weight. Similarly, a suitable corrective algorithm has to be added to the basic FCS.

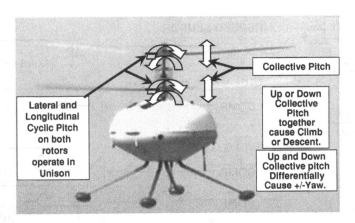

Figure 10.6   Plan-symmetric helicopter controls

c) To effect sideways flight from the hover, lateral cyclic pitch must be applied. The tail rotor will exert a very strong 'weathercock' effect which has to be precisely corrected by an adjustment in tail rotor pitch, requiring yet another addition to the FCS.

Although a SMR helicopter has a lower response to gusts in most modes compared with an equivalent-sized HTOL aircraft, its response to side-gusts is high, due to the very large fin effect of the tail rotor. It is difficult to see how that can be practically overcome, given the need for it as a powerful anti-torque measure.

10.2.2   Coaxial-rotor Helicopter

A coaxial rotor helicopter has symmetry in its rotor system and, in the case of the plan-symmetric helicopter, complete overall symmetry. It is therefore even simpler than for a HTOL aircraft to configure its FCS. Furthermore it is inherently less sensitive to gusts than any other configuration. Its method of aerodynamic control is shown in Figure 10.6 and a block diagram of an appropriate FCS is shown in Figure 10.7.

'Directional' Airframe Coaxial-rotor Helicopter (CRH)

'Directional' implies that it has an airframe having a preferred axis of flight, i.e. along which it has the lowest aerodynamic drag. A typical arrangement is shown in Chapter 3, Figure 3.9. With its rotor symmetry, it has none of the complex mode couplings of the SMR helicopter.

On the command to climb, the torque from each rotor remains sensibly equal so that little, if any, correction in yaw is required. In that event, it is achieved by a minor adjustment in differential collective pitch which removes any imbalance at source. Hence, unlike the SMR, there is no resulting side-force to balance. Similarly, entry into forward or sideways flight occasions no resultant side-force through rotor flapping. The flapping motion on each rotor is in equal and opposite directions thus the system is self-correcting. For these reasons, pilots flying crewed versions of the coaxial helicopter configuration, report on its ease of control compared with a SMR helicopter and, for the same reasons, the electronic flight control system is easier to develop.

There is a possible downside to the coaxial rotor helicopter. Its control in yaw relies upon the creation of a disparity in torque between the two rotors. In descent, less power is required to drive the two rotors and therefore less disparity in torque can be achieved, thus reducing the control power available. However, for all rates of descent short of full autorotation, the control available should remain adequate.

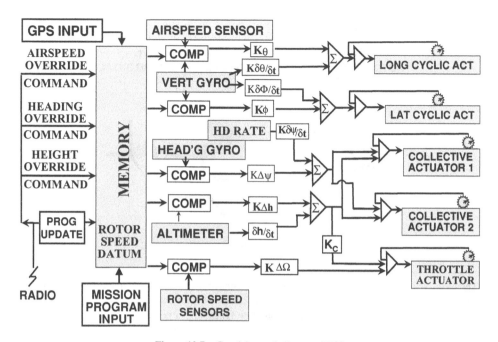

Figure 10.7 Coaxial-rotor helicopter AFCS

In full autorotation, calculations show that a small control power is available, but it is in the reverse direction. To overcome this problem, manned CRH are usually designed to be aerodynamically stable and incorporate rudders in the fin(s). In the event of total loss of engine power, unless very close to the ground, the pilot is required to put the aircraft immediately into forward flight where he has rudder control and conduct a run-on flared landing. This measure could be programmed into a UAV FCS.

Symmetrical Airframe Coaxial-rotor Helicopter

Otherwise known as a plan-symmetric helicopter (PSH), this is a special case of the CRH and, as explained elsewhere in this book, has several advantages over the directional CRH other than in aerodynamic drag of the fuselage. These advantages include a more compact aircraft for transport, more versatile operation of the payload (see Section 10.4), lower gust response and lower detectable signatures for stealth operation. It cannot be made aerodynamically stable in yaw, but is inherently neutrally stable. In normal flight conditions, it is stabilised spatially by the FCS. In full autorotation, unless corrective algorithms are added to the FCS to take account of the reversal in the control direction, the FCS would actually destabilise the aircraft.

A run-on landing, however, is unlikely to be practical for this type as it would require an undercarriage capable of such a landing and so is probably unsuitable for the configuration. However, it has the least response to gusts of all aircraft configurations, the response being zero in some directions and with no cross-coupling into other modes.

10.3 Convertible Rotor Aircraft

As discussed in Chapter 3, Section 3.5.3, convertible rotor aircraft may exist in two main variants – tilt-Rotor and tilt-Wing. Their means of control are similar.

COLLECTIVE PITCH AND LONG. CYCLIC PITCH ARE REQUIRED
PLUS ELEVATOR, RUDDERS AND AILERONS IN CRUISE FLIGHT

HOVER FLIGHT-	CRUISE FLIGHT-
ROTOR SHAFTS VERTICAL	ROTOR SHAFTS HORIZONTAL

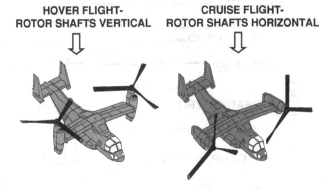

Figure 10.8   Tilt-rotor aircraft controls

The most basic approach is for each rotor to have control of collective pitch and longitudinal cyclic pitch control only (Figure 10.8) as opposed to helicopters which normally have cyclic pitch control in both longitudinal and lateral planes. In addition, both types have a powered means of tilting the rotor shafts (and usually engines) from the vertical forwards to the horizontal. In the case of the tilt-wing aircraft, the wing tilts as well. Thus the control strategy is as follows:

Hover Flight

Mode	Control
Climb or descent	Collective pitch change on both rotors
Fore and aft translation	Fore and aft longitudinal cyclic pitch change
Lateral translation	Differential collective pitch change
Heading change	Differential longitudinal cyclic pitch change

Cruise Flight

Mode	Control
Climb or descent	Elevator deflection
Speed change	Collective pitch change
Heading change	Rudderdeflection

The table shows the simplest solution to the means of control. There are further options:

a) By accepting the additional complexity of adding lateral cyclic pitch to finesse lateral translation in the hover and reduce the amount of roll incurred in the manoeuvre.
b) Differential collective pitch can be applied in cruise flight to assist in heading change especially in the transition between hover and cruise.

Transitional Flight

In any hybrid aircraft, the transition between hover flight and cruise flight is the most difficult regime to achieve and convertible rotor aircraft are no exception to this. Not only does it require the additional channel of control, i.e. the control of the actuators which tilt the rotor shaft axes at a controlled rate, but the FCS must phase in and out the control means for each flight mode in concert with, and appropriate to, the shaft tilt angle. Thus, during transition flight, both sets of controls (hover and cruise) must be operative in the correct ratios and in the correct phasing to ensure that the correct wing incidence is achieved. Any ancillary systems such as wing flaps must be phased in or out appropriately,

Hence the FCS for this type of aircraft is far more complicated than for either 'pure' HTOL or VTOL aircraft and is open to many different interpretations. Therefore no attempt is made here to show a FCS block diagram for these types.

10.4   Payload Control

In addition to maintaining control and stability of the aircraft, it is just as important to achieve that for the payload. Control of the aircraft is needed to get the aircraft over the target area, but will be useless unless the payload is properly controlled. The latter may be achieved using a system which is part of the aircraft FCS or by using a separate module. The choice will probably depend upon the degree to which the payload operation is integrated with the aircraft operation.

Control of the payload will include, for most imaging payloads, the means of bringing the sight-line accurately onto the target and keeping it there. This is probably carried out initially by 'manual' direction using actuation of the payload mounting about two axes. A gust-insensitive, spatially stable aircraft will facilitate this.

Subsequent maintaining of the sight-line on target will rely upon gyro-stabilisation of the sight-line, possibly aided by a lock-on pattern recognition system in the E/O sensor and/or use of the differential GPS with computation involving the GPS coordinates of the target and the aircraft at each moment in time. Other control will include switching to release dispensable payloads; bring payload sensors 'on-line'; changing E/O settings; making adjustments to FOV; initiating, for example, scanning programs, etc.

The integration of the payload and aircraft control and stability systems is at its greatest in the PSH configuration which is, in effect, a flying payload turret. The same set of heading and vertical gyros, for example, support the control and stabilisation of both aircraft and payload.

The FCS operates two sets of coordinate axes, those of the aircraft and those of the payload, even though the latter is fixed within the aircraft. Thus the payload sight-line may be pointing in one direction whilst the aircraft may fly in a totally different direction.

This facility enables the operation of a range of useful manoeuvres. For example in traversing a large expanse of sea or terrain, the payload sight-line may be programmed to scan at a range of frequencies over a range of amplitudes on either side of the aircraft line of flight to encompass a large field of regard in a reconnaissance mission. Another program may call for a continuous 360° rotation of the sight-line as the aircraft emerges vertically from the depths of a wood to scan for 'items of interest' in both near and far fields.

10.5   Sensors

Sensors, as shown in the FCS diagrams of Figures 10.2, 10.3, 10.5 and 10.7, include vertical attitude gyros, heading gyros, angular rate gyros when necessary, height and altitude sensors and airspeed sensors. Linear accelerometers may be used in some applications. Individual sensors may be used as described above or the sensors may form part of a 'strapped down' inertial measurement unit. It is not intended here to cover sensors in any detail as information is readily available from a number of different suppliers from whom specifications may be obtained. Some suppliers offer complete FCS 'ready-made' or at least custom-built for individual applications. In each case their qualities of accuracy, reliability, life, power supply, environmental protection and mass will be of importance to the UAV systems designer. Usually, though not always, the cost increases as the performance specification increases.

Some general comments may be in order.

Height and Altitude Sensors

Sensors for measuring tape height, that is height above ground, include those measuring distance by timing pulses of radio, laser or acoustic energy from transmission to return. These vary in their accuracy, depending upon their frequency and power, but are usually more accurate than pressure sensors measuring altitude. Radio altimeters vary in their accuracy and range depending upon their antennae configuration.

Laser systems may have problems in causing eye damage and precautions must be taken in their selection and use. They may also lose function when operating over still water or certain types of fir trees when the energy is either absorbed or deflected so that no return is received. Acoustic systems usually have a smaller range capability and must also be separated in frequency from other sources of noise.

Barometric (or pressure) sensors for measuring pressure altitude are less accurate than the tape height sensors and have to be adjusted to take account of the atmospheric changes which take place hour by hour and from area to area. However in transitional flight at altitude this does not constitute a real problem and can be backed up by GPS data. These sensors are not suitable for accurate operation at low altitude, especially in the case of VTOL aircraft. The static air pressure measurement from a VTOL aircraft is greatly affected by the induced airflow around the aircraft, the direction of which also changes with vertical or lateral manoeuvres.

Airspeed Sensors

For HTOL aircraft a standard pitot-static (PS) system is acceptable provided that it is suitably positioned to read accurate static pressure either as part of a combined unit ahead of any aerodynamic interference or as a separate static vent elsewhere on the aircraft. The compensating PS head developed by Bristol Aircraft in the 1950s improves the accuracy of the former type of installation.

In the case of VTOL aircraft the difficulty of measuring an accurate static pressure at different airspeeds, referred to above, also affects measurement of airspeed using a PS system. Apart from the inaccuracy of the classic PS system in measuring airspeed, and its inability to record speeds below about 15 m/s, fluctuating values from it can cause instability in the control system. Hence it is better to rely on data from a system integrated with GPS or better still from an omnidirectional air-data system that does not require knowledge of ambient static pressure.

Hover-position-hold Sensing

Holding station in a hover or near hover is often a requirement for a VTOL aircraft for take-off or for landing and also for several types of operations, current or projected, where surveillance from a fixed-point is required. If this is required at an established base, the task is solvable by means such as hovering over a beacon.

If the operation is required away from base, then options include the engagement of integrating accelerometers, pattern-recognition or, possibly in the future, photon-flow measurement on the E/O sensor or possibly Doppler interrogation of the radio altimeter, etc. These sensor inputs would be integrated into the FCS to operate the appropriate controls.

10.6   Autonomy

The 'jury' in the unmanned aircraft community seems to be 'still out' for the verdict on the definition of autonomy. Some suppliers of UAV systems claim that an aircraft has operated autonomously in carrying out a mission when it has flown a pre-programmed flight from take-off to landing without further instructions from outside. Others would label this type of activity as merely automatic and would say that to be autonomous the system must include an element of artificial intelligence. In other words the system must be able to make its own decisions without human intervention or pre-programming.

The main systems drivers for autonomy are that it should provide more flexible operation, in that the operator tells the system what is wanted from the mission (not how to do it) with the flexibility of dynamic changes to the mission goals being possible in flight with minimal operation replanning. This is coupled with reduction in reliance on time-critical communication and communication bandwidth, which in turn reduces the vulnerability of the system to communication loss, interruption or countermeasures. The goal is for the operators to concentrate on the job rather than operating the UAV.

The author is not aware of any fully autonomous system existing outside of science fiction (shades of the computer HAL in Arthur C. Clarke and Stanley Kubrick's '2001: A Space Odyssey'). It is conceivable that this may yet come about but would it be desirable?

References

10.1. W. J. Duncan. *The Principles of the Control and Stability of Aircraft*. (The classic work), Cambridge Aeronautical Series, Cambridge University Press, 1952.

10.2. Bernard Etkin and Lloyd Duff Reid. *Dynamics of Flight: Stability and Control*. John Wiley & Sons, Ltd, 1996.

10.3. Alfred Gessow and Garry Myers. *Aerodynamics of the Helicopter*. (An introductory text), Macmillan, New York, 1952.

10.4. A. R. S. Bramwell, G. Done and D. Balmford. *Bramwell's Helicopter Dynamics*. (More advanced reading) Butterworth – Heinemann, 1976.

10.5. Gareth D. Padfield. *Helicopter Flight Dynamics*, 2d edn. Blackwell Publishing, 2007.

10.6. Eduard Petrosyan. *Aerodynamic Features of Coaxial Configuration Helicopter*. Deputy Chief Designer, Kamov Company, 2009.

11

Navigation

It is necessary that the UAV controller knows the position of the UAV and, for automatic operation, for the UAV to have that knowledge on board. The currently popular method of position fixing and navigation between points is by use of the Global Positioning System (GPS).

11.1 NAVSTAR Global Positioning System (GPS)

GPS was developed by the United States' Department of Defence and officially named NAVSTAR GPS. It was initially limited to use by US military forces until 1982 when it was made available for general use. A receiver calculates its position using the signals transmitted from four or more GPS satellites selected from a constellation of 24 (nominal) satellites. The satellites orbit the Earth at an altitude of approximately 20 000 km and the satellites used for the measurements are selected by the GPS receiver on the basis of signal quality and good fix geometry.

Each satellite has an atomic clock and continually transmits its radio signals. The signals which contain the time at the start of the signal, travel at a known speed (that of light). The receiver uses the arrival time to calculate its range from each satellite and so its position on Earth. Radio frequencies used by the GPS lie within the L Band, from about 1.1 GHz to about 1.6G Hz.

GPS is available as two services, the Standard Positioning System (SPS) for civilian users and the Precise Positioning Service (PPS) for military users. Both signals are transmitted from all satellites.

The SPS uses signals at GPS L1 frequency with an unencrypted coarse acquisition (C/A) code. It is understood that this will in future be supplemented by an additional L2 service. SPS gives a horizontal position accuracy in the order of 10 m.

The PPS (also known as P code) uses both GPS L1 and L2 frequencies in order to establish a position fix. These signals are modulated using encrypted codes (into Y code). The Y code will be supplemented by a new military (M) code currently in development. The military receivers are able to decript the Y code and generate ranges and hence position. The PPS horizontal position accuracy is of the order of 3 m. The accuracy of GPS position fixes varies with the receiver's position and the satellite geometry. Height is available from GPS, but to a lower accuracy.

The accuracy of both GPS services may be improved by use of Differential GPS (DGPS). This provides an enhancement to GPS using a network of fixed, ground-based, reference stations that broadcast the difference between the positions indicated by the satellites and their known fixed positions. These differences are then used by each receiver to correct the errors manifest in the raw satellite data.

The accuracy of DGPS reduces with the distance of the receiver from the reference station and some measurements of this indicate a degradation of about 0.2 m per 100 km. A UAV system may use the

Unmanned Aircraft Systems – UAVS Design, Development and Deployment   Reg Austin
© 2010 John Wiley & Sons, Ltd

available network of reference stations or use its own ground control station as a reference station. This latter may be less appropriate for air- or ship-based control stations.

Although the US-provided GPS is the most extensive system currently operating, other systems are emerging. Such other similar systems include the European Galileo system and the Russian GLONASS with the proposed Chinese COMPASS and Indian IRNSS systems.

The above presents an over-simplistic coverage of GPS and the reader is referred to publications such as References 11.1 and 11.2 for more detailed information.

The availability of GPS has permitted UAV operation to be vastly extended in range compared with their capability of 25 years ago, enabling MALE and HALE systems, in particular, to be operated. There is continuing concern, however, that in the event of hostilities, GPS signals may be jammed. GPS signals at the receivers tend to be rather weak and therefore relatively easy to jam by natural emissions such as geomagnetic storms and by unintentional or intended radio emissions. It is therefore possible for unsophisticated enemies to jam GPS signals with a closer, stronger signal. In the, hopefully unlikely, event of hostilities with a sophisticated opponent, the satellites, themselves, could be destroyed, but it is more likely that they would wish to maintain the benefits of being able to use the system themselves.

GPS is basically a 'fixing' system, in that the measurements provide a sequence of discrete positions or 'fixes'. It is normal to integrate the GPS with a dead reckoning (DR) system. DR systems work on the basis you know where you are at the start of the mission and you then use time, speed and direction measurements to calculate your current position. The process of dead reckoning is probably the oldest form of air navigation (clock, airspeed and compass) but it has the disadvantage that position errors will grow with time due to inaccuracies in the measurement of the basic parameters.

GPS on the other hand provides a series of largely independent position measurements with some position error (noise) however it is not continuous and in some circumstances can have local errors. For example in urban areas the satellite signals can be subject to obscuration or reflection from structures which give GPS 'multipath errors' (erroneous range measurement). Also, as stated above, GPS can be subject to local radio frequency interference or deliberate jamming. Prudent GPS integration therefore combines GPS with a dead reckoning system to provide an element of smoothing to the raw GPS and a means of providing a continuing navigation capability in the event of GPS signal loss. This combination is normally undertaken in a mathematical filter such as a Kalman filter which not only mixes the signals, but provides an element of modelling of the individual sensor errors. The modelling enables the filter to give improved navigation during periods of GPS signal loss/degradation.

Numerous techniques continue to be proposed to reduce dependence on GPS, and users should consider a fall-back plan in the event of GPS loss. It is believed that some military operators currently retain, as a fall-back, systems which were in use before they were replaced by GPS. These systems are principally TACAN, LORAN C or inertial navigation. Each has disadvantages compared with GPS.

11.2   TACAN

Like LORAN C and GPS, TACAN relies upon timed radio signals from fixed ground-based transmitters to enable position fixing. The fix is based on range measurement from multiple transmitters or range and bearing from the same transmitter. The signals, being terrestrially based, are stronger than GPS signals and can still be jammed, although not as easily. For military operations, a major disadvantage of TACAN was that emissions could not be controlled to achieve stealth, and an enemy could track an aircraft equipped with the system.

11.3   LORAN C

This long-range radio system based on ground transmitters uses even stronger signals than TACAN and is less easy to jam though it does suffer serious interference from magnetic storms. Although funding is limited, enhanced development of LORAN, known as E-LORAN, is continuing as it is seen as a fall-back to the perceived vulnerability of GPS. It is principally used in marine service.

For military UAV application, its major drawback is its very limited availability. It is available principally in the populated areas of Europe and in North America and not in the areas of likely world trouble-spots. It is virtually nonexistent in the southern hemisphere.

11.4   Inertial Navigation

An inertial navigation system (INS) does not rely on external inputs. It is a sophisticated dead reckoning system comprising motion sensing devices such as gyroscopes and accelerometers and a computer which interrogates the data from them and performs appropriate integration to determine the movements of the aircraft from a starting set of coordinates to calculate the aircraft position at any subsequent time.

Past systems have been based on platforms gimballed within the aircraft to remain horizontal as determined by pendulums and attitude gyroscopes. The main disadvantage with them has been their need for many expensive precision-made mechanical moving parts which wear and create friction. The friction causes lag in the system and loss of accuracy. The current trend is to use what are termed 'strapped down' systems. The term refers to the fact that the sensors (accelerometers and rate gyros) operating along and around the three orthogonal aircraft body axes, are fixed in the body of the aircraft.

Lightweight digital computers are able to interrogate these instruments thousands of times per second to determine the displacement and rates of displacement of the aircraft at each millisecond during the flight and to compute the attitude, velocity and position changes. The sensors are usually solid-state and the accuracy of the overall system depends upon the accuracy of the individual sensors. Greater accuracy is obtained at greater cost but, inevitably, 'drift' away from the actual spatial position occurs with time.

Problems of accuracy therefore are more critical for long-endurance operations with HALE and MALE systems unless some form of update is possible during the flight. As stated above it is increasingly common for long-range/endurance systems to integrate the INS with GPS.

Heading updating is possible, of course, using magnetometers which point to magnetic north. Height or altitude can be sensed.

Developments in Doppler radar sensors provide good prospects for geo-speed measurement, although their use would have to be limited if the aircraft was to remain covert. A problem remains in sensing pitch and roll angles adequate for accurate navigation in the absence of IN and GPS, however pitch and roll accuracy sufficient for flight control is available. Developments which sense the horizon may come to fruition for operations at high altitudes where a horizon is distinct.

11.5   Radio Tracking

This is a well-established and ready solution for aircraft operating at shorter ranges, of the order of 80–100 km. It is particularly applicable to over-the-hill battlefield surveillance and ground attack operations or shorter-range naval operations such as over-the-beach surveillance missions where a line-of-sight radio contact can be maintained between the ground/sea control station and the aircraft.

The narrow-beam up and down data-links carry timed signals which are interpreted by both control station and aircraft computers giving their distance apart. Parallel receiving antennae at the control station (CS) enable it to lock onto the aircraft in azimuth and transmit that information to the aircraft.

In the event of loss of radio link, the aircraft and CS will be programmed to scan for the signal in order to re-engage. The aircraft will also carry a simplified INS in order for it to be able to return to the neighbourhood of the CS should there be a failure to re-engage. At the estimated arrival time, two options for recovery are available. Either an automatic landing program is brought into operation or a low-frequency omnidirectional radio system activated to re-establish contact and control the aircraft to a safe landing.

11.6   Way-point Navigation

Using any of the above technologies to ascertain its position, the UAV controller may direct the UAV to any point within its range by one or more of three methods.

a) Direct control, manually operating panel mounted controls to send instructions in real time to the UAV FCS to operate the aircraft controls to direct its flight speed, altitude and direction whilst viewing its progress from an image obtained from the UAV electro-optic payload and relating that as necessary to a geographical map.
b) Input instructions to the UAV FCS to command the UAV to fly on a selected bearing at a selected speed and altitude until fresh instructions are sent. The position of the UAV will be displayed automatically on a plan position indicator (PPI).
c) Input the coordinates of way-points to be visited. The way-points can be provided either before or after take-off.

Methods (b) and (c) allow for periods of radio silence and reduce the concentration necessary of the controller. It is possible that, depending upon the mission, the controller may have to revert to method (a) to carry out a local task. However, with modern advanced navigation capability and the introduction of 'autonomous' technology within the systems the trend is strongly towards pre-planning missions or in-flight updating of flight plans so that the operators are more focused on capturing and interpreting the information being gathered by the UAV than managing its flight path. Future systems with increased use of autonomy are likely to be based on the operators 'tasking' the UAV to achieve aspects of a mission with the UAV system generating the routes and search patterns.

References

11.1. Ahmed El-Rabbany. *Introduction to GPS – The Global Positioning System*. Sci-Tech Books, Artech House Publishing, ISBN 1580531830. 2006.
11.2. Ian Moir and Alan Seabridge. *Military Avionics Systems*. John Wiley & Sons, Ltd., Aerospace Series, ISBN 13 978-0-470-01632-9. 2007.

12

Launch and Recovery

12.1 Launch

The method of launching the aircraft may be considered within three types, each with an appropriate means of recovery:

a) a horizontal take-off and landing (HTOL), on a wheeled undercarriage, where there is a length of prepared surface (runway or strip) available;
b) a catapulted or zero-length rocket-powered launch when the aircraft has no vertical flight capability and where the operating circumstances or terrain preclude availability of a length of runway;
c) a vertical take-off and landing (VTOL).

12.1.1   HTOL

The length of run required to achieve launch is dependent upon the acceleration of the aircraft to lift-off speed. The thrust available to accelerate the aircraft is a major issue which depends upon the power available for thrust and the efficiency of the propulsor from stationary to lift-off speed.

On the assumption that the wing incidence is kept such as to generate no lift during the run until lift-off speed is reached, the profile drag of the aircraft will be negligible compared with the propulsor thrust. The drag will build up from zero to typically only 3% of available thrust at lift-off speed.

The efficiency of the conversion of engine power into thrust for propeller-driven aircraft is determined basically by the diameter of the propeller. A graph showing this for engine power up to 50 kW is presented in Figure 12.1. The justifiable assumption is made that only 85% of the power available from the engine is converted to thrust, the remainder being the power lost to propeller blade profile drag.

The speed at which the aircraft is safe to lift off is a function of the wing's aircraft mass/area ratio (wing loading) and lift coefficient. The graph in Figure 3.3 (Chapter 3) shows the variation of minimum speed for flight with wing loading. It is based upon the wing generating a lift coefficient of 1.0. This allows a margin before stall of only about 0.2 and could be optimistic in certain circumstances which would require a longer run before a safe lift-off is advisable. Using these criteria to calculate the acceleration to be expected, Figure 12.2 indicates the shortest take-off run likely to be achieved, in still air, by a sample of different UAV having different values of wing-loading.

The design of an aircraft involves many compromises and includes a compromise between design for good (short) take-off performance and good cruise efficiency (long range at high speed). A good take-off performance requires a large wing area (low wing loading) while a good cruise performance requires a minimal wing area (high wing loading) to minimise wing profile drag. It is not surprising, therefore, to

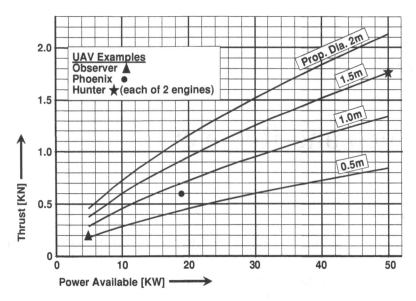

Figure 12.1 Thrust available for take-off

see the marked trend from the low wing loading of the close range aircraft to the high wing loading of the long range aircraft.

The Global Hawk, (see Figure 4.1 in Chapter 4) representing the HALE type, has the highest wing loading and the highest speed required for lift-off. It is powered by a turbo-fan engine whose nominal thrust of just over 40 kN gives it an estimated 0.3g acceleration to achieve a best run length to lift-off of about 600 m. However, a turbo-fan engine, though effective at higher speeds, does not develop its maximum thrust at very low speed. Therefore its actual acceleration may be less than that estimated

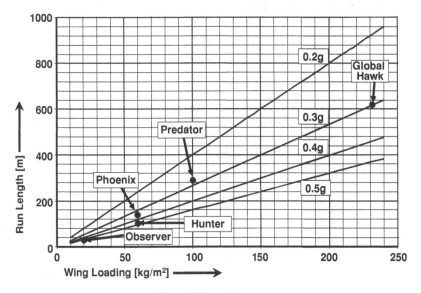

Figure 12.2 Run length

figure and therefore requiring a longer run. As this type of aircraft operates from fixed bases away from a combat zone, a suitably long runway will be available.

The Predator B (see Figure 4.1, Chapter 4), on the other hand, uses a variable pitch propeller which will convert engine power efficiently over a wide range of airspeed from zero to cruising speed. With its much lower wing loading of 100 kg/m² it should achieve lift-off in about 300 m and will therefore have a greater range of take-off strips available to it.

The Hunter UAV is designed to operate as a medium range system and, as such, will often be required to operate out of shorter, improvised airstrips. With full power from its two power-plants, calculations indicate that, under favourable conditions, it should be capable of lift-off after a run of only 100 m. It has the option of a rocket-assisted take-off (see Figure 4.11. Chapter 4) which will shorten the take-off run and steepen its climb-away.

Catapult Launch

The Phoenix and Observer UAV (see Figure 4.17, Chapter 4) are examples of close-range battlefield systems where no airstrip is available and the system is required to be mobile on the battlefield to move its operating site as expediency demands.

If using only thrust from the propeller the aircraft would require runs of about 150 and 40 m respectively, neither of which would necessarily be available under battlefield conditions. Further, the propellers, being of fixed pitch, would have to be designed for best performance for take-off. This would severely compromise the performance of the aircraft in flight.

The solution adopted is to catapult the aircraft into the air with sufficient acceleration and catapult throw (ramp) length for it to achieve flight speed on release. Although this method achieves the aim of relative mobility on the battlefield, it does require the transport of a catapult system in addition to the transport of the UAV and its control station. It also has the limitation of launch direction. Almost inevitably the catapult must be positioned where it can launch the aircraft sensibly into the wind. This is not always easy to achieve, especially in wooded or hilly terrain. A shift in the wind direction may mean a repositioning of the launcher.

Launchers range from the hand-launching of the micro-AV through the simpler, bungee-powered to the more complex pneumatic-powered launcher, depending upon the mass and wing loading of the UAV to be launched, Figure 12.3.

Typical characteristics are shown in the following table though it must be noted that some of the figures are necessarily approximate, but sufficiently accurate to indicate the trends. The values of wing area and thus of wing loading have not always been available to the author, in which case they have been estimated from drawings or photographs. The aircraft self-sustaining speed calculated on the basis of a lift coefficient of 1.0 is therefore also approximate, though in the case of Phoenix the manufacturer's figure has been quoted (indicated ** in the table) and is 5 m/s higher than otherwise estimated. This higher margin is probably necessary to prevent the aircraft stalling on release from the ramp if hit by an up-gust. This aircraft may be more vulnerable than others, in that respect, due to the higher drag of its under-slung pod.

UAV type	Shown in figure	Launch Mass (kg)	Wing loading (kg/m²)	V_{min} (m/s)	Ramp length (m)	Mean aceleration (g)	Catapult method
Wasp	4.25	0.275	4.9	8.8	2*	2.5?	Hand-launched
Mosquito	4.25	0.5	6.7	10	2*	3?	Hand-launched
Hawk	4.24	3.2	13	14.4	?*	?	Hand-held/bungee
SkyLite	4.24	6.0	7.5	11	2.6	2.9	Ramp/bungee
Observer	4.15	36	19	17.9	7	2.9	Ramp/bungee
Phoenix	4.15	177	55	35**	9	8	Hydropneumatic

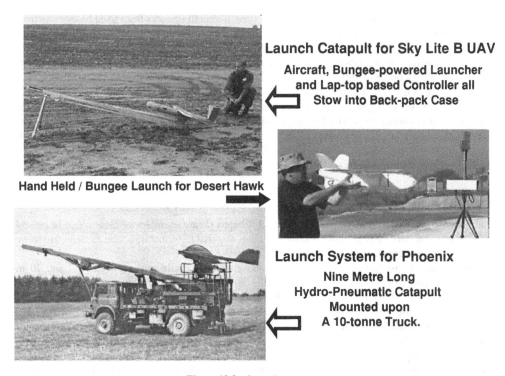

Figure 12.3   Launcher systems

Another aspect to note is that it is not possible to know, other than to postulate, the effective launch lengths of the first three of the examples shown in Chapter 4, Figures 4.23 and 4.24 (indicated *). In the case of the hand-launched micro-AVs, the length is dependent upon the technique of the thrower. The length of stretch of the bungee used to launch the Desert Hawk is unknown and probably varies with conditions and the physique of the operator.

All three systems will be able to benefit by launching the aircraft into the wind, if available, thus achieving flight speed over a shorter distance.

The acceleration required to be imposed upon the aircraft is seen to be no more than about 3g for all except the Phoenix. All of the on-board equipment within the aircraft, including sensitive camera equipment and control and navigation sensors must withstand this acceleration. Below 3g the amount of hardening and support needed for the equipment may be little. To withstand 8g, however, will require more substantial hardening and will add to the cost and probably weight and proving of that equipment if its reliability is to be assured.

12.1.2   VTOL

A controlled vertical take-off is the most elegant mode of launch. It enables the UAV system to be operated from almost any type of terrain, requiring no runway, airstrip or cumbersome catapult equipment. The launch is totally independent of wind direction and the aircraft can be airborne within minutes of the system arriving on site.

It is merely required, after initial start-up and check-out via the UAV on-board monitoring system display in the GCS, for the controller to input height (or altitude) and rate of climb and press the 'go' button.

12.2   Recovery

Recovery of the UAV requires not only a safe landing, but the return of the UAV to its base or hangar.

12.2.1   HTOL

Recovery involves the return of the aircraft to make a controlled touch-down onto its undercarriage at the threshold of runway or airstrip, deceleration along the runway, followed by the aircraft taxiing or being towed back to its base point.

For long-range aircraft such as Global Hawk, which operate from secure, fixed bases, a form of approach control with guidance, for example, along a radio beam may be used. Initial positioning onto the radio beam will be by using the Global Positioning System (GPS). Thus both take-off and landing may be completely automated, so reducing the occurrence of accidents which have happened during the use of manual control.

For UAV systems required to operate from small airfields or, in the case of medium-range systems, often from temporary airstrips with no landing aids, the take-off and landing must be manually controlled. This implies the controller acting as a remotely placed pilot, using the aircraft aerodynamic and throttle controls, aided by a view ahead of the aircraft from a forward-looking camera or by his having direct sight of the aircraft.

Although the take-off is relatively easy, the landing approach and touch-down at the correct position and airspeed require considerable judgement.

The initial positioning may again be by use of GPS and possibly the more accurate Differential GPS (DGPS) but, until suitable mobile systems are developed, the actual landing will be under manual control. This limitation currently presents considerable risk at night. It may therefore reduce, or totally prevent, operation after dark or in poor visibility.

After touch-down the aircraft must be decelerated and braking on unmetalled surfaces can present a problem. Arrester wires are sometimes used or the aircraft may release a drogue to increase its aerodynamic drag. After landing, depending upon the size and mass of the aircraft, it may be taxied or towed or man-handled back to its base point.

Catapult Launch System Recovery

As with its launch, this type of system presents the more complicated provision for recovery. There are a number of alternative solutions:

a) a skid or belly landing,
b) guided flight into a catchment net,
c) deployment in flight of a parachute with provision of an energy absorbing system to reduce the severity of ground impact,
d) guided flight onto an arresting pole.

Skid or belly landing. Applicable only to the very small UAV such as micro- and mini-UAV, e.g. Wasp, Mosquito and Hawk, which are rugged enough to have a survival rate upon this form of 'arrival' and cheap enough as to be considered expendable.

Guided flight into net. This was first tried with the Aquila UAV system in the USA in the late 1970s with disastrous results as on most occasions the UAV was severely damaged or completely destroyed. It contributed in large measure to the termination of the programme. The problems encountered were two-fold.

The first was ensuring that the UAV was directed into the centre of the net rather than striking the supports or missing the net altogether and striking the ground. The second was the abrupt deceleration, even if the UAV did contact the net centre, which severely damaged the UAV. No UAV system has since adopted this method, but development of a better system is now being undertaken by Advanced Ceramics Research Inc. for the US Navy. Using the improved methods of guidance available today, an accurate contact is now possible. Reduction of the deceleration is being addressed by two different means.

The first is to mount the net above a ground vehicle which traverses at a speed somewhat less than that of the recovery speed of the UAV so that the UAV catches up with the net at a reduced speed differential and therefore imposes a less damaging deceleration upon the UAV. This method, however, still requires a length of straight and level track for the ground vehicle and a sophisticated means of timing to ensure the UAV and ground vehicle meet as planned.

An alternative approach is to suspend the net above ground below a kite, parasail or perhaps balloon. This provides a degree of compliance for the impact, thus again reducing the acceleration.

Neither of the methods is without problems but have some promise.

Parachute deployment. Currently, this is by far the most usual method of the four. As introduced in Chapter 4, it requires the UAV to carry a parachute and an energy-absorbing airbag, both of which, with their release mechanisms, reduce the mass which can be invested in payload or fuel.

The parachute must be released by remote control over or near the point at which it is required the aircraft to land and the airbag(s) must also be released and inflated well in advance of impact.

One disadvantage of this method is that the aircraft/parachute combination is at the mercy of the wind, and its precise point of touch-down may therefore be unpredictable. The impact of the aircraft may be at an inconvenient angle and with a translational velocity with the probability of damage resulting. In a worst-case scenario, the aircraft may be swept into trees or other obstacles.

A variation of the method, under consideration, is for the parachute to be replaced by a parafoil which will achieve a controlled glide onto a pre-planned (D)GPS position or pre-positioned radio beacon. The problems then to be solved are the release of the parafoil in the correct orientation and its subsequent control to the point of impact.

Security on the battlefield may also be an issue.

A second disadvantage is that both parachute and air-bag(s) must be repacked or replaced prior to further flights constituting an operational delay.

A third disadvantage is that the aircraft must then be returned to the launcher from its point of impact which may be a considerable distance and probably requiring the addition of a further transport vehicle to the system.

The launch-to-recovery sequence is shown for the Phoenix system in Figure 12.4. The aircraft was designed to turn upside-down beneath a parachute for recovery in order to best protect the payload/avionics module on impact. A frangible, expendable, energy-absorbing element was mounted above the wing of the aircraft which, with the aircraft inverted, was the first to impact the ground.

The complexity of achieving this manoeuvre and the adverse influence of the drag of the pod are believed to have been but two of the reasons for its protracted development period. Sadly this approach did not prevent serious damage being suffered by the aircraft and the necessary repairs plus the replacement of the frangible elements caused an unacceptable logistics problem. Subsequently the frangible element was replaced by a number of airbags which is the more usual solution. As an example, a SkyLite aircraft is seen in Figure 12.5 in descent with a large airbag inflated.

More recently, a breakthrough seems to have been achieved with the Scan Eagle system where the aircraft is captured onto a section of rope which is suspended vertically between two boom arms with bungee shock absorbers. Clips on the Scan Eagle's wing-tips catch the rope, hooking it in place, with the bungees rapidly decelerating the aircraft to a halt. The aircraft then dangles on the rope before being reeled in and recovered by the ground crew. The launch-to-recovery sequence is shown in Figure 12.6.

All four of these methods impose considerable deceleration on the airframe and ancillary systems, often in excess of 10g; they must be sufficiently rugged as to accept the resulting loads without failure.

Figure 12.4   Phoenix – launch to landing

12.2.2   VTOL

The recovery of a VTOL UAV is simplicity itself. The UAV will have been commanded to a coordinate position using DGPS or other navigation system and at a given height above ground. The position-hold system will be engaged and the 'land' button pressed. The ideal control arrangement will automatically descend the aircraft at a vertical rate of descent proportional to its height above ground to an arrest at a nominal, for example, 0.1 m beneath undercarriage contact. This ensures that the aircraft is firmly on the ground before engine shut-down.

12.3   Summary

In summary, an example of launch by catapult and recovery by parachute is shown by the Phoenix system in Figure 12.4; whilst examples of a wheeled (rolling) take-off and recovery and a vertical take-off and recovery are shown by the Galileo Falco UAV and the M L Aviation Sprite UAV, respectively, in Figure 12.7. The latter illustration graphically indicates the ability of a VTOL UAV to 'land on a spot' which also enables it to take off from and alight onto a land vehicle or a small ship.

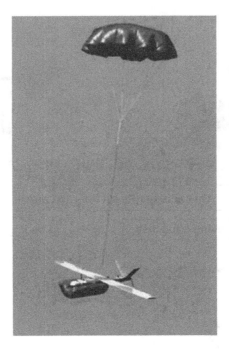

Figure 12.5   SkyLite UAV recovery

Figure 12.6   Scan Eagle launch and recovery. *Source*: Insitu

**Take-Off Run and Landing
on Wheels**

**Vertical Take-Off and
Vertical Recovery**

Figure 12.7   Launch and recovery methods

Of all the methods for UAV launch and recovery, the VTOL one is the most accurate, operationally convenient and the most gentle, imposing at most 2g onto the payload and the other aircraft equipments. VTOL UAV are, however, not appropriate to all operations, not being able, for example, to achieve the range and altitude of a HALE operation as readily as can a HTOL UAV, or the speed required of a UCAV UAV. VTOL UAV are therefore more appropriately used for close- or medium-range operations.

13

Control Stations

13.1   Control Station Composition

As introduced in Chapter 1, the control station (CS) for UAV(s) may be based on the ground (GCS), aboard ship (SCS) and possibly airborne in a 'parent' aircraft (ACS).

It may be simply the control centre of a local UAV system, within which the mission is pre-planned and executed, but may also be part of a still larger system, or 'system of systems' when it is also interfaced with other components of a network-centric system, sharing information with and receiving information from other elements of the larger system. In this latter case the mission planning may be carried out in a central command centre and 'retailed' to the individual CS for execution.

The architecture of a typical unmanned aircraft system is shown in Figure 13.1. 'Architecture' refers to the arrangement of the interfaces and data-flow between the sub-systems of the CS.

The CS is the man–machine interface with the unmanned air vehicle (or air vehicles) system. From it the operators may 'speak' to the aircraft via the communications system up-link in order to direct the flight profile or to update a pre-flight-entered flight programme. Direct operation of the alternative types of mission 'payload' that the UAV carries may also be required.

The aircraft will return information and images to the operators via the communications down-link, either in real-time or on command. The information will usually include data from the payloads, status information on the aircraft's sub-systems (housekeeping data), altitude and airspeed and terrestrial position information.

The launching and recovery of the aircraft may be controlled from the main CS or from a satellite (subsidiary) CS which will be in communication with the main control station by radio or ground-based cable (wired or fibre-optic), but have its own direct radio link with the aircraft.

The CS will usually also house the systems for communication with other external systems. These may include:

(a) means of acquiring weather data,
(b) transfer of information from and to other systems in the network,
(c) tasking from higher authority, and
(d) the reporting of information back to that or other authorities.

Therefore a CS will contain a number of sub-systems required to achieve its overall function. These will depend in detail on the range and types of missions envisaged and the characteristics of the UAV(s) that it will operate. In general these sub-systems may be structured to include elements as follows:

Unmanned Aircraft Systems – UAVS Design, Development and Deployment   Reg Austin
© 2010 John Wiley & Sons, Ltd

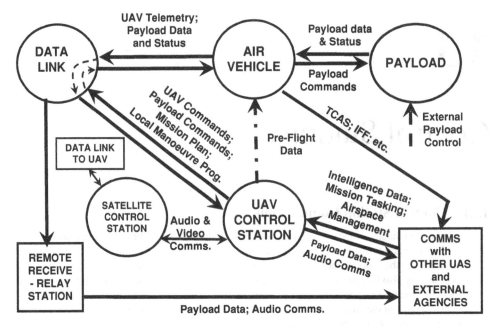

Figure 13.1 UAS architecture

a) The UAV flight controls; display(s) to present UAV status and recording equipment. The controls will interface with the UAV automatic flight control system (AFCS) either for manual real-time control of the UAV or to select and engage stored on-board flight sub-programs. One such program may, for example, control the UAV to orbit at a given radius and speed around a to-be-designated terrestrial grid reference in order to survey a site of interest. The grid reference will be inserted by mission planning, taking account of the terrain surrounding the site and the position of the sun relative to it at that time of day in order to obtain the most effective images from the most favourable direction.

Other programs, appropriate to rotorcraft UAV, may direct the UAV to hover over a selected point for surveillance or other purposes; also to hover over a point and descend at a controlled rate to a landing; or descend to ground level to take an NBC atmospheric sample and then to climb again to the operating altitude.

b) A sub-system to recognise the type of payload installed in the UAV and to suitably adjust the operation of the payload controls for that payload. Also included will be display(s) showing the payload status and data (imagery and/or other types of data) with recording media.

c) The ground elements (encoders, transmitter and receiver) of the communication link between the CS and UAV, and controls for their operation, for example raising or lowering radio antenna masts, steering them manually or automatically, changing frequencies as necessary, etc. A display may be included to show communication status. Whilst it is always desirable to have the radio antennae mounted on the CS to minimise conductor lengths, it is necessary to situate the antennae in a position to obtain good transmission and reception of the radio waves. In undulating or hilly terrain this may mean on or near a crest. For some military operations, unless the GCS is small and readily concealed, its position may become obvious to hostile forces and so vulnerable. A solution sometimes adopted is to mount the antennae on a small, mobile and less-detectable platform which can be located in a suitable position. This is often referred to as a remote ground terminal (RGT). The downside is that it will require its own levelling system and ground-deployed communication (probably fibre-optic) to the GCS. For many operations the difference in position coordinates between the GCS and RGT will have to be determined and entered into the mission computers.

d) Navigation displays for monitoring the position and flight-path of the UAV and the necessary computers to process the data.

e) Terrestrial map displays with computers to enable mission planning and the necessary calculations for it to be carried out. Records of previous mission flight paths will be retained for future repetition if required. Listing of selectable UAV flight programs will be held.

f) Communication systems with other players in the system of systems to enable data to be obtained for weather conditions, to receive mission requirements and to send data received from the indigenous UAV to other players.

Depending upon the complexity of the UAS, these subsystems may be housed as separate units within the CS or integrated as sub-modules within fewer units. Hence control stations, like the aircraft, come in all shapes and sizes and are staffed appropriately to the number and speciality of the tasks which they are required to perform.

13.2   Open System Architecture

The term open system architecture (OSA) implies that the system and most of its sub-systems are designed to accept the addition of new elements without redesign of the existing elements. Hence a new payload might be added to the UAV without fundamental change to the electronics or mechanics of UAV or the CS. Indeed an entirely new type of UAV might be added to the system and be readily operated in parallel with the current UAV from the same CS. To achieve this, the design standards of the electronic equipment and computer operating formats, etc. must be specified and be common to all potentially cooperative systems.

One recommended design standard is the open systems interconnection (OSI) architecture of the International Organisation for Standardisation (ISO). The OSI will call up the more detailed standards of, for example, RS-232 and MIL or STANAG standards.

STANAG 4586, in particular, as discussed in Chapter 9, defines a NATO interoperability standard that establishes a common protocol to facilitate the interoperation of various types of UAV and different control stations. STANAG 4586 is seen as being the adopted interface control definition (ICD) standard for both military and future civilian systems and defines a number of common data elements for two main interfaces. These are:

a) the data link interface (DLI) between a control station and UAVs,
b) the command and control interface (CCI) between the CS and other networking systems

CS range from a simple laptop computer control with a tripod-mounted antenna for micro- or mini-hand-launched UAV, through mobile stations in all-terrain vehicles for close- or medium- range systems, to comprehensive central command stations for MALE and HALE systems operating at longer ranges from fixed bases. A few typical examples are given below:

13.3   Mini-UAV 'Laptop' Ground Control Station

The Desert Hawk III, Mini-UAV, system uses a man-portable GCS, as shown in Figure 13.2 with communication system and antenna. The latter is mounted upon a tripod, which can be adjusted for more height range or placed on top of a building at up to 100 feet distant from the GCS.

The GCS incorporates the graphical user interface (GUI) and features a touch-screen laptop, allowing operators easily to input way-points onto a map base, and providing easy access to key and frequently used features. The Desert Hawk III software also uses digital terrain elevation data (DTED) with terrain contours for additional mission assurance. DTED information is uploaded to the unmanned vehicle at the time of launch, warning of lost-link or non-line-of-sight communication situations. Mission information is immediately updated when the unmanned vehicle is re-tasked to support a changing tactical environment.

Desert Hawk Back-Packed Control Station

**Both UAV and Control Station can be back-packed
or carried in a small suitcase**

Figure 13.2   Desert Hawk GCS. (Reproduced by permission of Lockheed Martin Corp)

An optional feature is a remote viewing terminal (RVT) which can operate in parallel with the main GCS and is back-packable. It allows imagery from the UAV also to be downloaded by forward military units.

13.4   Close-range UAV Systems GCS

The aircraft of these systems usually will be either ramp-launched or VTOL. The requirements and capabilities of both systems will be generally similar, except for the specifics of the control during launch and recovery.

This example is generally based on the Sprite VTOL system GCS, chosen principally due to the ready availability to the author of its pictures and details. It is merely a typical example and other systems will differ to some degree, depending upon their roles and sophistication. Differences between the VTOL and ramp-launched systems will be described later.

The GCS for close-range systems will usually be mobile and housed within an 'all-terrain' vehicle. The Sprite system is no exception to this and the 4×4 vehicle selected by the Swedish Government for their Sprite systems is shown in Figure 13.3. A pneumatically raised steerable radio mast is seen at the rear of the vehicle which carries antennae for the two radio frequencies used to communicate with the aircraft and also inter-network communications as required. A DGPS antenna is an optional roof-mounted fit. As with all mobile GCS, the vehicle is fitted with chassis-mounted jacks or stays which are lowered once the vehicle is on-site. These are necessary to stabilise the vehicle to prevent it 'rocking' under the influence of wind or the operators moving around within the vehicle. A stable base is required for narrow-beam antennae to be maintained accurately in azimuth and elevation.

Figure 13.4 carries a view of the rear of the vehicle which has accommodation for the modular control consoles shown in Figure 13.5. The two crew members are seated in the front cab en route and at the consoles when the system is in operation. In the case of the Sprite system, two aircraft are also accommodated in the GCS together with support equipment which includes a fuel supply, system spares, tools and operating manuals.

The Sprite system normally requires only two crew members to operate. The left-hand station is occupied by the aircraft controller who is usually the Commander (see definitions in Chapter 5). The right-hand station is occupied by the payload operator who controls the orientation of the imaging sensors and interprets the images. The payload operator also controls the several alternative payloads that Sprite

Figure 13.3   Swedish Army vehicle containing Sprite GCS

may carry, dependent upon the mission. He may also assist in the navigation and condition monitoring of the whole system. Different users may prefer an alternative assignment of tasks and may require the addition of a third crew member who may then adopt the role of Commander. Depending upon the degree of integration of the system within an overall network of cooperating systems, the Commander is most likely to be responsible for the inter-system communication. Whatever assignment of tasks is preferred by individual operating organisations, the automation of the flight profile and the ability to call up flight,

Figure 13.4   Internal view from rear of GCS vehicle

Figure 13.5   Sprite system control modules

imaging, navigation or housekeeping data on any of the three monitor screens enables a flexibility of tasking and easy cooperation between crew members.

It is of practical advantage for one crew member to be able to perform at least some of the tasks of another since that gives each member a better appreciation of the skills needed by the other and the pressures that the other may have to bear. It also may help one crew member to carry out another's tasks in an emergency.

The GCS also incorporates electrical power generation and air-conditioning which is required not only for the comfort of the crew, but for the climatic control of the computers, radio equipment and the monitors. This equipment also has its own controls and monitoring systems provided in a separate rack. Radio communication equipment for net-working with other agencies is provided. Figure 13.6 shows typical images on the mission monitors at the top of each console module.

The left-hand monitor carries a video image from a daylight colour TV camera which, on this occasion, is seen over-looking a riverside industrial area in heavy mist. When necessary, time is superimposed

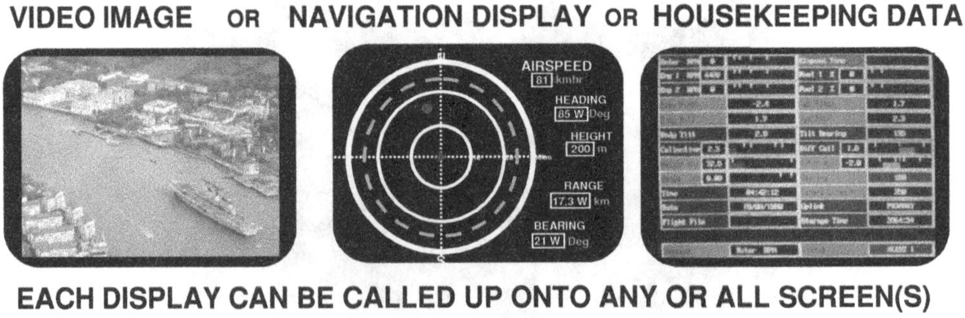

VIDEO IMAGE   OR   **NAVIGATION DISPLAY** OR **HOUSEKEEPING DATA**

EACH DISPLAY CAN BE CALLED UP ONTO ANY OR ALL SCREEN(S)

Figure 13.6   Sprite System. Ground Control Station Displays

digitally upon the video image for later reference and possible legal evidence. Controls beneath the right-hand display allow for camera lens and elevation and azimuth field of regard adjustment. The displays may be used for infrared images sent down from a thermal camera, and controls beneath the displays are available to select 'black hot' or 'white hot' images.

The centre monitor carries the navigation display which shows an icon representing the aircraft position in bearing and range relative to distance rings centred on the position of the GCS. In the Sprite system, an orange concentric ring shows the limits within which the aircraft must be kept for it to be able to return to base with the fuel then available. The ring obviously continues to contract during the flight as fuel is consumed.

To the right of the geometric display are digital data showing the aircraft height, speed and direction, and coordinates of its position at any time. If required a digital map of the area may be overlaid onto the display.

The right-hand monitor carries the housekeeping data. In the case of Sprite, this data is on two pages, either of which can be selected as required. The data includes items such as:

time,

fuel content of the tanks as percentage of full,

engine cylinder-head temperatures,

engine bay temperatures,

rotor speed and engine speed,

engine failure,

electrical power supplies,

control functioning and positions. i.e. elevator, aileron, rudder, engine throttle(s) for fixed-wing aircraft or collective and cyclic rotor pitch angles for rotorcraft,

functioning and temperatures of critical components, such as altimeters and gyros,

the type of payload that is mounted,

camera settings such as field of regard, lens settings, etc.,

coordinates of aircraft position,

commanded and actual aircraft height, speed, etc.,

radio transmission frequency option selected by FCS logic,

other data which is dependent upon payload and mission requirements.

Each parameter is allocated a condition bar which, where appropriate, indicates the value of that parameter. The bars change colour, dependent upon parameter condition. If the parameter is operating well within its range, the bar is green. If it is nearing the limits of normal range, the bar changes to orange. If it is outside limits the bar becomes red.

If any bar becomes red, a flashing red central warning light alerts the operators and indicates on which housekeeping page is that parameter. The operator can then select that page onto any screen to decide what action to take. As previously stated, for the Sprite system, each of the displays, video image, navigation or either of the two house-keeping pages can be called up onto any or all of the monitors.

Beneath the monitors are the control 'decks'. Here reside the keypad controls for inputting mission data or way-point updates to the aircraft. 'Control sticks' also are provided for optionally taking direct operator control of the aircraft in flight. Controls are provided for starting the aircraft, selecting pre-flight test data, ejecting any umbilical cord, unless this is done manually, and for activating the aircraft launch.

In racks beneath the control decks, are recording systems for recording the video images and house-keeping data. The power supplies for the computers, monitors, recorders and radios are also contained here.

Prior to launch it is necessary to perform a check on the GCS itself. The controls and displays for this may be mounted in the same racks as the aircraft control system or separately. It is a priority to check-out the efficiency of the GCS power supplies, communication systems performance and its operation of the aircraft control system.

The aircraft check-out process will be commenced similarly for both aircraft types, with the fixed-wing aircraft lifted onto the launcher ramp, and the rotorcraft placed on the ground near to the GCS. Electrical power is usually supplied to the aircraft by umbilical cord. An umbilical cord may also take check-out data from the aircraft to the GCS or it may be done via a radio link.

Any flight programmes may be inserted into the aircraft FCS at this stage or prior to it, and basic pre-flight electrical and electronic check-out performed on both airframe and payload operation. Engine(s) will next be started and control functioning carried out. If everything is in order, any umbilical cables will be jettisoned.

Some differences between ramp-launched or VTOL aircraft now occur. The former will be released with the launcher facing into the wind and the engine(s) at full throttle to carry out a programmed climb-out to a proscribed altitude, flight-path and speed.

The latter will sit stationary on the ground beside the GCS with rotors stationary and the engine(s) running at idle speed. The aircraft will be fully under GCS radio control. The engine throttle(s) will next be opened to 'flight' position, engaging the rotor(s) which will accelerate to a governed flight speed and confirmed as such. The aircraft can now be lifted-off to a low hover and, if deemed desirable, control and video checks can be made again in-flight. With the aircraft now airborne and ready to go, either the pre-programmed mission instructions are enabled or a vertical rate of climb and height command is inserted to send the aircraft to hover at altitude, awaiting further instructions.

For recovery, following completion of the mission, the fixed-wing aircraft will be brought over a suitable clear landing area, the engine shut down and the parachute and cushioning airbags released. Provision will be made at the operators' console for the controls of this phase.

The VTOL aircraft will be brought to a hover above a nearby landing 'patch' and then descended to ground either by direct manual control or by an auto-land program. After landing, the engine(s) will be closed down and the rotors brought to a halt.

13.5   Medium- and Long-range UAV System GCS

A medium or long-range UAV will usually be launched horizontally on wheels along an airstrip or hardened runway. Provision has therefore to be made to control the aircraft during its take-off run to flight speed and lift-off. This and subsequent recovery, unless made by parachute, is currently accomplished under direct operator control, usually with the aircraft in direct view of the operator. Progress is now being made to provide some degree of automation for the airstrip launch and recovery phases.

A medium-range system will normally employ a mobile GCS, either in a road vehicle or in a trailer so that it can be suitably positioned relative to the airstrip. A vehicle-based GCS for the Hermes 450 UAV System, for example, is shown in Figure 13.7, and the interior of the station is shown in Figure 13.8.

The equipment will be generally similar to that contained in the close-range system GCS but, taking account of the longer duty time of the operators, may provide more spacious accommodation and up-graded system–operator interfaces. Additional crew members may be required, especially for the more complex payloads carried. A specialised image interpreter may be included and a system Commander would then probably be obligatory in overall command and integrating role.

Figure 13.7   Hermes UAV GCS vehicle

Figure 13.8   Hermes GCS interior

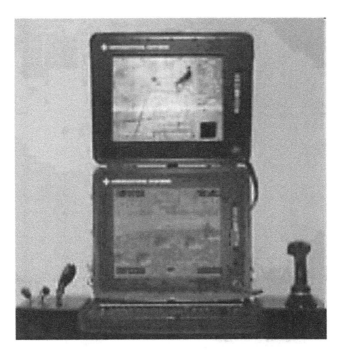

Figure 13.9   Predator GCS for launch and recovery phases

In addition, as the longer-range and larger aircraft may carry a wider range of imaging sensors to include optical, infrared and synthetic aperture radar imagers. These may be displayed independently or fused together. This capability will require greater computing capacity, more monitors and recording equipment.

It will be noted also that, with the extended range and endurance of this type of system compared with close-range systems, further and more powerful radio equipment has to be accommodated. This will include satellite communication in order to relay commands and data to and from the aircraft during beyond line-of-sight operation.

In turn, all this extra equipment will generate heat which will add demand for more extensive air-conditioning facilities.

The specifically long-range system, however, is probably operated from a fixed base on an existing airfield with metalled runway surfaces. An additional mobile control station may therefore be provided for the launch and recovery (Figure 13.9).

As with the medium-range UAS, development is currently being made for the automation of the launch and recovery, easing the task of the operators and possibly removing the need for the auxiliary control station.

As an example of the even more comprehensive range of equipments installed within a long-range, long-endurance system GCS, two views of the interior of the Predator System GCS are shown in Figures 13.10 and 13.11. More, and longer-ranging, radio equipments will be installed for wider networking. Crews will operate in shifts during the long flight-times of the aircraft which will often exceed 30 hours. Provision will therefore be made for 'handing over' responsibility for the mission.

If the aircraft carries armament, then a further crew member, the Weapons Systems Operator, may be required to select, monitor, release and guide the weapons onto target.

Figure 13.10   Interior view of Predator GCS

Obviously it is desirable not to over-crew the system and different operating forces will have their own ideas on this.

Currently, systems such as Predator and Global Hawk may launch their aircraft from a GCS on airfields relatively close to the theatre of operation but, after launch, be controlled from a command centre which may be up to two thousand kilometres away, or more. This enables the more forward GCS to be less comprehensive in equipment and more mobile than it would otherwise be. It reduces the number of personnel committed in the forward area.

Figure 13.11   Another interior view of Predator GCS

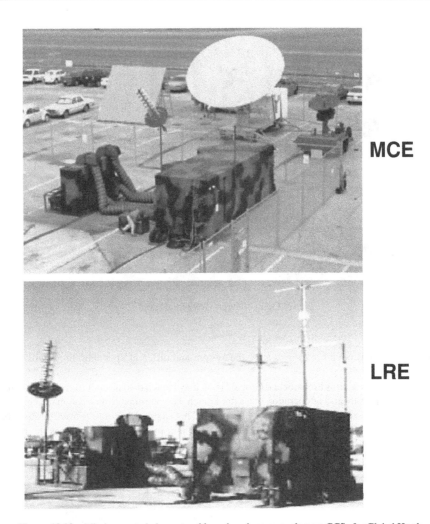

Figure 13.12   Mission control element and launch and recovery element GCSs for Global Hawk

The Global Hawk system employs two ground-based command and control elements. These are the launch and recovery element (LRE), otherwise known as a satellite control station, and a mission control element (MCE). An external view of both is shown in Figure 13.12.

The LRE contains equipment similar to that in the control stations of close-range systems, though with longer-range radio, to enable direct operator control of the aircraft during launch and recovery and to direct and monitor the flight of the aircraft for up to 200 nautical miles (370 km) towards its planned area of operation. At, or about, that distance the aircraft is acquired by the MCE for control during the majority of the mission. The LRE is designed to be rapidly deployable and fits onto two pallets which can be carried within a tactical transport aircraft.

The MCE uses satellite communications as the UAV will usually be out of its line-of-sight, i.e. beyond line-of-sight (BLOS). Unlike the LRE, the MCE can control more than one UAV at the same time. It also contains more communication equipment as it can act as a command centre, receiving and transmitting information from and to other intelligence agencies. The MCE is installed in a 20 m-long trailer and, although deployable, is not as readily so as the LRE.

13.6   Sea Control Stations (SCS)

A comparison can be made with the various 'grades' of ground-based control stations. Naval operations may be conducted either using a UAV system based aboard the ship or by taking over control of an aircraft initially launched from a land site. In either case the UAV system control may be completely or partially integrated within the ship's control centre and using the ship's power supplies. Integration of the UAV system radio antennae alongside the ship's radio and radar antennae may require more consideration.

In the former case, the UAV may be a close-range or medium-range system rather than a MALE or HALE system which would be problematic in its accommodation and launch and recovery for all but aircraft carrier vessels. The on-board UAV system therefore is more likely to operate a type of fixed-wing aircraft which is more readily launched and recovered, such as Scan Eagle (see Chapter 4, Figure 4.18) or rotary wing types such as Firescout, Sea Eagle or Sprite (Figures 4.13, 4.14 and 4.19 respectively), that are even more small-ship compatible. The sea control station would in each case be similar to, and probably use the same modules from, their ground-based counterparts. Support equipment might be a little different (see Chapter 14).

13.7   Air Control Stations (ACS)

Development has been proceeding to assess the logistics and practicality of controlling a UAV from an airborne platform. The UAV might be launched from the controlling fixed- or rotary-winged manned aircraft, off-ground or off-board launched, with the manned aircraft subsequently taking over control.

Such an arrangement would offer a more covert observation of targets, such as moving vehicles or fleet-shadowing. In attack operations it would offer a more covert penetration to a target to release weapons or to direct fire from the airborne platform or ship itself.

A number of problems have yet to be resolved before such a system reaches fruition, not least controlling the spatial juxtaposition of UAV and manned machine, antenna mounting and operation of the latter and stowage of the ACS within the tight confines of all but a large manned aircraft. One of the drivers for UAV autonomy (Chapters 10 and 27) is that it could be a major contributor to the facilitation of control of UAV from manned airborne platforms.

14

Support Equipment

Any electromechanical system will require equipment to support its operation. The equipment items may be available at various levels. For a UAV system many of the items will be at first-line availability – i.e. immediately available and carried with the system. Other items will not need to be immediately available and will be in a base-store. We deal here with the first category which will almost certainly include:

a) operating and maintenance manuals,
b) consumables such as fuel and oil, recording discs, etc.,
c) components required for first-line servicing,
d) critical components which may have an occasional need for replacement due to a vulnerability to damage, or components which are replaced 'on condition',
e) tools,
f) subsidiary equipment.

14.1 Operating and Maintenance Manuals

The manuals will usually include a System General Specification and a System Operating Time Log, either separately or as part of the Operating Instructions and Maintenance Manual respectively. The System General Specification will not be as detailed as the System Design Specification, but sufficient to define the major structure of the system and its components, the limitations in its operation and the reasons for them. It will refer to the System Design Specification.

The Operating Instructions will include directions for setting up, checking out, preparation for and carrying out the missions and, of course, the retrieval of the aircraft at completion of the mission and system stand-down. It may also include directions for training unless there is a dedicated System Training Manual.

The Time Log will keep the operating history of the system. It will record the identity of the operating crew, the time and duration of each mission or test and the conditions under which it took place; and any significant technical observations and recommendations. It also serves as a record for alerting imminent scheduled maintenance.

The Maintenance Manual will call up the servicing and item replacement intervals for each module of the whole system. Directions will be given for checking the condition of specified components, carrying out necessary cleaning, lubrication and adjustment, etc. and component replacement on a 'lifed' basis. A record will be made of the completion of scheduled maintenance and also any corrective maintenance required and completed.

Unmanned Aircraft Systems – UAVS Design, Development and Deployment   Reg Austin
© 2010 John Wiley & Sons, Ltd

The manuals will be drafted in the design stage and amplified in the development stage. They will be regularly reviewed for updating. Copies, both hard and on disc, will be kept with the system, usually within the control station.

14.2   Consumables

Depending upon the size of the system and the quantities required, lubricants, cleaning materials, batteries, CDs, fuel may be carried within the CS, especially if the CS vehicle uses the same type of fuel as the aircraft. Otherwise, for larger systems or for safety reasons, fuel will be carried in a separate vehicle or bowser.

14.3   Replaceable Components

The lifed components will be listed in the Maintenance Manual and the replenishment of these to the UAV system's operating base will be required from the logistics support organisation. If the UAV system is a mobile one then the operators must ensure that the necessary components are carried, dependent upon the operating hours which are expected to be made during the time that the system is away from its base or other source of supply.

14.4   Vulnerable and On-condition Components

Vulnerable components may include those which may be damaged, for example, on landing during adverse weather conditions. These could include the detachable wing-tips or propeller of a fixed-wing aircraft, or elements of the undercarriage in fixed- or rotary-wing aircraft. They might possibly include the rotor blades of a rotary-wing machine.

On-condition replaceable components might include spark-plugs in ignition engines. The control station equipment may also contain components that are in these categories. These components will be identified during system development, and proving trials and will be listed as such in the Maintenance Manual.

14.5   Tools

Tools will include those required for routine operation and for maintenance, and will cover electrical, electronic and mechanical elements. They will range from electronic test meters and battery-chargers to torque spanners. There may also be jigs or rigs for testing sub-system functioning.

Tools required in routine operation may include start-up or check-out equipment. Jigs may include tools for checking, for example, control settings and ranges. Rigs may include those required for a payload functioning check.

The number and types of tools carried will depend, as before, on the type of UAV system deployed.

The requirement for tools will be called up in the design phase and confirmed or modified in the development phase of the system. An aim should be to minimise the number of tools required, particularly special-to-type tools, and employ only standard, international tools. Depending upon their category, the tools will be listed in the Operating or Maintenance Manuals or in both.

14.6   Subsidiary Equipment

Possibly considered as being part of the UAV system, especially if integral with the CS vehicle, the necessary means of generating electrical power may, instead, be carried in a trailer. It may therefore be considered as subsidiary equipment and may also require its own dedicated fuel supplies and maintenance equipment.

A UAV system operating from a fixed base may normally be able to utilise supplies from a local grid, but the precaution of having a stand-by indigenous generator is essential.

15

Transportation

Transportation of a UAV system will range widely from the one or two man-carried backpacks of a micro-UAV, through to a VTOL close-range system carried entirely within a single small 4×4 vehicle, to a mini-UAV HTOL system with a 4×4 vehicle towing a single trailer, up to the multi-vehicle/trailer transport system of a close, medium, or long-range HTOL UAV system.

15.1 Micro-UAV

Both HTOL and VTOL micro-UAV systems may be backpacked and will be operated in the field. An example of the former is the Lockheed Martin Desert Hawk. The system, comprising the laptop control station and the aircraft is shown in Figure 15.1.

15.2 VTOL Close-range Systems

As with the micro-UAV system, these systems are generally compact, with the aircraft being man-portable and operable by two personnel in the field. A coaxial-rotor or ducted-fan VTOL aircraft with its control station and support equipment can be accommodated in one small 4×4 vehicle, as demonstrated by the Sprite system and shown in Figure 15.2. Other, less compact, VTOL aircraft such as the EADS Scorpio or the CybAero AB Apid will require a trailer in addition, or a much larger single vehicle.

15.3 HTOL Close-range Systems

This type of system is also required to be operated in the field. A HTOL close-range system aircraft is most likely to be too heavy and dimensionally too large to be launched by hand or by a simple bungee system. It will require a launching ramp system. The recovery may be either by skid landing or by parachute and airbag. The gross mass of the aircraft will, however, be within the limit of a two-man lift and should require no additional vehicular means of retrieval.

Both aircraft and ramp, with much of the support equipment, can be carried in a small trailer towed behind a single small 4×4 vehicle which is then able to house the rest of the system, including the operators.

Examples of such systems include the Flight Refuelling Ltd Raven, the Cranfield Observer and the EMT Luna. The trailer of the Raven system, shown in Figure 15.3, carries two disassembled aircraft within it and the launch ramp is mounted above.

Figure 15.1   Desert Hawk system

15.4   Medium-range Systems

These systems will be generally operated from forward area bases with a temporary airstrip available. In both HTOL and VTOL form, the aircraft will be heavier and require a larger supply of fuel which may have to be transported unless the systems are operated from fixed bases. The aircraft will no longer

Figure 15.2   Sprite VTOL close-range system

Figure 15.3   Raven UAV launch from trailer

practically be man-portable. Some will have a wider range of payload types than the close-range systems and may require increased accommodation for a multiplicity of monitors and an extra member of the crew.

The HTOL aircraft may be ramp-launched or make a wheeled take-off to begin the mission. In the former case it is probable that a crane will be required to position the aircraft onto the ramp. A further vehicle, possibly with an integral crane, will be needed for retrieval. For a wheeled take-off/landing vehicle, unless the system operates from a fixed base and the aircraft can be taxied into its compound it will also require a further vehicle for retrieval. For changing sites, either system will require its aircraft to be partly dismantled and possibly carried on the retrieval vehicle. In practice, however, that may require a larger vehicle which may not be suitable for the retrieval operation.

Thus the HTOL system may comprise three vehicles, or four if one vehicle cannot be adapted to fulfil two of the four rôles. Clearly this is an area in which the initial system design needs to address the vehicle footprint and to strive to minimise the number and types of ground vehicles needed for deployment.

A single, though larger, GCS/tow-vehicle with a trailer should suffice for the VTOL aircraft, even for changing operating sites. Unless the aircraft can land onto the trailer, provision will have to be made to winch the aircraft onto the trailer.

As an illustration of the role and size of vehicles Figure 15.4 shows the number of vehicles used for a Phoenix system. It should be noted that every additional vehicle increases the number of personnel required to operate it and requires extra provision for fuel. That, in turn, adds to the further expansion of the system.

15.5   MALE and HALE Systems

Both of these types of system will generally be operated from fixed bases with metalled runway surfaces and electrical power available. Unlike the previous systems, they will be generally static installations with their movement seldom required.

The control stations will usually completely occupy a large trailer, as shown in Chapter 13 (Figures 13.11 and 13.12) and will require a substantial sized vehicle if it is required to be moved.

Phoenix Troop:-
5 Vehicles, 11 Personnel, for 1 Air Vehicle

Figure 15.4   Phoenix system

Maintenance of the aircraft will generally be carried out in a hangar with the usual facilities and that of the GCS on site. Supplies will be land- or air-freighted in.

Currently there are no VTOL aircraft operating in this rôle, but the situation may change as indicated in Chapter 27.

16

Design for Reliability

Some newcomers to the UAV scene imagine that the aircraft component of the system, at least, can be cheap since it has no need to be reliable. This misguided thought rests purely on the idea that, if the aircraft crashes, then no on-board life is lost.

This is far from being acceptable. The reliability of a UAV system must be assured for the following reasons:

a) If a UAV system fails whilst on a mission, then that mission has failed. In a military operation, this lack of information, etc., could result in loss of initiative or worse, hundreds of deaths. In both military and civilian operation a failure could result in injury or even loss of life to the operators.
b) If the aircraft crashes, injuries or fatalities might be caused to the over-flown population.
c) Any loss or malfunction of the system can result in loss of the service provided, loss of the facility and costs of repair or replacement. Unreliability is a major driver in whole life costs.

Although reliability is the core consideration, the wider aspects of availability and safety also enter the equation and are defined.

Reliability

This is measured as the mean time between failures (MTBF) in hours of a system or sub-system and obviously the greater the value, the more reliable the system or sub-system. It is often more convenient to express this as the reciprocal, i.e. the degree of unreliability, and is usually presented as the number of failures of a system, or sub-system, in 10 000 hr.

Availability

Of particular importance to an operator is whether a system is 'ready to operate' when needed. It is measured as the number of times when the overall system is ready (available) compared with the number of times that it is required. This is presented as a percentage and, obviously, 100% is the aim.

Availability is affected by the frequency of failure of a system and the time taken to repair or replace (and check-out) the failed component or sub-system. Based upon a theoretical continuous availability requirement, the overall time of unavailability will be the number of failures in a period (say N), e.g. 100 000 hr, multiplied by the average time to repair and return to service (say T). The percentage availability is then $[10^5 - (N \times T)]/10\ 000$ expressed as a percentage.

Poor design for ease of maintenance can significantly worsen availability by increasing the time taken for repair or replacement.

There may be periods during which a UAV system is not needed for operation and is in store. It is therefore important for the designers to ensure that the system will not deteriorate when in store so that, at least for relatively short periods of inactivity, it is immediately available when required. Failures might be caused, for example, by batteries discharging or humidity and temperature changes causing damage through condensation. Such failure will reduce the system's availability figure.

For longer periods of inactivity it is acceptable that a 'time to reinstate' will be required and specified since it may be necessary to inhibit engines and remove, then subsequently replace, batteries or other components which may deteriorate with time.

Safety

Failures of the system, and this more often applies to failures within the aircraft, which are 'catastrophic' – e.g. cause the aircraft to crash and cause injury or even fatality to the over-flown population or to the operators themselves – constitute a safety hazard.

Failures which may, or may not, cause further damage or loss of the aircraft whilst it is on the ground may be a hazard to operators. This may result from the failure of critical components such as helicopter rotors, propellers or rotating engine parts such as turbine discs. Particular precautionary measures are applied to these components which are extensively tested in the development phase of the system.

16.1   Determination of the Required Level of Reliability

It is necessary first to define the drivers for the need for reliability in aircraft, and the level depending upon the result of failures (see also discussion of this in Chapter 18).

a) There are those catastrophic failures which will cause the aircraft to crash and possibly cause injury or loss of life to people on-board or on the ground.
b) Those failures, termed class A which will cause severe, probably irreparable, damage to the aircraft.
c) Class B failures which will debase the aircraft function, may cause its mission to be aborted, but not cause it to be severely damaged.

Failures of Type (a)

Although any loss of life for whatever cause is a tragedy, in the real world accidents do happen. In the case of aviation, aircraft are designed to a system which aims to minimise the occurrence of failures that may lead to loss of life. However, realistically, the achievement of 100% reliability is an impossibility and an element of risk must be accepted.

Civilian airliners are expected to achieve a high level of reliability such that the probability of a crash causing death or serious injury is accepted to be no more than 10^{-9} per flying hour. This takes into account the fact that the probability of a passenger aboard the aircraft sustaining serious injury in the event of a crash approaches unity. In the case of the UAV, of course, it is zero for passengers so relates only to injuries or fatalities suffered by people on the ground.

The probability of death or injury happening to persons on the ground, if those persons are connected with UAV operations, and so obtain benefit from them, might be considered to be acceptable by them at the same level as airline travel of 10^{-9}. People who view UAV operations as providing no benefit to them might expect a lower probability of their being affected. Therefore an acceptable probability of a UAV crash causing death or serious injury of 10^{-10} per flying hour has been suggested by the author.

It may be argued that the probability of a person on the ground being struck by a crashing UAV is the product of three probabilities – the probability of UAV failure type (a) occurring multiplied by the probability of the UAV penetrating to a victim multiplied by the probability of a person being in the vicinity. i.e. $P_d \times P_p \times P_b$.

Comparing the probability of injury caused by the crash of a typical, averaged-size commercial airliner with that of an average-size ubiquitous type of UAV might involve the comparison of the characteristics of a Boeing 757-767 type with an amalgam of a close- to medium-range type of UAV as follows:

Aircraft type	Civil airliner	UAV
Mass m (kg)	100 000	200
Impact velocity v (m/s)	70	30
Wing-span b (m)	40	6
Product $m \times v \times b$	2.8×10^8	3.6×10^4

This makes the assumption that the likelihood of aircraft components striking a person is proportional to the momentum of the aircraft and the swathe-width that it would carve out. On that basis, the probability of a typical fixed-wing UAV causing injury is 1/10 000 less than the airliner, i.e. 10^{-4}.

The probability of a person 'being there' is in some way proportional to the population density of the area of operation. An airliner, at least in its departure and arrival phases of operation, will be flying over densely populated areas. Apart from micro-UAV designed to provide surveillance in urban areas, UAV will generally operate over more sparsely populated areas. The ratio of over-flown population density during critical phases of operation of airliners to UAV may be as high as 1000 to 1. i.e. 10^{-3}.

Hence, it may be argued that, depending upon the size and mass and the area over which it operates, the acceptable probability of a catastrophic failure for the UAV might be reduced as follows:

$$P_d = 10^{-10} \div 10^{-4} \div 10^{-3} = 10^{-3}$$

That is to suggest that the occurrence of one critical defect, which could cause the aircraft to crash, is acceptable every 1000 flying hours for the small- to medium-size UAV operating over sparsely populated areas. The probability of such a defect must be lower for larger UAV and if operating over more highly populated areas.

The above analysis is merely conjectural and is provided to illustrate the factors and issues involved. As far as the author is aware there is no formally accepted analysis nor percentages. Clearly more soundly based criteria are needed and demonstrated to have been met before more general operation of UAV in unrestricted airspace can be expected.

Failures of Types (b) and (c)

The other criterion for assessing a required level of reliability for UAV systems, is economic. Herein lies a dilemma. The achievement of reliability in a system is expensive. The costs arise from the extensive testing necessary to develop and prove a given level of reliability in the system, and also from the costs of maintaining the quality standard in continuing production. The former is virtually impossible to predict but is nonrecurring and so is mitigated if a large number of systems are produced. The latter is a recurring cost and will be applied to each system sold.

There is therefore a balance to be achieved between initial affordability and operating costs. Other complications, of course, in the equation are:

a) No matter how attractive the system is in terms of costs, if it fails to achieve a mission, then its whole purpose is in question. This applies to both military and civilian operation.
b) The philosophy of regarding the UAV part of the system as expendable is mooted for some military operations. Consider the UAV operating, for example, a surveillance mission and is covering a scenario where it is imperative to maintain that surveillance. It is suggested that rather than discontinue the surveillance to return the UAV for refuelling, the option of keeping it on station until it crashes through lack of fuel should be taken. If the UAV is expensive, the option to expend may not be taken, especially if the purchase cost has limited the numbers available to the military operators.

It is necessary to know how often scenario (b) will occur compared with scenario (a). For a system of very low reliability, (a) will occur more often. For a system of very high reliability, (a) will seldom occur, but nor may (b). In these extreme cases, the cost of systems losses and repair, together with the resulting loss of business is likely to far outweigh the extra cost of the reliable systems. Somewhere between these extremes must lie a practical answer, but it is a difficult equation to formulate, let alone to solve.

The costs added to achieve a specified level of reliability depend upon:

a) the complexity of the total system,
b) the level of reliability specified under specified ambient conditions – temperature, altitude, humidity, precipitation, day/night, type of operation, etc.,
c) the availability of components with a known level of reliability,
d) the success of the design phase in 'designing-in' reliability.

Therefore it is impossible to generalise what the total costs will be. Even if a history of the development costs of previous systems is known, no method has yet been devised to use that information to predict the costs of a future system. Therefore the determination of a required level of system reliability on an economic basis does not appear to be currently feasible.

Historical Levels of Reliability

The defect rate or the loss rate in aviation is measured as the number of occurrences per 100 000 flying hours. This information is seldom released by operators or manufacturers so that only a limited amount of data is available.

According to the US National Defense Magazine, quoting the US National Defense Department's 2002 UAV Roadmap report, the initial program reliability goal for the Predator was a loss rate of less than 50 per 100 000 hours. In the same report, US Pentagon announced that the Predator RQ1B suffered a loss rate of 31 and, though it had met the initial goal, it was considered to be unacceptable. The Defense Department set a target rate for large UAV of less than 15 to be met by 2015. No target was set for small UAV.

What is unclear is whether either the reported losses, or the target values, include losses due to enemy action or operator errors. The implication is that the losses due to defects and losses due to enemy action cannot be separated as it is not known what has happened if a UAV fails to return.

16.2   Achieving Reliability

Reliability has to be designed-in from the beginning of the design of the system. It may be either impossible, or very expensive to correct any shortcoming in the later development or operational stages. Poor design can lead to premature wear or fatigue of materials.

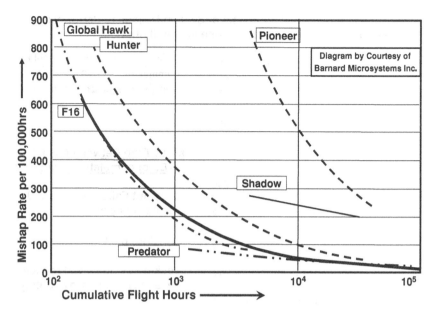

Figure 16.1   Mishap rate comparison (Barnard Microsystems Inc.)

Engineers have developed techniques to ensure that a system achieves a given level of reliability and these will be explained. However, it is first necessary to define and quantify the level of reliability that is required.

Historically, UAV reliability has been seen to improve as the systems accumulate flying hours. Figure 16.1 shows the trend for five different types of UAV systems. Interestingly the trend for the Global Hawk almost matches that of the F16 jet fighter aircraft. The statistics shown here, however, are open to some questions. The UAV data relates only to the aircraft and no data is presented for faults in the control station which may have contributed to or even caused the loss of the aircraft.

The Predator and Global Hawk were, until recently, manually flown on take-off and landing. The 'pilots' had very few physical clues and limited peripheral vision compared with pilots flying manned aircraft, and many mishaps occurred in these phases. These incidents may not be included in the data shown.

Other aspects worthy of considering in viewing statistics are:

a) Commanders may be more prepared to risk the loss of a UAV compared with a crewed aircraft.
b) It might be expected that a more complex aircraft would suffer a higher defect rate than a less complex one.

The figure indicates that the smaller (and simpler) aircraft suffer a greater mishap rate than the larger ones. However, it may be that this is not purely as a result of technical fault, but that they are operated in a more hostile environment and that they are seen as more expendable in the field than the more expensive aircraft.

Nevertheless, the trend indicates a strong reduction in defects as flying hours are amassed due, presumably, to fault identification and modification action being taken. The necessity for the development of reliability during the service of the aircraft is, however, disruptive and expensive. It is better to achieve reliability before the system enters service.

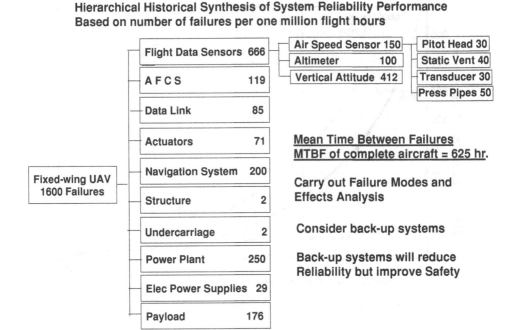

Figure 16.2 Reliability synthesis

16.3 Reliability Data Presentation

The probability of a defect in a system is the sum of the probabilities of defects of all of the components of which it is composed.

Figure 16.2 presents a synthesis made in 1993 by Reliability Consultants Ltd of a notional UAV aircraft based on averaged data obtained from established military aircraft data banks for fixed-wing aircraft. It suggests that using selected equipment a UAV aircraft *entering service* would have an overall failure rate of 160 failures per 100 000 flight hours. In light of the records of Figure 16.1 the 1993 analysis is seen to be very optimistic, but all the aircraft may be using new, relatively unproven, equipment, especially the smaller aircraft which often have been forced to use power-plants assembled using chain-saw engine components, control equipment from model aircraft sources, etc.

However, the synthesis is valuable as an example of how an aircraft designer will assess the probable reliability of a new design of aircraft. For UAS the total system failure rate is important. Therefore a similar synthesis of other major elements of the overall system, principally the control station, and its interfaces with the aircraft, will need to be added to the air vehicle analysis.

All the sub-systems synthesised (or analysed in the case of records from operating UAV systems) should be reduced to component level as shown for the flight data sensors in Figure 16.2. The power-plant, as a further example, will be decomposed to show defect rates for basic engine, fuel supply, carburetion or injection, cooling system, etc.

Twelve years later (2005) than the Reliability Consultants data referred to above, some data on the modes of failure of UAV aircraft were released and are summarised in Figure 16.3. Unfortunately no data on the performance of the control station is available to the author.

Figure 16.3 presents the failure modes as a percentage of the total aircraft failures attributed to five areas – power-plant; flight controls; communications; miscellaneous and that resulting from human

Aircraft	Power-plant	Flight Controls	Communi-cations	Human Errors	Miscellan-eous
Predator A	23%	39%	11%	16%	11%
Predator B	53%	23%	10%	2%	12%
Pioneer 2A	29%	29%	19%	18%	5%
Pioneer 2B	51%	15%	13%	19%	2%
Hunter 5A	38%	5%	31%	7%	19%
Shadow	38%	0	0	38%	24%
Average of above UAV	38%	19%	14%	17%	12%
Average of IAI UAV Fleet	32%	28%	11%	22%	7%

Figure 16.3   UAV failure mode summary

errors. The data is that obtained from six different aircraft. At the bottom of the figure the average for all aircraft is shown and also the averaged failure modes for the Israeli Aerospace Industries (IAI) UAV fleet.

It is of interest to see how the 1993 prediction compares with the realisation, although it would have been more meaningful if the recent results had been released showing the data reduced to component level. Also the 1993 prediction did not consider the human operators as part of the system and this omission will skew the comparison of results.

Assuming that the flight control sub-system in the 2005 data includes the flight data sensors, the AFCS, the actuators and the navigation system, the 1993 prediction indicated that the majority (66%) of the failures would occur in the flight controls and that 16% would occur within the power-plant.

In reality, Figure 16.3 shows that the greater number of failures occurred within the power-plants (38%/32%) with the flight controls (19%/28%) in second place. The communications proved less reliable (14%/11%) than predicted in 1993 at 5% of the total. Looking at the data for the individual aircraft indicates some flaws in the form of presentation shown in Figure 16.3. A comparison of the results for the Pioneer A and B models could indicate that the power-plant in the B model was substantially less reliable than that in the A model. However, the two aircraft are substantially the same with the same power-plant, the difference only being that the B model is so called to signify that the aircraft was later fitted with a greater range of payloads than that originally carried.

Inevitably, as portrayed in Figure 16.1, as the aircraft amassed flying hours, modifications were made to rectify the faults in the flight controls, the communications and the other miscellaneous sectors. Understandably it was more difficult to make changes to the power-plant so that that sector is now the major contributor to defects on a percentage basis, even though its failure rate is unlikely to have changed.

Another surprise is seen in the comparison between Predators A and B. Direct comparison is less meaningful here since the two aircraft are somewhat different, the latter being virtually a scaled-up version of the former, but built by the same company and with similar technology except for the engine. The surprise is that the well-developed aircraft turbo-prop engine of the B Model is shown to be responsible for over half of the total failures whereas the piston engine of the A Model is responsible for less than one-quarter of the total aircraft failures. There has to be another hidden factor here.

Hence the only meaningful presentation of data which can be realistically used to identify areas which are the most critical in the search for reliability is the detailed reduction of the system into the failure rates of its components as illustrated in Figure 16.2.

16.4   Multiplexed Systems

There is a distinction to be made between reliability and safety. A system can be unreliable but safe if the failures are such that no catastrophic crash ensues.

Duplication of the less reliable elements may offer a higher probability of safe operation, but will increase the overall failure rate, thus making the system less reliable. A higher failure rate will be expensive in repairs and the inconvenience will not endear the system to the operator. Thus, ironically, duplicating troublesome elements to improve safety is no panacea in terms of reliability and will make the search for reliability even more urgent. In some airliners and sophisticated combat aircraft, AFCS electronic channels are triplexed or even quadriplexed.

This is done since in a duplex system it may not always be possible for the aircrew to determine which of the two channels has failed. With three or more channels, a voting system will disable the single channel which 'disagrees' with the other two or three. To the author's knowledge no UAV has adopted such a system and, with increased reliability of computer-based AFCS systems, may not be merited.

The US RTCA (Radio Technical Commission for Aeronautics) committee has recommended (Ref. RTCA SC-203 - 2003) that any UAV which is intended to enter national air space must have two engines, as the available statistics seem to show that the power-plant is the least reliable part of a UAV.

Note, however, that three years later, Northrop-Grumman announced that their Global Hawk UAV System had received Military Airworthiness Certification on 25 January 2006 and a National Certificate of Authorization from the FAA which enables the aircraft to be flown routinely within National Airspace. The aircraft has only a single turbo-fan engine (see Chapter 4, Figure 4.1), but the engine is manned-aircraft certificated with known high reliability. It is understood that Global Hawk mission planning for each sortie has contingency for possible, albeit unlikely, engine failure at any time throughout the sortie. The system back-up capability, including emergency electrical power, would enable gliding flight to be made to pre-planned landing/crash areas. The provision of this contingency may also be a factor in the certification of single engine operation.

Installing two engines, however, will increase the weight and cost of the UAV and reduce its reliability. It will only increase its safety if the aircraft, following the failure of one engine, is able to retain the performance which it would have had as a single-engined design. It is likely to require that the engines be installed as totally separate and independent power-plants, complete with their own fuel supply, ignition systems and alternators (if engine-driven), etc.

An aircraft having two engines with a full one-engine-failed performance means that each engine will normally be operating at half power or less. That is good news for a piston engine as it can then be expected to have a longer life and be more reliable, thus possibly saving the investment cost of the second engine. It may also reduce the fuel consumption of the aircraft, as the sfc of piston engines are at their worst at high powers, unless it is negated by extra nacelle drag.

Gas turbine engines, however, are at their most efficient and have their lowest sfc at high powers so may not give as much benefit in a twin-engine configuration.

Of course, there are other benefits from having an "over-powered" aircraft to offset the extra weight and cost of a second power-plant. These are an enhanced rate of climb and a shorter take-off run.

The received data tends to indicate that, except in the case of the Hunter UAV, the communications systems ranked only third in the unreliability stakes. However this must depend upon the electromagnetic environment in which the UAV must operate.

It is reported that at least one type of UAV system has suffered many losses through interference from external radiation sources. In addition to other means of protecting the integrity of the communication links, as discussed in Chapter 9, there is an advantage in having two parallel command links, provided that they operate at widely separated radio bands. The implementation of such a system in the Sprite UAV, with a logic system incorporated to select the better signal at all times, certainly proved its worth. The air vehicle was able to fly through a wide range of radio and radar transmissions, which included ships' radars and missile guidance radars, with impunity.

This example also illustrates the advantage of having the alternative (back-up) system take a different form from the primary system. Direct duplication of systems runs the risk of common failures. Another example of the provision of a back-up system which uses a different *modus operandi* from the primary system is again taken from Sprite. The rotor speed was measured from a magnetic pick-up on a rotor shaft and the signal fed to the AFCS, An output from the AFCS then adjusted the throttles of the two engines to maintain a constant rotor speed. This system was designed so that any failure left the throttles fully open. An engine speed pick-up counted the ignition spark frequency and in the event of that exceeding a fixed value, the ignition system reduced the frequency of sparking until the engine speed stabilised at the correct equivalent rotor speed. This provided a rough but effective back-up method which had no possibility of a common failure with the primary control system.

16.5   Reliability by Design

At the commencement of the system design it is necessary to prepare a system specification which, together with system performance estimates and general arrangement drawings of the major elements of the system, will form part of the system type record. The specification should include:

a) The ambient conditions in which the system will operate, i.e. the temperature, precipitation and humidity ranges, wind speed and turbulence level and, for the aircraft, the altitude range.

b) The accelerations and loads which are estimated to be imposed on ground and in-flight operation. The ground loads for the control station and its equipment can arise as it is being driven over rough terrain. This can give rise to very high accelerations in all directions and it will be necessary to provide suitable protective suspension for the sensitive components of the control, communication and display equipment.

   For the aircraft, too, ground loading must not be ignored. If the aircraft is to be carried to its operating site in a ground vehicle, attention should be paid to the accelerations imposed on it in that phase. Also for HTOL aircraft, in taxiing or being towed into position or during the take-off or landing run, very high loads may be experienced by and through the undercarriage should wheels run into drainage gulleys or pot-holes in the surface. These loads have often been ignored or severely underestimated in the past. The vibration specification should also take into account the levels and frequencies for all phases of operation including taxi and take-off from less-than-perfect airstrips.

   A flight envelope must be prepared for the aircraft, indicating its range of operating speeds, acceleration limits in manoeuvres and estimated accelerations imposed by air turbulence at its range of operating weights. A different flight envelope may be required for different altitudes, especially for the HALE and MALE system aircraft. The flight envelope for VTOL aircraft should extend to appropriate values of rearward and sideways flight. Impact velocities and resulting accelerations on landing must be specified for all aircraft.

   The failure to recognise the significance of these several factors has often led to inadequate design and subsequent failures of structures and equipment which have been expensive, difficult or even impossible to rectify at a later date.

c) A system hierarchical chart which should take the form shown as an example in Figure 16.4.

   The overall system is reduced in line 2 of Figure 16.4 to the prime elements of the system and in line 3 to sub-system modules of those elements. The sub-systems are then reduced to sub-assemblies

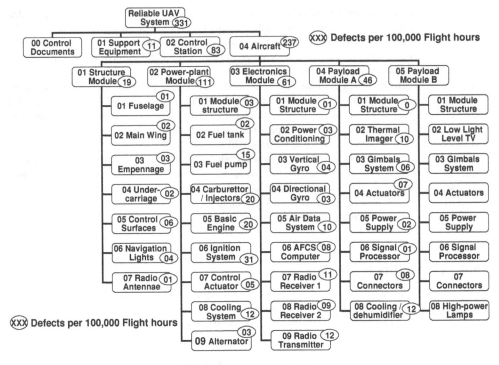

Figure 16.4   System hierarchy

and then (not shown in the diagram) to component level. Each element of each stage is given a reference number.

This system has the merit of allowing each subsystem or component to be identified with the reference numbers incorporated into the drawing system and reliability records of the manufacturing and operating organisations. Note that only the airborne element of the system has been expanded here in order to save space, but each of the other elements will be similarly treated.

As examples of this referencing system, the thermal imager payload module would be referenced as 'Reliable 040402-0000' and an actuator operating a starboard-side aileron might be referenced as 'Reliable 040105- 0201', with the last four digits referencing the starboard aileron as 02 and the actuator as 01. Each component, wherever possible, will be marked with its reference number.

The system hierarchy document will also be used in the synthesis, and later in the recording, of failure rates for the system. An example of such use of the document is shown in Figure 16.4 with artificial defect rate numbers in place.

d) An advisory document which will specify such matters as any particular cautionary measures to be taken in the design, such as for safety reasons; any specific materials or substances to be used or avoided; reserve factors to be applied to particular areas of structures or mechanisms; and concerns with electromagnetic compatibility standards, etc.

Any system as complex as a UAV system will be designed, built and developed by a team of people. The information contained in the specification will serve as essential guidance to the designers and ensure consistency.

The system hierarchy document will allow a systematic synthesis to be made in the allocation of acceptable failure rates to the several system elements in the design to achieve the required overall level of system reliability. It will assist in the choice of 'bought-out' components. If the components are already approved aerospace equipment, the supplier's compliance documents will show the mean time between failures (MTBF) of their components tested under specified conditions. This is readily converted into a failure rate per 100 000 hours for inclusion in the reliability synthesis.

With the UAV industry being a relatively young one, there will probably still be areas where there are no approved components available or where available aerospace components are too large or otherwise inappropriate for UAV use. In that case the UAV system manufacturer will have to look for other sources of supply, i.e. to arrange the testing and approval of equipment currently available for automotive or agricultural use or to develop in-house or contract out the development of suitable equipments. This is discussed further in Part 2.

Meanwhile listed below are areas to note from the author's experience where problems of reliability have occurred. This is not meant to be all-inclusive but just a snapshot of some of the more common problems experienced.

Structure

Properly considered and designed UAV structures have had a good record of reliability, unlike those from model aircraft makers who generally have little knowledge of air-loads, etc. or understanding of the design of lightweight structures. Where defects have occurred it has usually been at joints in the structure where the diffusion of stresses across metallic joints or from metallic lugs into a plastics composite structure, has been inadequate. The extra cost and weight of applying a more generous reserve factor in these areas is well justified.

Failures by metal fatigue in both HTOL and VTOL aircraft should be eliminated by rigorous design and testing as discussed in Chapters 6 and 19.

Power-plant

Defects may occur within the power-plant or in other sub-systems but caused by the power-plant. The former include inadequate consideration in the initial design of the ambient conditions in which the aircraft will operate. Hence, unless sufficient cooling is provided, a piston engine, in particular, may suffer overheating in hot conditions and seize, or fuel may become vaporised. This is particularly true on long-duration missions.

If the system is operating near equatorial conditions, then the heat from solar radiation must be considered as well as the ambient temperature. Temperatures as high as 100 °C have been experienced on the upper surfaces of wings and fuselage, even with the aircraft in flight. This heat may then be conducted into the internal elements of the aircraft, including the engine.

Under cold conditions, any water in suspension in the fuel may form droplets or, worse, freeze and cause icing of the carburettor, fuel metering system or injectors. In heavy precipitation, water may be ingested into the fuel unless diversionary and/or filtering means are employed. In cold conditions, snow or ice may be ingested and provision must be made to counter this by suitable heating.

The latter, i.e. defects caused by the engine, can result from vibration damaging sensitive equipments or heat transfer causing overheating of other components. Where possible a low-vibration engine should be chosen in preference, but also the engine should be mounted on anti-vibration and heat-insulating mountings. Coating the inside surfaces of the engine compartment with a heat-reflective material is also a sensible measure.

Another source of engine-related defects is an inadequately earthed or screened ignition system causing interference with the communications, payloads and other electronic components.

Electronic Components

Many electronic components produce heat. Therefore they should not be too tightly packed and should be well supplied with heat sinks extending into fan-induced cooling airflows. Electronic boxes should be screened and sealed against moisture ingress caused by rapid temperature changes in environments of high humidity.

Payloads

It is impossible to generalise since, as described in Chapter 8, there are a wide range of types of payload appropriate to the many different UAV missions. However, a factor common to many is the inherent problems of quick-release electrical connectors and, again, the damaging effects of humidity. Humidity can also degrade the performance of camera lenses and payload windows. Provision for some form of dehumidifying should be made.

External Structures and Mechanisms

The main problem here is icing of wing surfaces and jamming of control surfaces, both of which can cause loss of control. For all but the more expensive HALE, MALE and UCAV systems, heater-mats are too expensive and complex. The application of anti-icing pastes may be a more practical answer.

Connections

Whether electrical or mechanical, all connections should be designed so that it is impossible to connect them incorrectly in orientation or to connect them to other connectors by mistake. In the aerospace industry in general, the making of incorrect connections has been the cause of many tragedies through controls being reversed in direction or emergency systems not working when called upon to do so. The reliance on colour-coding alone is not recommended. Adjacent connectors should be so shaped that their incorrect orientation or inadvertent connection to adjacent, wrong connectors is impossible.

Control Station

In terms of the environment, much of the discussion above is also relevant to the control station. Humidity allied with temperature change and overheating of electronics (and personnel) may account for the majority of the problems.

Possible interference between systems due to radiation is a further factor to be addressed, usually eliminated by adequate screening and earthing.

As mentioned earlier, soft-mounting of the various assemblies is advisable as a precaution against damage caused by driving a land-vehicle-based control station across rough terrain.

16.6   Design for Ease of Maintenance

As previously suggested, difficult access to areas requiring regular maintenance can lead to reduced system availability due to pressure of time on technicians. The design of the aircraft and the control station as demountable modules and sub-modules is a major step towards ensuring accessibility. Where this is not possible, resort to removable panels has to be made, but it is preferable that the panels can be hinged in such a manner that they remain attached to the aircraft or control station or module so that they are not dropped, damaged, lost or incorrectly replaced after maintenance.

All tools should be of a common standard wherever possible and 'operator friendly' to ease the technician's task, especially when maintenance is carried out under adverse conditions.

17

Design for Manufacture and Development

During the main programme of design, thought should have been given to the design of components in order to facilitate their manufacture, and also to which components would be manufactured in-house and those which it would be more appropriate to source from specialists. A detailed programme plan would have been constructed and regularly reviewed and updated as the programme progressed.

Those sub-systems and components which are particularly critical to the success of the programme and those which may be expected to have a long development time will have been identified at an early stage so that the design, manufacture and testing of these would be carried out first. Several actions may have to be carried out sequentially and several others may be carried out in parallel as best fits the time-scale and cash-flow into the programme. These actions will be planned in what is usually known as a PERT (program evaluation and review technique) diagram which especially identifies those actions which are interdependent and may form a critical path to completion, particularly if the estimated time of any action is found to be longer than its prediction.

During the main design phase, the selection of a power-plant (and, desirably, alternatives) will have been made. Evidence of its performance and other characteristics will have been required from its manufacturers or suppliers. However, the performance of the unit may be affected by its installation in the UAV. It is highly desirable, therefore, that an engine unit be installed into a power-plant assembly, representative of that of the UAV, and tested using a dynamometer or thrust measuring system, as early as possible in the programme. As most UAV systems are propeller, rotor or impellor driven, the discussion below will cover this means of propulsion. For propulsive jet driven systems a similar argument applies, but with means of measuring the thrust from the engine.

The process therefore will require the design and manufacture of a representative engine installation with instrumentation and the acquisition or design/build of a suitable dynamometer. It is recommended that, unless the engine manufacturer is of very high regard, the UAV manufacturer always has the ability to check the engine performance before installation into the UAV, both during development and on a selective basis during subsequent production.

Dynamometers may take three basic forms in terms of means of accepting the load:

a) electro-brake dynamometers using generators or alternators to convert mechanical energy into electrical energy which is then dissipated or stored by various means,

b) water-brake dynamometers which pump water to convert mechanical energy into potential energy and heat energy, and

---

c) fan-brake dynamometers which convert mechanical energy into an amalgam of kinetic, pressure and heat energy.

The first tend to be most appropriate to the testing of the more powerful engines and the latter two for the smaller units. Of the latter two, the author, from experience, favours the fan-brake system as, although larger than the water-brake, it is less complex and less prone to problems such as cavitation.

The design of moulds, jigs and fixtures required for prototype component manufacture, adjustment, assembly and testing will have begun, as will the preparation of specifications of the out-sourced elements and the liaison with potential suppliers.

It is likely that some components will be selected to be manufactured in a different form for the prototype system(s) than for later systems in production (series manufacture). For example, initially, some metallic components may be machined from solid or fabricated rather than being cast or forged. This process will speed the availability of those components and delay the need to invest in expensive tooling until the design has been proven to be satisfactory. Similarly, electronic systems may initially be 'breadboarded' and the investment in specialised chips delayed until the parameters required of them are confirmed. Developmental design will be required to issue preliminary drawings and specifications to cover these components/sub-systems. This approach is less likely to apply to the design of the GCS where, in the main, standard equipment is more likely to be adopted, with particular system requirements being reflected in dedicated software design.

Rigs for mounting and testing the static strength and stiffness of, for example, spars or torsion-boxes or root fittings of wings or rotor-blades, as appropriate to the UAV will be designed for early availability. In the case of rotor-blades, a balancing rig will be required to check and adjust their chord-wise and span-wise balance.

Take-off and recovery can be critical parts of the UAV flight spectrum. Landing on wheeled or skid undercarriages for fixed and rotary-winged UAV will require the undercarriage to absorb and damp the energy of impact. A drop test of the undercarriage, to ensure that it does not allow deceleration greater than that specified, should be carried out. This will require a rig, with appropriate instrumentation, to be designed and manufactured in good time.

Catapult-launching will impose a high acceleration onto the UAV. It may be that, in addition to the design of the catapult system, a 'dummy' UAV may be required to assess the catapult performance, in terms of controlled acceleration to the necessary speed, in advance of subjecting the real thing to its rigours. This acceleration can be as high as $8g$, as previously recorded and, although some thrust, and therefore acceleration, will be provided by the aircraft propulsor, its contribution will be small, probably of the order of $0.2g$.

Although three-dimensional computer-based drawings will initially be used to assess the correct juxtaposition of components, a full-scale wooden three-dimensional mock-up of the aircraft, and possibly of the control station, may still be thought to be necessary to confirm the results. It would also be used to layout electrical and electronic components and wiring looms.

Preparing for later ground tests of the aircraft is a further design function. It is likely that a rig will be required to mount a HTOL aircraft during ground running tests. Instrumentation to measure temperatures, vibration levels, fuel consumption, noise, radio reception, etc., together with displays and recording equipment will be needed. All of this equipment must be forward-sourced.

VTOL aircraft of whatever configuration offer a significant advantage over HTOL aircraft, in this phase of the programme, in that they can be more fully functioned without ever leaving the ground than can the latter. However, that does mean investing in an up-front expenditure on rigs. This investment will be repaid handsomely in saving cost and time whilst providing greater technical assurance in the development and trials phase. A progressive programme of static and dynamic testing the rotor(s), controls, transmission and power-plants individually and then as an assembly may be prepared and the necessary rigs designed. With careful planning much of the same equipment may be employed in each case and for the total assembly, thus minimising investment capital.

At this stage the designers should also be giving further thought to the aspects required to be tested during the development phase, not only to confirm that adequate performance and integrity has been achieved, but to obtain the data necessary for the later certification of the system.

Consideration should already have been given to the siting of the test facilities for ground testing the UAV system (see Chapter 19).

It is important now to begin negotiations with the management of a suitable site for flight testing the system. Agreement will have to be reached as to the projected time-scale and extent of the test programme, and, of course, the appropriate fees. The site operators may require specific provisions for safety to be met, including in particular, installation in the aircraft and GCS of equipment to interface with their controllers. It is important to maintain liaison with the site management for both parties to be updated should there be any change made in the site operating requirements or change in the projected development programme (see Chapter 20).

Outline test schedules and test requirements (TReq) should be prepared, calling up the parts of the system to be tested, and the appropriate test equipment, including instrumentation, method of conducting the test and results required. The TReq should include pro-forma charts or tables as an aid to alerting the test engineers as to what is required. Many tests are performed outside the comfort of an office environment and any means of easing the tester's work is to be recommended.

An outline of the operating and maintenance manuals will also have been prepared at this stage on the basis of the proposed/predicted or specified *modus operandi*. The manuals will be consulted and fleshed-out during the development and trials phases.

It is not proposed to discuss the production methods of UAS in this volume since they are much the same as any other aeronautical venture so far as the UAV component of the system is concerned and using similar materials. The major differences lie in the generally smaller sizes of components and this may require closer tolerancing in manufacture of mechanical parts, but alleviated by refined design appreciation, as indicated in Chapter 6. A greater use of composite materials may be evident in the UAV due to the effect of scale on both mechanical and structural components. A greater use of modular construction may also be a characteristic of UAS production, with closer integration between the electronic and electrical and mechanical/structural elements.

The manufacture of the control station may be more in line with general commercial or military practice, rather than aviation practice, with the exception of critical elements which could cause disruption of the system in the event of their failure.

Part Two

The Development of UAV Systems

18

Introduction to System Development and Certification

18.1 System Development

The overall 'life' of a significant aircraft-based programme may be considered to be divided into up to ten phases, depending upon the origin of the system.

These phases are listed in chart form in Figure 18.1 against a time frame which indicates the length of time which each phase may occupy for a significant new system. 'Significant' implies a programme which may require an investment of perhaps several hundred million or more dollars, euros or pounds, such as for a manned ground-attack aircraft or a UCAV. Smaller, less complex, systems will require less time in each phase, with a proportionately lower investment cost. In practice some of the phases may be allowed, with caution, to overlap one another.

The term 'development' of a high-technology system can imply either the route from an idea to the entry of the system into operation (phases 1–9) or simply that part of the programme in which the pre-production systems are constructed, tested and proven ready for series manufacture (phases 5–8). For this work, the latter view, that 'development' is phases 5–8, has been adopted. The activities of the four phases are covered in Chapters 18–24.

The development phase may be extended, at some cost to the programme, if it is found that corrective modifications to the design have to be made at that stage. Careful attention in the earlier design phases should reduce the need for such action, as will company experience in the field and good teaming between design, test and manufacturing staff.

A manufacturer's responsibility, however, does not cease with the supply of his product to the customer. The responsibility continues with the support of that product for the remainder of its useful life. In many cases, especially that of aeronautical systems, that support phase may be longer in time than the sum of all previous phases.

To ensure an acceptable level of safety in the operation of any airborne system, including UAV systems, the airworthiness of the system and its safe usage of the airspace in which it operates must be addressed at an early stage in the design, and pursued in the development and operational trials phases. There is no reason that the proven principles long applied to manned aircraft should not also apply to UAS. It should be emphasised, however, that the UAS has to be treated as a complete system and must take into account the reliability of all elements of the system, particularly of the control station and communication links.

Unmanned Aircraft Systems – UAVS Design, Development and Deployment   Reg Austin
© 2010 John Wiley & Sons, Ltd

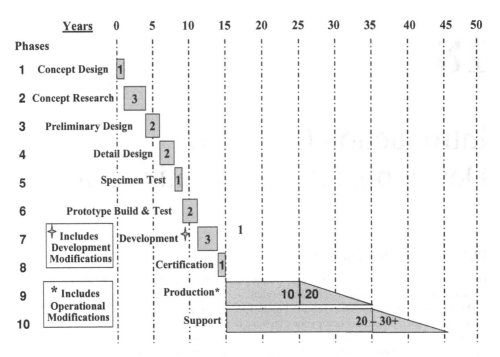

Figure 18.1   Phases of UAS development

18.2   Certification

A statutory duty of the national Regulatory Authority is to ensure the safe design and operation of the system and to issue a Certificate of Airworthiness for the system once it is satisfied that the system meets its requirements for safety. Successful system development must therefore be directed to and culminate with the achievement of certification.

Advisedly, the developer will have been liaising with the Authority from the early stages of design to welcome its advice and to build up a rapport with its officers. A draft of the proposed Certification Programme should be prepared and submitted for comment early in the development phase, together with a draft Compliance document. The latter will indicate the way in which the emerging system will comply with the airworthiness requirements of the Authority. The compilation will begin of a Type Record, containing the design calculations and the results of the component, ground and flight testing as they become available.

18.3   Establishing Reliability

As discussed in Chapter 16, 'designing-in' reliability is costly. Proving a specified level of system reliability and life through testing and further development of the system as necessary is even more costly. It makes economic sense, therefore, to define the life required of the system and the level of reliability required of it over that life, and not to over-specify it. The specified level of reliable life may be achieved by replacing specific components at shorter periods within the overall system life. In particular, a given component of the system may be 'lifed', and replaced, at a shorter period than the system or sub-system containing it.

In a particular military operation, for example, it may be deemed that the aircraft is overly vulnerable and through attrition will be expected statistically to survive only for a few hours. It may also be decided that under certain circumstances, the aircraft should be expendable. Economically, therefore, it may be that the aircraft should be made as cheaply as possible within the requirement for it to achieve a necessary performance and maintain a given level of reliability, but for only a limited number of hours until it is lost or withdrawn from service.

On the other hand, the economic analysis of a system deployed in a civilian operation with an aircraft carrying expensive payloads over populated areas is likely to conclude that a higher level of reliability for a longer life is a necessary solution. In this latter case, a greater investment in the design and development phases of the system is justified.

These are decisions which should be made pragmatically before the design work is even begun. Early consultation as to the airworthiness and airspace usage requirements for the intended operations should be made with the Regulatory Authorities (military and civil) referred to in Chapter 5.

The procedures outlined in the following chapters are appropriate to the more stringent requirements and, with the approval of the Authorities, may be reduced.

The airworthiness of the whole system is dependent upon its reliability to continue operating under the conditions to which it will be subjected. This will be assessed and improved as necessary during testing.

The environmental conditions under which the system as a whole may be required to operate must be specified at the outset of the design and provision made to enable it so to do. It may not be possible to realistically assess the success of the provisions until quite late in the development programme. Any shortfall could require extensive changes to elements of the system and so cause significant delay and add extra cost to the programme. It therefore behoves the designers to address the possible problem areas and provide 'escape routes' for later solutions to be incorporated without imposing lasting penalties upon the design, should they subsequently be found to be unnecessary.

Depending upon the specified conditions, problems which may arise may include overheating of components, especially those deep within the system. It is not only the ambient air temperature which may be of concern, but heat transfer from other components and, something which is easily overlooked, the effect of solar radiation. This may occur not only during flight of the aircraft, but in soak conditions on both aircraft and control station. On the other hand, operation at altitude or in arctic climes may bring very low temperatures to adversely affect items such as batteries or other electronic components. Thus some heating may be required. Cold soak during periods between operations may make starting difficult.

For some usage a measure of the system's vulnerability, in particular that of the aircraft, might be included in the testing. For general operation, this might include vulnerability to icing, precipitation or sand ingestion as well as electromagnetic interference and possibly bird-strikes. The smaller aircraft such as mini-UAV and MAV may be particularly vulnerable to high rates of precipitation, and bird-strikes as well as high winds. Indeed these factors may severely limit their operations.

Military operation might require the demonstration of the aircraft level of stealth and the protection of critical areas against, for example, small arms fire. In both military and civil operation, unduly fragile elements of the system may be subjected to accidental damage by operators. Any elements at risk should be designed to be adequately rugged or shielded from inadvertent abuse.

For civil operation, in particular, limits may have been placed upon the system's environmental impact arising from noise and toxic emissions. These values may have to be measured and demonstrated to comply.

It is desirable to phase in testing of these aspects as soon as is practically possible though, unfortunately, it may not be possible for many until the system is largely complete. Initial developmental testing of the aircraft can be carried out in general using the well-tried methods adopted for manned aircraft, but taking account of any effects of scale and with the exception, of course, that the UAV will have no sub-systems relating to human occupation. In the later stages however, before in-flight testing begins, integration of the complete system with the inclusion of the control station and any launch and recovery sub-systems, etc. is likely to require approaches more specific to unmanned systems.

Condition measuring and monitoring instrumentation will need to be installed in the aircraft. Recording equipment will have to be installed, either in the aircraft or in the CS after downward transmission of the data from the aircraft. Measuring and monitoring equipment will also need to be installed in the ground station to obtain and record the condition and performance of the CS elements. This may include automatic on-board check-out equipment or built-in test equipment (BITE) to ensure that all system parameters are within specification before the aircraft is launched.

Early flight testing will take place on designated test ranges and the safety rules may require additional equipment to be installed in both the aircraft and control station. For example, the Range Safety Officer may require that he will have his own means of tracking the aircraft and an independent means of terminating the flight should an emergency arise. Safety, in terms of means of collision avoidance with other occupants of the airspace, may have to be demonstrated before the UAV system is allowed to enter full operation.

Testing, especially of life-testing of components and assemblies, costs much money. A trade-off, balancing the worth of demonstrating an assured life in operating hours of components and assemblies against the cost of so doing, must be considered. The worth of longevity will be measured in terms of saving the costs of replacement parts, but more so in terms of reducing the cost of the down-time when the system is not available for operation and the labour costs involved in the replacement.

There may be inspection and servicing periods determined by factors other than wear or fatigue of parts, for example legislation or lubrication. It is usually desirable that replaceable component life-spans coincide with these periods or are multiples of them. Cheaper, readily accessible and easily changeable components may be acceptable with shorter lives than the more expensive less accessible components.

In military service a system, and particularly the aircraft, which by nature of its operation is subject to a high attrition rate, may not justify large expenditure on establishing longevity. An investment in designing-in stealth to the aircraft at the outset of the design might well reduce, considerably, its rate of attrition and readily justify investment in life extension. It is generally recognised that stealth features must be incorporated in the initial design and cannot effectively be 'bolted on' afterwards. Although this approach would result in a more expensive aircraft, far fewer may be required, thus resulting in a less expensive and more reliable system overall. It is an interesting equation.

The cost of establishing the life of a system involves the testing for fatigue and wear of a number of specimens, usually three. Apart from the cost of the specimens, the design and construction of test rigs and the labour costs of monitoring them have to be included in the equation.

Development testing will usually begin with the testing of critical components and will progress to the testing of assemblies and sub-systems until it is possible to assembly the system as a whole for both ground and then in-flight testing. However, as previously emphasised, the engine is probably the most critical of all of the elements in the development of a UAV and is probably the most closely integrated. Should it subsequently fail, it will not be easy to find a suitable replacement and even if that were possible, it may require considerable design changes to integrate a new and different engine.

Other elements which may subsequently be found to be inadequate should be more readily modified or replaced by available alternatives. Even if it is necessary to initiate a replacement element from the design stage, the time of so doing is likely to be considerably less than finding and integrating a new engine at a later stage. Therefore, the author advises early selection and testing of the engine and subsequently integrating it into a power-plant. This might even be carried out to advantage in a time-frame parallel to the early phase of preliminary design.

The comments above and in the following chapters apply principally to close- and longer-range UAV systems in terms of performance, size, complexity and cost. MAV and even mini-UAV systems may not need to undergo quite as rigorous a development and certification as will be described. They may be exempted on the grounds of their being limited in operation to benign environmental conditions and that the UAV remains within sight of the operator. The UAV may be expendable due to its low cost and, because of its low mass and flight speed, will cause no serious damage following loss of control. It may be required to carry no instrumentation and only optical images on the communication down-link. However, any such exemption must be with the agreement of the regulating authority.

19

System Ground Testing

19.1   UAV Component Testing

It is often difficult to define exactly what is meant by a component as one may range from the very simple to the complex involving more than one part and bordering on being an assembly. Component testing may be required to establish the ultimate strength, fatigue life or wear of a mechanical component or the correct functioning within limits of an electrical component or circuit. In any case, specimens will have to be manufactured or purchased and means of their testing prepared.

Electrical or electronic items will usually have been bought-in with appropriate certificates of conformity. However, for example, if it is proposed to use COTS (commercial off-the-shelf) components for a military application it may be necessary for the main contractor to certificate those items under his own acceptance as an Approved Manufacturer. It therefore behoves him to ensure that the COTS equipment meets and maintains the required quality standards. The testing of such components may involve their being subjected to specified temperature and acceleration cycles, but otherwise measurement of their performance under such conditions will be made by standard instruments.

Critical mechanical components to be tested for ultimate strength, fatigue life or wear, for example, are joints between composite and metallic materials where diffusion of stress into metallic transport joints is required. These typically occur at wing and other aerodynamic surface roots. Other components might be undercarriage flexures, mechanisms such as control bell-cranks or complete electromechanical actuators. For rotorcraft these may include gears, transmission shafts, rotor hub components and blade root attachments.

For tests of ultimate strength, it may be possible to use standard tensile or compression testing machines. Estimates of the proof or ultimate loads to be applied will have already been calculated with the required reserve factors superposed. It may be decided to test each of three components to destruction to determine the value of the reserve factors actually achieved and the degree of scatter in consecutively manufactured specimens. This information may be particularly valuable should it be decided later to increase, for example, the design gross mass of the aircraft. An excessive scatter in component performance may point to a need to improve production control, both in terms of manufacture and inspection.

Special rigs may have to be constructed to test for the fatigue life of a component. Oscillatory loads of the appropriate amplitude and frequency must be applied on top of the specified mean (steady) load. Such tests will usually be automated as they may be required to run continuously for very long periods.

These comments on production control are possibly of even greater significance where fatigue is concerned. The maintenance of a good standard of surface finish is important in achieving a predictable fatigue life, especially of light alloy components. Wear tests will also require rigs to operate moving components under load and again surface finish control is similarly important.

Other long-term tests of a different nature may be required to be carried out in parallel with the mechanical and electrical/electronic tests. An example is the proving of the long-term integrity of composite plastic fuel or oil tanks to ensure that they are inert to their contents.

This may only be achieved by constructing a small number of vessels, possibly of alternative materials, filling them with the appropriate fuels, etc. and keeping them at a representative temperature for several years.

Reporting of the progress and results of the tests must be carried out on a structured basis to ensure that the tests furnish the information required. They must be recorded in a form in which they can be included in documentation leading to certification of the system for civilian or military operation as appropriate.

The Test Requirements (TReq) will have been prepared initially during the design phase, as outlined in Chapter 17, but will be formally finalised at this juncture. They will carry the signature of a senior engineer approved by the certificating authority and responsible for the system.

The test equipment called up may be standard and already available or may have to be manufactured to drawings prepared in-house. The results and conditions logged during the testing will be recorded in a Test Report (TRep) using, where applicable, the pro-forma supplied in the Treq. The TRep will be signed by the senior test engineer in charge.

It is convenient for co-identifying the two reports if the TReq and the TRep carry the same number, e.g. TReq UAVS X003 and TRep UAVS X003. This formality is necessary to ensure that the information is acceptable to and traceable by the certificating authority. The process will be applied to the testing of the complete system, both on the ground and in flight.

19.2   UAV Sub-assembly and Sub-system Testing

These elements of the system may range from the aircraft undercarriage, through the mechanical and electronic flight control system, the payloads to the complete power-plant, the communication system and the aircraft recovery system.

For the control station (CS) they may comprise the levelling/stabilising system of the CS vehicle, the deployment of the radio antennae/mast(s), the air-conditioning of the crew working space and operating equipment, the aircraft starting system, the system check-out system, launch system, and communication system, etc.

19.2.1   Undercarriages

Undercarriage configurations range from the very simple to the very complex, as discussed in Chapter 6. They, therefore, may be categorised as components or, more probably, assemblies and require testing, from basic structural and energy absorption tests up to full functioning tests.

All of the types will have to undergo the former testing which will usually take the form of 'drop-testing'. Here a rig will be constructed incorporating the undercarriage and its mounting structure. At its simplest it will mount a platform above it, containing recording equipment and instrumentation to measure deceleration/acceleration and time. Other instrumentation may include strain-gauges on selected areas of structure and flexures in order to confirm that the working stresses are as predicted. Masses are added to bring the whole assembly to a total mass equal to the maximum design gross mass of the aircraft.

To carry out the test, the rig will be raised to a height from which, following release, the undercarriage will impact at the specified velocity. Before release, the timing device, instrumentation and recorders will be actuated. Subsequently the results will be analysed to determine the maximum deceleration and the form of oscillation that takes place following impact. The stresses will be compared with the design values. Further drops may be made to confirm the results. Component fatigue tests may then be made or repeated on local elements to determine the number of landings which should be allowed before scheduled replacement.

If the undercarriage incorporates plastic composite materials, creep tests of those elements should be made under load and in adverse temperature conditions to determine if the undercarriage as a whole or any elements of the assembly may distort with time and have an elapsed time-critical replacement life.

If the undercarriage is designed to be retractable, the retraction mechanism must be tested, but may have to wait until it can be installed into a prototype airframe for full sub-system functioning.

19.2.2 Flight Control System

The flight controls in both HTOL and VTOL aircraft are most likely to be electrically powered, with electrical actuators directly operating the aerodynamic control surfaces. As described in Chapter 10, the FCS logic component with its attitude, speed and other sensors coupled with the computing core may be functioned independently from the actuators. Quantified inputs may first be made to the computing core and the outputs recorded to check for correct responses. Subsequently the full FCS with attitude sensors, but without the actuators may be mounted on a gimballed table to check the response of the system. Finally, the actuation system may be added, complete with simulated aerodynamic surfaces, inertia and resistance. At this point, a prototype airframe may be available in which to install the complete control system for later testing on the aircraft ground rig.

19.2.3 Power-plants

As recommended in Chapter 17, a suitable dynamometer will have been obtained if the engines are sourced from other than well-established approved aircraft engine manufacturers, and probably of power less than about 50 kW. The supplier should have been required to conduct tests to determine the engine performance before delivery and to have run the engine for, say, at least 10 hours to determine any change in performance over that period. The performance, especially of piston engines, may improve over a 'running-in' period and thereafter suffer deterioration over time. It is desirable that this rate of loss of performance is determined at an early stage as the aircraft warranted performance, in terms of power and fuel economy, must take that feature into account.

It may also be found, depending upon the effectiveness of the production control, that there is a scatter in performance levels between individual engines. Therefore a minimum must be established, and agreed with the engine supplier. No engine should be accepted that does not meet the agreed minimum. For the approved aircraft engine manufacturers, of course, this is established practice, but may not necessarily be so for the manufacturers of the smaller engines entering the UAVS scene.

The bare engine will be coupled to the dynamometer and this may require the provision of a mounting rig. Instrumentation to measure the mean and oscillatory torque of the engine, its shaft speed and fuel consumption at various throttle openings should be supplied with the dynamometer. Other measurements to be made, such as engine temperatures, ambient temperature and pressure, and generated noise will require specific test equipment.

The test results will yield information which will be normalised to represent values at standard temperature and pressure. The normalised values will then be presented as a carpet graph showing engine power and fuel consumption as a function of throttle opening and output shaft speed. Subsequently these tests must be repeated with the engine built into an installation representative of the power-plant in the aircraft in order to determine any effect that this may have on the engine performance, operating temperatures, etc. and noise levels. If any adverse effects are found, it may be necessary at that stage to consider remedial modification to the installation.

19.2.4   Payloads

As noted in Chapter 8 there is a range of payloads, from simple lightweight video systems to more capable and complex multi-spectral systems, available from specialist suppliers. Alternatively the UAV manufacturer may have specified a system to meet his, or his customer's, particular requirements or even developed his own system.

Whichever route is taken, it is desirable that the payload can be functioned as far as is possible as a stand-alone subsystem, i.e. bench-tested, before being mounted into the aircraft. There are, of course, a large range of potential payloads, all performing for different purposes. It is therefore not possible to generalise in a discussion of their test procedures.

Power supplies will, in general be connected as will all the interfaces needed between the sensor, the communications and control station.

Functioning of camera performance, direction of field of regard, lens control and stabilisation of electro-optical systems should be possible.

It may not be practical to completely test functionality in some systems such as radars, magnetic anomaly detection (MAD), NBC etc systems until they are airborne in the UAV.

19.3   Testing Complete UAV

The Operating and Maintenance Manuals will have been prepared as previously discussed. They will be used and modified as experience is gained during this phase. The pre- and post-flight checks will be carried out including checks for wear, fluid leaks, signs of overheating, security of connectors, etc.

At some stage, communications testing must be carried out on a prototype airframe to prove radio transmission and reception. This will include confirmation of acceptable antennae positioning to achieve adequate gain at all aircraft to CS orientations. It may therefore be necessary to mount the airframe on an elevated platform. It also becomes necessary to ground-test a complete aircraft with subsystems operating.

From this point the testing of HTOL and VTOL aircraft may take different routes.

19.3.1   HTOL Aircraft

At some time the largely complete aircraft must become airborne under control from the CS. To reduce the risk attendant on that first flight, as many as possible of the subsystems will be progressively integrated into a complete airframe and tested for correct functioning and to ascertain if there is any undue adverse interaction between them. These unwanted interactions may be due to electro-magnetic interference, vibration or inter-system heating, etc.

It may be necessary to construct an 'aircraft ground functioning rig' upon which to mount the complete aircraft attached to the rig at strong-points on the fuselage or wing attachment points. An interpretation of this type of rig is shown in figure 19.1. The rig should permit undercarriage retraction and extension if relevant.

Appropriate instrumentation will be set up to measure and record commands, responses and conditions. The more critical values will be displayed. This instrumentation will probably include linear and/or angular potentiometers to measure control surface and throttle displacements, ammeters, voltmeters, temperature measurement, accelerometers, strain-gauges and engine speed measurement, etc. It may also be possible to include means of measuring propeller thrust which would be of particular advantage in future in-flight testing. Most of the instrumentation will be carried on to in-flight testing and, in addition to performing its task on the ground rigs, will be proven for the later operation.

It may be considered prudent to assess the structural characteristics of the wings, in particular the torsional stiffness and position of the flexural axis in order to assure the non-occurrence of aerodynamic

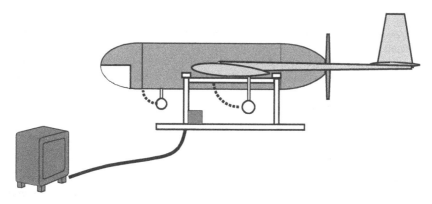

Figure 19.1   HTOL aircraft ground test rig

flutter. Determination of the flexural axis position is often accomplished by mounting the aircraft upside-down in the rig and loading the wings with weights along the line of the aerodynamic centre, representative of the span-wise lift distribution at maximum lift. The maximum lift is that specified by the Flight Envelope in the Type Record. Measurement of wing deflection at the leading and trailing edges will enable the position of the flexural axis to be determined and, if found to be too far aft of that calculated in design, correction can be made by adding mass forward or adding suitable stiffening. It may also be opportune to measure the stresses at critical points such as the wing root fittings.

A more sophisticated, pivoted, rig would enable the aircraft to be set and functioned at different pitch and roll attitudes. However, it would be debated whether the extra cost of this extra facility is warranted on the basis of cost/benefit.

The installation and functioning of the several sub-systems may be made in total from the beginning or more usually added as the programme of testing is developed.

Power-plant functioning should be an early element of the tests, with vibration levels measured at various points on the airframe to establish modes, and in particular to record vibration levels at sensitive positions such as the payload and electronics systems mounting. This should be carried out over a range of engine speeds and thrust levels.

If environmental noise may be an issue, opportunity should be taken to measure that at several distances from the aircraft.

Although temperatures should be measured at the several critical areas during engine and electrical system operation, they may not be particularly representative of airflow cooling during flight, but only during ground-running.

With the flight control system installed it will be possible to function and measure the response of the control surfaces to the inputs from the AFCS. Unless the control surfaces are aerodynamically balanced, it may be desirable to load the surfaces to represent the aerodynamic loads at V_{NE}.

The size of a small UAV may permit a complete aircraft to be mounted in a wind tunnel to measure the aerodynamic derivatives to confirm or modify those previously calculated. In some cases, for example given the availability of a blower-tunnel, it may even be possible to function a complete aircraft with all systems operating, including with the engine running.

The use of a wind tunnel is expensive, but will come closer to reproducing the conditions found in flight and so reduce later-stage risk and expense.

If the aircraft is to be launched by catapult, the complete aircraft will be mounted upon the catapult 'trolley' or relevant interface and fully functioned short of actual launch.

The recovery equipment, the parachute and airbags, if used, is best tested on the aircraft ground functioning rig. The parachute can only be tested for its release and ejection system functioning. Obviously its deployment will require the airflow of flight to inflate it. The airbag(s) system, however, can be tested for both release and inflation whilst on the ground, provided that due safety precautions are taken.

If the aircraft is launched on a wheeled undercarriage, and sufficient clear space is available, then taxiing trials up to just below lift-off speed may be carried out, exercising the undercarriage and coming closer to reproducing flight conditions. In particular, acceleration rate and yaw control can be confirmed. The 'first flight risk' of a catapult-launched aircraft is therefore greater than that for an aircraft which takes off from a wheeled undercarriage.

19.3.2   VTOL Aircraft

VTOL aircraft, especially rotary wing, have an advantage in both their ground and flight testing. It is possible at an early stage to exercise the aircraft from component to full integrated aircraft level under equivalent full hover-flight conditions without it ever leaving the ground. Thus it does not incur risk in the later, in-flight testing to the same degree as does a HTOL aircraft.

Inevitably, one of the more critical elements in a rotary wing UAV is the rotor system. It is fundamental to ensure that all blades match in terms of radial and chord-wise balance and torsional stiffness and are free from aerodynamic flutter. Therefore it is advisory to construct initially two rotor blade sets to measure and confirm that their pre-adjusted radial and chord-wise balance and torsional stiffness are within the specified design range.

The rigs required to do so will have been designed and made in advance. The chord-wise balance is usually determined by swinging the blade at right-angles to the chord-line whilst it is hung from a point suspension at the root end. The point of suspension is adjustable chord-wise. The suspension point is moved for successive 'swings' until the blade swings without twisting. This is the point where the centre of mass of the blade is directly outboard of the suspension. If this does not coincide with the design point, then adjustment is made to masses provided for the purpose in the leading edge of the blade. If any of these are found to be outside an adjustable range then either the production method or detailed design should be reassessed and corrected.

The blades will be subjected to a tensile test, possibly to destruction to determine the realised reserve factor on the centripetal loading at the over-speed condition. The design rotational over-speed is a speed higher than that of the top end of the governed operational range. It is usually set at 15% above the maximum normal governed speed and will be called up in the system Type Record.

Rotor hub components, drive shaft(s) and mechanical controls will have been manufactured in parallel with the blades and all assembled. The usual procedure would then be to test the complete rotor system, driven by an electric motor, to measure drive torque at several speeds up to the over-speed value and rotor thrusts up to a maximum as defined by the flight envelope. Typically the latter might be up to twice the maximum design gross weight of the aircraft.

For the single-main-rotor configuration helicopter, the tail rotor must also be similarly tested and this is likely to require a gearing change if the same rig is to be used. The maximum thrust at which the tail rotor is to be tested is conventionally twice the thrust required of it to balance the torque reaction of the main rotor at maximum transmission power. Note that for multi-engined aircraft, the power transmitted is normally restricted to less than full power from all engines.

A rotor system test rig is shown in Figure 19.2 and, in this case, a second rotor drive has subsequently been added to measure intra-rotor reaction in a coaxial twin rotor system.

Testing will continue with a drive/reduction gearbox added to the system. The tasks will be to measure the level of power loss due to friction and any wear of the components over a period of typically ten hours. If the gearbox(es) are oil lubricated, oil temperature may be measured and the pattern of oil distribution will be determined using dye additive in the oil. The gearbox performance may be first tested as a unit separately from the rotor system or driving the rotor system. In the former method some other means will be required to simulate rotor loads.

Figure 19.2   Rotor test rig

On successful completion of these tests using electric drive, the actual power-plant(s) will be added and the test repeated as an integrated system. A cyclical variation of rotor thrust will be demanded to represent, as far as is possible, the thrust and power predicted in real operation. Performance, including fuel consumption and engine and gearbox temperatures will be measured over a more extended period of perhaps 100 hours.

Finally a complete aircraft, probably with a real payload, will be assembled and mounted on an aircraft functioning rig. Such a rig, as exampled in Figure 19.3 for a coaxial rotor system, will, with advantage, give the aircraft freedom in pitch, roll and height so that it is, for all purposes, freely hovering.

The rig should be constructed with the moving parts as lightweight as possible so as not to increase overmuch the effective inertia of the aircraft, although the measured responses may be corrected to take into account the increased inertias. A similar rig may be designed and built for a single main rotor aircraft.

The pitch and roll freedoms, through the use of pivots, can allow the full angular movement of the aircraft. The height freedom will, of course, be limited to probably less than one metre. Test inputs and results may be obtained by radio-link or hard-wired through an umbilical cable. The inputs may direct the aircraft in yaw and then add perturbations to check the response of the FCS in returning the aircraft to the commanded orientation. Similarly commands and perturbations may be input around the pitch and roll axes, with the responses recorded and compared with the design requirements. Adjustment can then be made to the values in the AFCS to produce the required responses.

Similar instrumentation as listed for the HTOL aircraft will be installed as appropriate, but with the addition of means of measuring total rotor thrust. This may preferably be accomplished by the inclusion of a load cell in the mechanism that allows the aircraft vertical freedom within the rig. Thus full rotor thrust can be demanded and measured up to the maximum of the flight envelope specification, or beyond if possible and desirable. Measurement of the several performance and condition parameters can then be made under hover-flight manoeuvre conditions, giving even greater understanding and confidence in advance of free-flight.

Further, a spectrum of estimated flight manoeuvres in terms of rotor power and thrust can be sequentially applied to the UAV in the rig to prove the UAV reliability over a given period of, for example, an initial 100 hours, before it is committed to free flight. Having effectively carried out hover-flight on the rig, the aircraft can be taken into actual hover-flight with considerable confidence.

Figure 19.3 Rotary-wing aircraft functioning rig

As with the HTOL aircraft, if the functioning rig is placed into a blower tunnel, forward-flight conditions could also be more accurately reproduced. However, the expense of this is not considered to be warranted for a VTOL aircraft since it can be taken in small steps from actual hover-flight into forward-flight.

Ground testing of both HTOL and VTOL aircraft, unlike flight testing, does not need to occupy a great amount of space. It can, however, take many hours and it is recommended that the test facilities be housed in a hangar with suitable provision for airflow into and out of the hangar whilst engine and/or rotor testing. This will allow testing to proceed in inclement weather, and also protect sensitive instrumentation and improve security. The test hangar used for Sprite aircraft testing, which encompassed engine dynamometer, rotor rig and aircraft functioning rig, is shown in Figure 19.4. The hangar had the facility of a sliding roof and adjustable side walls. The operators' cabin which contains power supplies, controls, displays, recording equipment, etc. is alongside.

19.3.3 UAV Environmental Testing

Although both the control station and the UAV must be able to function in a range of specified environments, which include weather conditions, the demands upon the two may differ. For example, the UAV may be required to operate at altitudes which are far higher than those to which the CS will be exposed. The testing of each are therefore considered separately with the CS testing being discussed in Section 19.4. The ultimate weather and climatic testing of the UAV will inevitably await flight, but it is prudent to conduct as much of this testing as possible on the ground during the ground test phase or in parallel with the early flight testing.

Weather testing will principally involve response to wind strength with gusts and precipitation. The former is difficult to reproduce realistically on the ground and results must await flight testing. Initial exploration of the latter may be carried out more safely on the ground.

Figure 19.4 Sprite system test hangar

The Particular Specification of the UAV system, and of the UAV itself, will state the degree of precipitation in which it must satisfactorily operate and the climatic conditions pertaining. For testing the HTOL aircraft in rain it will be appropriate to set up a water spray rig, emitting water onto the front of the aircraft at a density, droplet size and velocity as far as is possible commensurate with the maximum speed of the aircraft. All systems, including most importantly the power-plant, must be operating. The actual temperature at the time of the test will be of little significance, provided that it is not so low as to incur icing. The test will not exactly reproduce conditions as encountered during flight, even if the rain tests can be made in a blower tunnel as, of course, the UAV will be stationary, but it is likely to indicate any particularly vulnerable areas of water ingress and any erosion of propellers.

For the VTOL aircraft, the test will be of greater significance since the test, using the aircraft functioning rig, will reproduce hovering conditions very well. For this the spray rig will be positioned above the UAV and emit the water spray downwards. Extending into representation of forward flight, a horizontal component should be introduced and the aircraft rotated in the rig to cover lateral and rearward flight.

As with the HTOL aircraft propellers, the testers should look for evidence of possible rotor blade erosion.

Assessing the effects of precipitation leading to icing is less easy, and would be carried out in specialist low temperature facilities more readily available in Canada or Finland. It may be more practical to delay this until the flight testing stage, provided that provision of means of incorporating anti-icing/de-icing air intakes, propellers, sensor windows, etc. is anticipated.

Climatic conditions, whether of extreme heat or extreme cold, will require insulation and heating or cooling vulnerable systems as is appropriate. A factor that may sometimes be forgotten is the effect of solar radiation. The author recalls a sub-tropical operation when the ambient temperature was 'only' 5 °C, but the solar radiation resulted in a temperature approaching 10 °C on the upper surfaces of the aircraft. Environmental chamber testing of the UAV should include radiant heating as well as air temperature control.

Depending upon the type of operation expected of the system, screening against sand ingestion or electromagnetic radiation must be considered and incorporated if relevant. Whilst the testing of the former may again wait upon flight trials, the latter can be assessed as an element in the ground testing.

Electromagnetic radiation might be intentional in military operation, unintentional in many civilian operations and unavoidable in, for example, power-line inspection. Screening against all these threats is possible and laboratory facilities are available to test for appropriate screening performance.

19.3.4 UAV Testing for Reliability and Life

Having proven the correct functioning of the UAV as a complete system, by adding modifications as and if necessary, it is now important to ensure that it will continue to function reliably for its specified life-time, and that any lifed and replaceable components continue to meet or exceed their specified design life.

Failure may occur due to wear or fatigue, as discussed in Chapters 6 and 16. Critical components will have been subjected individually to wear and/or fatigue tests, but based on estimated rates of both. The impressed forces may have been underestimated and testing within the environment of the aircraft may now yield the truth. Strain-gauging of such components operating *in situ* will provide updated values, but demonstration by life-testing of the complete aircraft will be necessary to achieve certification.

This may be carried out by a sophisticated application of loads on a ground rig or, at greater risk, delayed until in-flight testing. It may be deemed necessary to carry out preliminary instrumented flight tests to measure the loads incurred by various manoeuvres and then to apply them in extended life-testing on the ground rig.

We now see another significant difference between the testing of the HTOL aircraft and the rotary-wing VTOL aircraft.

Short-period oscillatory loads in HTOL aircraft may arise from engine vibration and air turbulence. Whilst the former, if present, is largely predictable, the latter is not. A typical gust spectrum is assumed. Depending upon the operating scenario of the system, the actual loads may differ significantly.

In a loading test spectrum, the oscillatory loads due to turbulence will be superimposed upon the spectrum of steady loads which themselves will vary, depending upon the current manoeuvre. The most damaging manoeuvre is likely to be a 'maximum-g' turn under full power. Therefore the prediction of HTOL UAV system and component lives will require a margin to cover the variation in operating conditions.

The rotary-wing aircraft may also have oscillatory loading from the power-plant, but are far less affected by air turbulence. The principal oscillatory loads come from the rotor system itself and are predictable. Furthermore in a small UAV the frequency of the oscillatory loads is likely to be of the order 50 Hz or higher so that a million (10^6) cycles is completed within 5–6 hr.

If, for example, a component life of 100 hr is required, the components will be required to undergo 20 million (2×10^7) cycles without failure. However, for well-designed, well-finished components manufactured from good-quality steel, if the sample components when under component test (see Section 19.1) have survived for 6 hr under a rigorous spectrum of loads, they will have been operating at a stress level at which they should continue to operate for an infinitely long life. Therefore a 100-hr endurance test of the complete UAV can now be carried out on the aircraft functioning rig, applying a representative load spectrum, as discussed in Section 19.3.2, through the application of cyclic and collective pitch, with a high degree of confidence.

19.4 Control Station Testing

In parallel with the UAV testing, the control station will have been prepared. The several sub-system equipments, such as the communications, controls, displays, recording equipment, power-supplies etc. will have been separately checked out. Most of this equipment will have been out-sourced and come with certificates of conformity.

The CS vehicle or trailer will have been prepared with the necessary accommodation for the crew and equipments and UAV(s) if relevant. Radio antennae will have been installed and elevating means, if relevant, will have been functioned. Air-conditioning for crew and equipment, as appropriate, will have been installed and operated. Now comes the time to assemble and test the CS as a complete entity and then to integrate its operation with the UAV.

An on-board check-out of all sub-systems will be made to ensure their correct and continued functioning in their positions in the CS. Checks will ensure that the ergonomic interfacing with the operating crew is satisfactory, that there is satisfactory system integration and no adverse mutual electromagnetic or physical interference and that the air-conditioning system(s) maintain appropriate ventilation and temperatures for crew and equipment. Radio communications will be functioned and checked for performance by transmitting data and control commands to a slave radio receiver, preferably positioned at some distance away.

If all is satisfactory, the next step will have been the integration of the CS with the UAV on its ground functioning rig. The start-up procedure will be carried out and built-in-test-equipment (BITE) functioned.

The BITE addresses the state of the UAV systems to ensure that it is ready for flight. i.e. that the on-board power supplies, sensors, control systems, payload, fuel gauging, etc. are all operating within the correct limits and that housekeeping data and health and usage monitoring system (HUMS) equipment (if fitted) are registering.

During the ground testing of the UAV, the manipulation of controls and measurements made by the on-board instrumentation may have been transmitted by hard-wiring from and to a separate console for display and recording. This control must now be transferred to the CS and communicated by radio (or other communication system used). Similarly the results of the instrumentation should now be transmitted to the CS for display and recording, either separately to or as part of the aircraft housekeeping data.

19.5   Catapult Launch System Tests

Catapult systems are described and illustrated in Chapter 12. They differ principally in the length of 'throw' and the degree of acceleration required to launch the UAV. These factors, together with the mass of the UAV, will determine the power characteristics required and therefore the power transmission means, i.e. bungee-powered or hydropneumatic-powered for the heavier, higher-wing-loaded aircraft.

In the design of the catapult, the effect of any propeller thrust on the acceleration will probably have been ignored, since the propeller will usually be designed for efficiency in flight and will produce relatively little thrust from static condition. The effectiveness of the catapult, the system to release the aircraft from the trolley and the systems to arrest the mounting trolley itself, may therefore be tested using a 'dummy aircraft', having the same mass and interface fitments as the real aircraft. Recordings will be made of distance travelled versus time to ensure that the correct speed for launch is achieved. The launch is likely to be controlled from the CS and, if so, the dummy testing should be carried out under control from the CS, whether the link is by cable or radio.

19.6   Documentation

It is important that all these tests are accurately called up and the results recorded in the Treq and Trep documents so that they can be produced for subsequent certification of the UAV system. Any shortcoming in performance, ease of operation or reliability will be reported for modification action and subsequent re-testing.

The System Hierarchy document will also be contained in one of the control documents which will be held by the test engineers and subsequently by the operators. It will allow the testers and operators to identify faulty components for replacement and also to compile a record of failure rates for each sub-assembly. This information is used to determine which elements should be improved or replaced to give the most cost-effective increase in reliability.

With the ground testing satisfactorily completed and the integration of the system proven, the system should now be readied for the in-flight testing phase.

20

System In-flight Testing

20.1   Test Sites

Whilst testing a UAV system on the ground requires relatively little space, this is not true of in-flight testing. As recommended in Chapter 12, the location of a suitable and available site should be found early in the UAV programme.

Even mini-HTOL UAV such as Desert Hawk, hand or bungee-launched and capable of flight within a speed range of order 20–50 kt will require an area of about 800 × 800 m in order to provide contingency with sufficient safety margins under different wind strengths and directions. This will allow for only initial flight testing at moderate speeds to check out the instrumentation aboard the aircraft, and perhaps a limited confirmation of acceptable handling and aircraft stability. Extending the tests to cover maximum speed and sensor performance and more rigorous testing of the control and stability of the aircraft and effectiveness of the communications will require a field of order 1.5 × 1.5 km. A medium-range HTOL UAV such as the Denel Seeker II, runway-launched, with a speed range of about 50–120 kt may require a take-off run of order 300 m before it can lift off.

Initial flights will usually be made with minimum payload mass and moderate fuel load in the aircraft, but it will be carrying instrumentation. The aircraft may therefore enter first flight at less than its design gross mass, possibly at 80% DGM, and so may need a shorter take-off run than in service operation. However it is neither prudent to begin a first flight on a day with other than little wind nor to take off with full engine power applied so, in practice, a full run-length will be required.

The aircraft will then have to climb to a safe height of perhaps 100 m, taking another 1000 m before turning at a moderate rate of turn pulling about *1.1g*, assuming a minimum flight airspeed of 50 kt. This would take it a further 100 m forwards and see it returning on the 'down-wing' leg displaced at about 200 m laterally from its take-off path. Assuming that it makes a similar turn before descending back to the threshold of the runway, an oval of overall length 2500 m and width 200 m would have been covered.

Contingency areas must be planned all around the oval of 200 m on either side and on either end as a bare minimum. This would result in a field size of about 3 km long by 600 m wide for a very limited test at a low airspeed. Operating in the presence of wind from various directions could push the site requirement to about 31.5 × 31.5 km. Initial flight tests of catapult-launched aircraft could be carried out in a somewhat smaller site of possibly 2 × 2 km. VTOL aircraft are the least demanding on site area and initial, low-speed flights could be made within an area of about 1 × 1 km.

Tests at maximum performance of all types except mini- and micro-UAV will require the availability of sites having a length of between 2 and 4 km, depending upon maximum speed of the aircraft in order to allow measurement of performance at steady flight conditions. A further issue is the altitude at which it is necessary to test the performance of the aircraft systems. Many otherwise acceptable sites may have height restrictions imposed upon UAV operations due to their being over-flown by manned aircraft.

Only the largest aircraft companies have access to their own airfields where initial in-flight testing can safely be conducted, and even those are unlikely to be able to accommodate full performance tests, especially if height restrictions apply. Therefore a test range which would not have such limitations must be sought sooner or later. It is therefore best to make early provision for operation from approved sites and to book time well in advance.

If the development of the UAV system is under military contract, then most countries will be prepared to make a suitable military site available. These may be military airfields or army areas used for infantry or armour exercises or artillery ranges. If the development is purely civilian, it may still be possible to have the use of such facilities upon application.

However, there are a growing number of specialist sites being set up in several countries where not only time on site may be bought, but several have specialist personnel and equipment available for hire. Some of these sites are listed in Appendix A – UAV Organisations.

It may have been noted that no mention has been made here for provision of sites for the long-range and MALE and HALE UAV. This is because testing of the smaller, shorter-range, limited endurance systems as described above will be possible in dedicated airspace or, at least, in time-allocated dedicated airspace. Although it may be possible to begin some flight-testing of the larger systems on these mentioned facilities, it must be emphasised that their continued development will require a much wider use of the airspace in all three dimensions and, at least, transiting through civil airspace.

Therefore permissions for testing will have to be obtained at higher administrative level, involving more deeply Air Traffic Control, requiring the avoidance of other air traffic, and may also have inter-state political implications.

20.2   Preparation for In-flight Testing

Off-site Preparation

Before locating the system at the chosen test site, discussions have been held with the range management, including the Range Safety Officer (RSO), as to their requirements in addition to fees. A document laying down the range operating procedures to which the system operators will be required to conform will be presented.

The Range Manager may wish to have copies of the System Specification, the aircraft Build Standard, which will include the instrumentation that is fitted, the Operating and Maintenance Manuals and the system 'pedigree' as documented in the draft Type Record and Test Requirements and Test Results Documents. He will also expect to see the proposed programme of tests well in advance of the arrival on site of the system.

As noted in Chapter 18, the RSO may require the installation of range safety equipment in the form of a range-supplied transponder system to be fitted to the aircraft for range tracking purposes and equipment to enable him independently to terminate the flight should he judge that to be necessary for safety reasons. As far as is possible, this equipment should be checked out for correct functioning before the system is relocated to the test-site. Flight termination methods are likely to be an immediate cutting of the engine power by the most appropriate means and release of a parachute on fixed-wing aircraft. For VTOL aircraft this will most probably entail the engine power to be cut and a change in rotor pitch to ensure a rapid descent unless the aircraft has a built-in system for automatic entry into autorotation following power-loss.

Most of the instrumentation used and proven in the ground testing will be retained in the UAV. Further instrumentation may be added, for example, vertical and lateral accelerometers to record in-flight manoeuvre conditions. Other equipment, which may or may not have been included before, such as fuel-gauging may now be added. Recommendations, voluntary or obligatory, may then be received from the RSO to improve safety or to facilitate the tests. Offers may be made for the availability on loan of on-site test equipment for these purposes.

Test Crew Training

The task and capabilities of the operators of the system under developmental testing will be significantly different from that of the operators of the future users. The test crew that will be responsible for the initial flight testing of the UAV system will, at least in part, be drawn from the engineers who will have carried out the earlier ground testing. It is important that they are familiar with the working of the complete system and what is required of them and of the system in the in-flight tests.

The engineers responsible for the earlier testing will have knowledge at least of that part of the system on which they have worked. It is most probable that they were more widely involved and should have been encouraged, if encouragement was required, to gain an overall understanding of the total system. Selection will have been made of the more suitable persons to carry the system through into flight status.

Although some degree of automation of the control of the UAV in flight will be operative for the first flights, it is probable that most aspects of the flight will be controlled directly by the operator; 'piloting' the UAV within his sight. There should be an understood 'fall-back' to manual control to cover failure or inadequacy of those aspects which are automated.

Should an irrecoverable emergency arise during flight, a forced recovery to land would be made and the person whose responsibility it is to initiate this must be agreed. It would be expected that the responsibility would be that of the appointed engineer in charge of the test.

For the in-flight tests, there must be a clear delegation of tasks. More operators are likely to be employed in testing than in user operations since data is being acquired for the development of the system. Hence the team may consist of an aircraft controller, a payload operator, an engineer monitoring the instrumentation aboard the aircraft and another monitoring the recording equipment for both aircraft and CS data. In user operations only two, or even merely one, operator may be required.

Prior to flight a detailed Test Requirement will have been prepared calling up the Build Standard of the system and scheduling the flights within the initial programme. It will detail the required flight paths and profiles and the data required to be recorded from each flight.

If possible, it could be of advantage for the expected aircraft flight control characteristics to be computer-simulated, especially for the catapult-launched aircraft, so that the aircraft controller can have developed an understanding of what skill level may initially be expected of him.

If possible, a full simulation of the flight profile would be made so that all of the operators can play their expected parts in the simulated operation.

On-site Preparation

Having established the system on site, and positioned the aircraft as required by the site manager and appropriate to the weather conditions, a check-out of the system will begin with the aircraft and GCS powered. It may then be required that the aircraft be repositioned at a distance from the GCS representative of the furthermost extent of the proposed flight-path and the engine(s) started. Checks will be carried out to confirm the integrity of the communication link(s), and the satisfactory functioning of the range safety equipment.

If these checks are found to be satisfactory, the aircraft will be repositioned – the wheeled HTOL aircraft at the threshold of the airstrip or position from which it can be taxied to there; the catapult-launched aircraft onto its catapult; and the VTOL aircraft wherever it is required to be, probably in the vicinity of the CS. The system checks will be repeated and, if satisfactory, the aircraft will enter the next phase.

20.3   In-flight Testing

Basically there are three approaches to consider in the flight testing of a UAV, characterised by its method of launching as outlined in the previous paragraph.

UAV with Wheeled Undercarriage (See Chapter 12, Figure 12.7)

With this system, with the aircraft at a low AUM, a series of ground taxiing tests can be made up to speeds approaching that of lift-off. This will give the operator a better feel of the aircraft responses to control and confirm, or modify, the indication from any simulation previously made.

With pre-flight checks made, again with the UAV at low AUM, the system and crew are now prepared for that all-important first flight which may be a simple flight profile. After take-off, a circuit or two would be made at moderate speeds and low-rate turns followed by a landing. This should be sufficient to give the controllers confidence in the system and their interface with it. The take-off and landing performance at low AUM and under the prevailing ambient conditions will have been established and will enable this to be projected for other ambient conditions. Post-flight checks will be made and the flight data analysed.

Subsequent tests will progressively see the system become more automated both for the aircraft and CS. The flight profile will be gradually expanded in terms of aircraft gross mass, range, speed, manoeuvres and altitude. Other payload systems may be added as available. For each flight, pre- and post-flight checks will be made and any defects and signs of wear will be recorded and reported for modification action as appropriate.

Catapult-launched UAV (see Chapter 12, Figure 12.3)

This type of launch and recovery constitutes the greatest risk within a UAV system test. Although wind-tunnel testing and simulation may have been carried out, the moment of truth arrives as the UAV leaves the catapult. The speed margin between sustainable and unsustainable flight may be small, and the UAV systems will have been subjected to the highest accelerations of the whole flight spectrum, which may result in a power or control failure at the critical moment of departure.

If it is possible to configure the UAV so that rolling take-offs could be carried out before an attempt is made to catapult the UAV into flight, then this would considerably reduce the risk and give the operators greater confidence in the system. Apart from the possibilities of faults in the catapult system or faults caused by the high acceleration or faults in the deployment of the parachute or airbags, the level of risk would be reduced to that of the UAV designed for a rolling take-off.

Subsequent to the post-flight inspection showing no faults and the analysis of the recorded results showing satisfactory performance, the flight programme may then be continued, as for the HTOL aircraft, extending the parameters step by step.

VTOL UAV (See Chapter 12, Figure 12.7)

The VTOL UAV carries the lowest risk of all types when entering into free flight for the first time. In rig tests, the system will have operated the UAV in effectively hovering flight for many hours and proven its hover-flight characteristics and an acceptable level of reliability. Lift-off into free hovering flight should therefore produce no surprises. The operators will be able to explore vertical climb and descent and then venture into low-speed translational flight with considerable confidence.

Until the descent to land is automated, the operators should be aware of two aspects where care is required. The human eye tends to visualise distance relative to the size of a moving image. Therefore in manually controlling the descent of a VTOL UAV and assessing its height above ground prior to initiating a deceleration, he may believe that the UAV is at a greater height than it is in reality. The tendency, therefore, of an uninitiated operator is to leave the deceleration too late. This will result in a last-moment demand for an excessive decelerating thrust and/or a heavy landing.

The other misapprehension is for the operator to wish to demonstrate his skill in performing a perfect soft landing with the aircraft hovering with its undercarriage 'kissing' the ground or, worse still, scrub. An inadvertent lateral control application or a heavy side-gust, especially in the case of a single main/tail rotor helicopter, can then tip the aircraft over onto its rotor. It is always advisable to land the aircraft firmly. A correctly programmed automatic landing system will attempt to land the UAV at a small distance, for example 0.1 m, below the ground surface.

Following successful results from the vertical flight and low-speed tests, the flight speed will be increased in small steps, in all orientations, to reach the maxima called up by the flight envelope. Modifications may be found to be necessary in the control ranges or in the AFCS as the programme progresses. With these successfully made, as with the previous configurations, the flight programme will expand the flight profiles to the full extent of the flight envelope with increasing AUM, endurance, severity of manoeuvre, etc. and with the full relevant level of automation incorporated.

20.4   System Certification

For all of the configurations, after sufficient flight time has been accomplished to determine the system performance including endurance and reliability, and the results are accepted by the relevant certificating authority(ies) the System Design Certificate can be issued by the authority. The minimum number of flight hours to be completed in this phase, with specific tests made, may be a requirement of the relevant certificating authority.

The certificate will denote the limitations of the flight envelope of the aircraft/control station combination, such as the maximum AUM, speed, environmental conditions (air temperature, altitude, wind-speed, precipitation), restriction on range and limitation on airspace within which it may operate. The decision may then be made to enter the next phase, i.e. that of system trials.

Further Reading

A further source of information on flight testing UAS is the AGARDograph: RTO-AG-SCI-105 entitled 'Unique Aspects of Flight Testing Unmanned Aircraft Systems'. This relates to experience in the USA. It does not cover rotary-wing UAS testing.

Part Three

The Deployment of UAV Systems

21

Operational Trials and Full Certification

Once the fundamental testing of the system is complete, it is then necessary to prepare and assess the system for its performance under operational conditions. This may be seen as two types of trial which may be termed 'company trials' and 'customer trials'.

21.1   Company Trials

Company trials are those foreseen as necessary to ensure that the system will be capable of efficient operation under conditions in which future customers may have to operate the system and which may not have been met in the basic tests. These will almost certainly include an expansion of the operating temperature range and possibly of altitude. It may also mean operation from different types of base such as off-board ship.

If sufficient facilities, i.e. hot and cold chambers/wind tunnels are available within the country of origin to artificially reproduce other ambient conditions, then advantage may be made of them. If the facility allows for the power-plant to be operative then this should be done to check for engine over-heating or the possibility of fuel system icing or vaporising of the fuel. Unless the facility is a wind-tunnel, then it will not be possible to check for icing of fixed wing surfaces. Rotary-wing UAV have the advantage that the possibility of rotor blade icing can be explored. For off-board operation, ship-motion platform simulators may be available.

Depending upon the roles for which the system has been designed, the level of altitude for which the UAV will have to be cleared will vary. For close- or medium-range systems clearance of the UAV to operate at up to 2000–3000 m may suffice. Depending upon the location of the test site it may be possible to arrange with the local Air Traffic Control (ATC) for a limited number of short flights to be made.

Instrumentation aboard the UAV should measure and record ambient temperature and pressure and engine and flight performance over a moderate range of airspeeds. This would be carried out at a few step-changes in increase of altitude with a careful watch being kept on the measured values. Should any critical value approach a limit, the programme would be discontinued for remedial action as necessary.

Altitude and endurance testing of MALE or HALE UAV must be carried out from specialist locations and over designated test areas away from air traffic lanes and probably under the direct supervision of ATC.

Unmanned Aircraft Systems – UAVS Design, Development and Deployment   Reg Austin
© 2010 John Wiley & Sons, Ltd

If the appropriate test facilities are not available 'in country' then it may be necessary to hire such facilities in another country or to operate the system in trials in countries where the actual ambient conditions are met.

Either of these programmes would be expensive and an alternative may be to accept full certification, but on the basis of restriction to a smaller range of temperatures and altitudes until the expansion may possibly be done with the cooperation of potential customers.

21.2   Customer Trials and Sales Demonstrations

Potential customers will expect demonstrations of the effectiveness of the system before purchasing. Often the customer will be prepared to visit a site designated by the manufacturer. Others will expect the manufacturer to demonstrate the system on the customer's territory. Some potential overseas customers for the system may have requirements for different payloads and to use the system for other tasks than previously engaged. They may need to operate under more extreme environmental conditions than those in which the system has previously operated.

An assessment has to be made as to the commercial viability of transporting a system overseas for demonstration, noting also that a communication system of different frequency and/or format may have to be developed to operate in another region. To respond to requests for different payload types may also be expensive. However, provided that a successful sales outcome results from such expenditure (some of which the customer may be prepared to cover), advantage may be taken, with the customer's cooperation, to expand the flight envelope and certification of the system and also introduce a further payload type and rôle to the system inventory.

22

UAV System Deployment

22.1 Introduction

In Chapter 1, the reader was introduced to a list of roles for which UAV systems are now in use or expected to be so. Having, hopefully, gained a better insight into the working and capabilities of the several types of systems, it is opportune to look more closely into the possibilities for their application.

An earlier view of military forces was that UAV systems are appropriately used in DDD roles. That is rôles which are too dull, dirty or dangerous in which to engage human pilots. Whilst this philosophy remains true, the foreseen roles for UAV systems have expanded far beyond that boundary, as the following discourse is hoped to show.

Although many UAV systems will be seen to offer benefit in both military and civilian applications, the two will be viewed separately. Today, more UAV systems are in military use than in civil use. This is not because there are fewer applications in the latter, but rather because of the greater difficulties in their introduction to the civil market. The difficulties lie in two areas.

First, since civilian uses imply the operation in open airspace, rather than on a battlefield or within military enclosures, the regulating authorities have yet to accept their general operation. Obviously means must be in place to:

a) prevent injury to persons or animals and damage to property due to failures of the UAV; and also
b) prevent injury or damage caused by collisions between UAV and other airborne vehicles.

Requirements exist for the assurance of the airworthiness of the systems and meeting these requirements to ensure the achievement of (a) does not constitute a problem other than, perhaps, the cost of so doing.

The problem of (b) remains since the authorities look for a reliable means being in place of a UAV sensing the presence of another airborne vehicle and avoiding collision with it. This is currently required not only in open airspace, but in airspace dedicated to the use of the UAV system. In other words, should another aircraft stray into UAV-dedicated airspace, the onus is currently upon the UAV to avoid it.

A technical solution for 'sense and avoid' for the UAV is currently being sought and is expected to be achieved before long (see Chapter 27), but the cost of such a system may still inhibit UAV use in some applications on economic grounds. The other reason for the delay in opening a civil market for UAV systems is the anticipated cost of product support. One exception to this situation is that of powerline inspection. Utilities companies have a need for a significant number of UAS for that and possibly other duties. Generally they are large organisations with a technical background and maintenance facilities. They therefore can readily, and economically, interface with a UAS manufacturer in supporting their operation. Another exception is that of crop spraying which is widely carried out in Japan. Both of these operations are described in Chapter 26, Sections 26.2 and 26.6.

The responsibility of the manufacturer to the customer/operator does not end with the delivery of the UAS which may continue in operation for many years, depending upon its rôle. The manufacturer has a responsibility to provide technical advice and spare parts as scheduled and required together with updated information on maintenance of the system through additions to the operating and maintenance manuals. This will include any changes to the service lives of components and any precautionary inspections or modifications that may be found to be necessary following extended life-testing at the manufacturer's works or experiences by operators in the field. Modification 'packs' may be supplied to be applied by the operator or, with the customer's agreement, the affected element of the system to be returned to the factory for re-work.

The military scene is characterised by a smaller number of users requiring large numbers of UAV systems. The civilian scene, on the other hand, is generally characterised by a much larger number of potential users, widely dispersed around the world, operating only small numbers of systems, perhaps only one or two in many cases. The support, in terms of technical support and the supply of spare parts, of the former is easier to put into place and should be cost-effective. Appropriate military support and supply lines may already exist and can be accessed by the manufacturer via a central depot. The support of the civilian operator is much more difficult unless it can be sprung from a pre-existing military support system. It can be seen, therefore, that the combination of the two problems has delayed the take-up of the larger civilian market to be behind that of the military.

In the next four chapters we will look at the current and potential roles for UAV systems in the armed forces in the traditional order and in civilian and commercial operations. It is necessary to recognise how the manner of the operation will drive the choice of type of UAV which in turn will drive the design of the rest of the total system.

The operations which have received the greatest press reportage in recent years have been the tasking of long-range, long-endurance UAV conducting surveillance over theatres of war from medium to high altitudes and more recently being equipped with armament in order to immediately attack a recognised enemy rather than risk delay in calling up ground troops or attack aircraft. The types of UAV suited to these operations are large HTOL aircraft, such as Predator or Global Hawk, operating from established airfields at some distance from the theatre of conflict, but controlled from even further afield (see Chapters 4 and 13). These are very expensive UAV carrying sophisticated and expensive payloads and controlled from a complex centre of communications. They are not to be seen as expendable and so must be operated and controlled from secure sites.

At the other end of the scale are micro-UAV which have yet to see operational use. The smallest UAV systems in use employ mini UAV such as the Desert Hawk fixed-wing aircraft. They are used nearer to the 'front line' of army operations and must be mobile, simple, cheap and expendable. These UAV have their limitations, principally due to their low wing loading necessary for launching. They are unable to operate other than in quite benign environments and are especially vulnerable to high winds and air turbulence.

VTOL UAV, particularly of the rotary-wing type, are invaluable where very low speed flight and high manoeuvrability is required, where launch sites are very restricted in area and in conditions of air turbulence which often occur at the lower altitudes. The attributes of rotary-wing UAV have only recently come to be recognised by the general UAV community and the development and adoption of such systems are currently being more actively advanced.

22.2 Network-centric Operations (NCO)

Although the next four chapters address the operation of UAV systems in a number of specific roles, it must not be forgotten that, in reality, and especially in military operations, the UAV system will not necessarily be operating alone. More usually it will be receiving and using information from other sources which may include earth satellites, other UAV systems, manned aircraft, naval vessels and ground-based systems.

In its turn, it may disseminate information which it has self-acquired or received. Indeed, a UAV system, such as a HALE system may be a major supplier of information to a network. An illustration of surveillance network using airborne systems is shown in Figure 22.1.

Networks are not limited to airborne surveillance. A range of activities, air-, sea- and land-borne, covering reconnaissance, surveillance, support, defensive and attack operations, etc. may be coordinated through a network. This is illustrated in Figure 22.2.

For the network interoperability of systems to succeed, however, robust technologies must be developed to ensure that:

a) communications between systems are secure via multiple links;
b) interfacing between key systems is standardised;
c) interfacing with the human operators is adaptable and user-friendly;
d) radio frequency bandwidth is coordinated.

A possible weakness of NCO is if one of the major systems fails, then the whole network may fail. For example, if over-reliance is placed upon the Global Positioning System, and this fails due to sunspot

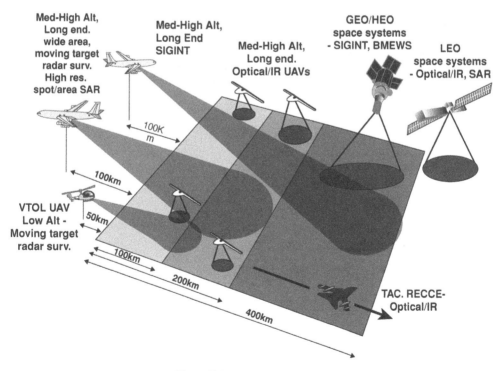

Figure 22.1   Surveillance network

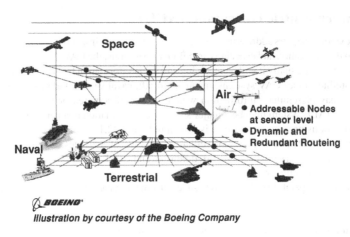

Illustration by courtesy of the Boeing Company

Figure 22.2   Network-centric architecture (Boeing)

activity or through its being neutralised by a sophisticated adversary or even through electromagnetic pulse weapons being used, albeit on a limited scale, the whole system could descend into chaos. It is therefore argued that individual systems should have a means of 'fall-back', even though that may mean a reduction in effectiveness in that event.

22.3   Teaming with Manned and Other Unmanned Systems

Teaming a UAV system with manned systems and/or with other unmanned systems may offer advantage to both. This may result in a local 'system of systems' or may form part of a larger NCO. Some such teaming arrangements will be included within the operational rôles discussed in the following chapters.

23

Naval Rôles

In naval application, UAV systems may be mounted from three different types of base:

a) off-board operation;
b) off-shore operation;
c) long-range support from airfields.

Several of the rôles may be carried out from each type of base, but generally employing different categories of UAV as befits the base.

In operations from off-shore, and especially from off-board, the UAV will generally be operating at low altitudes above the sea. It will come into contact with salt-laden atmosphere and it will have to be protected against corrosion at component and overall air vehicle level. Provision may have to be made for washing the air vehicle.

The longer-range operations from airfield bases are likely to be higher-altitude reconnaissance missions and so may avoid salt contact problems.

Naval operations generally have the advantage of cooler atmospheric conditions, due to the moderating effect of the sea, and are usually launched from lower altitudes compared with army systems. Naval UAV therefore have less demand for high power and cooling systems. However the UAV are more likely to encounter higher winds and turbulent air conditions than UAS being operated over land.

Another condition of concern in the design of ship-borne UAS is that they must be able to operate within a field of intense electromagnetic radiation so require shielding against this.

The provision of emergency flotation gear in the UAV may be considered prudent, if only to enable recovery of any expensive payload in the event of an inadvertent ditching. Hangar space on board ship, especially of small vessels is limited and this will require the UAV(s) and any support equipment to be compact when folded for stowage. The UAV should also occupy a minimum of deck space during operation.

Off-board operation of UAS will usually require the UAV control station to be integrated within the ship's control room and interface with some of the ship's sub-systems such as navigation and communication both internal and external, A separate, 'satellite', control station may be required in the vicinity of the launch-pad for the launch and recovery phases of the operation but during those phases must interface with the main control. The control station equipments will also have to be 'navalised' against salt corrosion. All these aspects have to be considered in the design and development of Naval UAS. UAS operated from off-shore or inland bases will not generally be controlled from ships at sea.

Unmanned Aircraft Systems – UAVS Design, Development and Deployment   Reg Austin
© 2010 John Wiley & Sons, Ltd

Rather, they will communicate with the ship(s) during combined operations and so facilities for that communication must be available within the ship's control centre.

Since the year 1960, navies have been looking into the possibilities of using UAVs in many different rôles. In the 1960s the US Navy put Gyrodyne DASH helicopter UAS into service. Their rôle was to operate from small naval vessels to drop torpedoes into areas of the sea in which sonar systems had indicated the presence of a hostile submarine. The results were not entirely satisfactory and several were lost at sea due to a number of reasons, believed to include electromagnetic interference problems and the loss of control of the UAV 'over the horizon'. In later years the US Navy has carried out trials with fixed-wing aircraft but these, of course, had the problem of being launched and even worse, recovered back onto ships.

Some success has been achieved by both US and UK navies in operating the Boeing/Insitu Scan Eagle fixed-wing UAV from off-board. The aircraft is launched by catapult and recovered by a 'sky-hook' device. (see Chapters 4 and 12 for details). However, in order to fly relatively slowly to engage the sky-hook, the aircraft has a low wing-loading and high aspect ratio wing which must render it vulnerable to high winds and turbulent air. Other disadvantages of the system must be the additional launch and recovery equipment deployed on deck. The current aircraft is of less than 20 kg AUM and is necessarily limited in payload. A heavier aircraft would be at risk of causing damage to the ship's antennas, etc. should it fail to engage with the sky-hook. Its progress will be watched with interest.

The US Navy has, however, recently decided that any new naval UAV for off-board operation must have a vertical take-off and landing ability and is assessing the performance in sea trials of the Northrop-Grumman Firescout VTOL UAS (see Chapter 4).

During the 1980s the British Navy carried out limited trials of the SPRITE VTOL UAS with considerable success.

23.1   Fleet Detection and Shadowing

Although the initial detection may be the province of satellite surveillance, the subsequent shadowing is likely to require a dedicated facility. This may be by use of a naval MALE UAV operating at moderate altitude, equipped with passive sensors, if it is required to remain covert, or radar if covertness is not required and it is necessary to penetrate cloud cover. The UAV system will be operated from an airfield and report to naval headquarters.

Should it be decided to counter an enemy fleet, the opposing force will probably need to launch an indigenous UAV from off-board covertly to shadow the enemy more closely and to attempt to intercept any electronic signals. Passive sensors would be deployed and data sent back on a directional communications link to the mother ship. A medium-range VTOL UAV would be employed to advantage as it would probably be more covert than a fixed-wing UAV and be better able to maintain station aft of the fleet.

23.2   Radar Confusion

Should it be decided to mount an attack upon the enemy fleet, a suitable UAV could be sent to approach the enemy fleet from a predetermined direction. It is possible that the mission could be entirely pre-programmed, otherwise radio communication would be relayed via a satellite or RUAV. The UAV would carry a payload capable of emitting radar signatures representing those of various types of vessels or an amalgam of several. After arriving on station, the UAV would be required to descend to fly at representative ship height and speed and begin emission at an appropriate time. By this means the enemy attention would be diverted from the real direction of attack.

A novel idea might be that the function could be provided by a fixed-wing UAV with a flying-boat hull, catapult-launched from a ship. After reaching position, it would descend onto the sea surface and motor at fleet speed whilst emitting its radar signals. This approach, however, would introduce a number of extra risks such as making it difficult to achieve a safe programmed take-off from the sea at distance or to survive the rigours of rough seas. It is unlikely therefore that the fixed-wing UAV, with its expensive payload, could be recovered.

23.3   Missile Decoy

A UAV fitted with a payload capable of emitting sources of detection used by anti-ship missiles can be deployed appropriately to attract the incoming enemy missiles. The loss of a number of unmanned UAV would be a small price to pay in saving the loss of costly naval vessels and their crews.

23.4   Anti-submarine Warfare

The following three basic techniques are available to detect submerged submarines although they are not mutually exclusive:

a) active sonar;
b) passive sonar;
c) magnetic anomaly detection (MAD).

The use of *active sonar* requires a buoy to be winched down from a hovering helicopter to a considerable depth in the ocean. The buoy emits a stream of acoustic signals which travel out from the buoy and are reflected back from any object in the water. The returning signals give information as to the distance and bearing of the object away from the helicopter and its geographic positional coordinates may be fixed. This enables a second helicopter to move rapidly to those coordinates to release depth charges or a homing torpedo.

However, a large helicopter transmits pressure waves downwards onto the water surface and these can be recognised by the submarine. Anti-helicopter submarine-launched missiles, such as Triton, are being developed which will increase the odds in favour of the submarine by homing onto the characteristic rotor noise and engine exhaust heat.

A counter to this may be offered by deploying VTOL UAV as shown in Figure 23.1. The active sonar buoys may be lowered from the UAV which would have much reduced signatures compared with the manned helicopter. Another decoy UAV could carry a payload designed to emit high levels of noise and heat signature to attract the submarine-missile. The appearance of this missile will give away the presence and position of the submarine for a counter attack launched from the heavier manned helicopter which is monitoring the situation from a safe height.

Passive sonar buoys, which merely listen for noise generated from the submarine's systems or wake, may be trailed from, or distributed by, one or more VTOL UAV. Data radioed from the buoys to the UAV, or directly to a ship, can determine the presence and position of an underwater vessel. A number of buoys may be carried by a UAV which would be smaller, lighter less expensive and more readily expendable than a manned helicopter.

Magnetic anomaly detectors, which can detect metallic objects beneath the sea surface by recognising any distortion of the earth's natural magnetic field, are now of a mass low enough as to be trailed from a small UAV of gross mass of no more than 300 kg. The task may be carried out using a HTOL or VTOL UAV although the latter has the advantage of being operable from a ship at sea and being able to remain more accurately above any target so discovered.

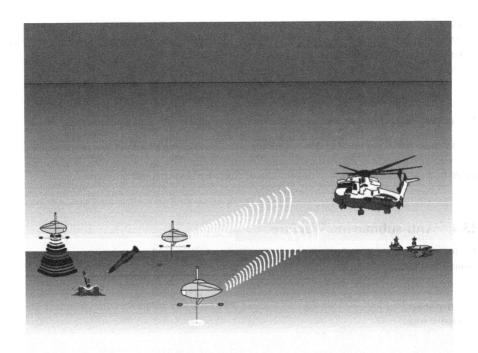

Figure 23.1 Future ASW operations

23.5   Radio Relay

Secure communication between ships at sea is limited in range by the radio horizon unless reflected by the Heaviside layer or relayed by satellite. These present problems of security or capacity. An indigenous relay capability can be provided by launching a UAV equipped with a transceiver system within its payload.

Figure 9.3 in Chapter 9 indicates that, for example, a relay UAV (RUAV) maintaining station at a height of 200 m above the mother ship can provide a radio link between its ship and another about 65 km away. If the RUAV climbs to 1000 m, the link range will be extended to about 135 km.

Similarly a UAV with a relay payload can extend the communication range to a distant operating UAV. Again viewing Figure 9.3, it can be seen that the RUAV from 1000 m height can communicate over 260 km with an operational UAV (OUAV) also flying at 1000 m. If cloud cover at 300 m requires the OUAV to descend to bring its electro-optic sensors to bear, then the communications link range will be reduced to about 200 km unless, of course, the RUAV moves towards the OUAV when each kilometre that the RUAV moves out will allow the OUAV to add that distance to its range from the ship.

23.6   Port Protection

Naval vessels in port are vulnerable to attack, and damage to the port facilities can deprive the fleet of vital operational support. UAV systems based at the port or aboard ship are able to maintain a watching surveillance for attacks mounted from the sea or from land. They may then be able to alert a defensive strike or initiate the strike themselves.

23.7   Over-beach Reconnaissance

Prior to the approach to a beach upon which a landing is to be made, a stealthy UAV may be launched from a ship whilst the vessel(s) are still out of sight from the beach. The UAV will survey the beach, while undetected, for defensive measures using a passive EO/TI payload. A decision can then be made as to where a landing may best be made. Stealth is important in this operation so as not to alert defensive forces to an impending landing.

23.8   Fisheries Protection

Fisheries protection implies ensuring that illegal fishing is not carried out within protected waters. This is customarily carried out by naval vessels patrolling the fishing grounds which may be quite extensive in area. It is an extended task, requiring the location of suspect vessels and approaching them to assess their origin and their intent. It is not illegal for an unlicensed vessel to be present within the grounds as long as it is not fishing. Therein lays the problem. The crew of an illegally fishing vessel will be alerted to an approaching ship as soon as that ship's mast appears over the horizon. They then have plenty of time to haul in their nets before the patrol vessel (NPV) is close enough to see the activity and its crew have no overt reason to board the vessel.

If the NPV has a UAV system on board utilising one or more stealthy UAV, the NPV can steer a steady course through the fishing grounds and send the UAV to view the fishing vessels. A stealthy UAV with passive sensors can approach closely to a suspect vessel without alerting the crew. It can then obtain evidence for future action by sending recordable photographic images of the fishing vessel's perceived activity, with its position and time, back to the NPV. A VTOL UAV can approach closely enough to photograph the name of the ship and its port of registration.

Therefore the NPV has no need to spend time pursuing an extended course to visit each suspect vessel, but simply visit only those vessels shown by the UAV to be illegally fishing and with full evidence to justify its boarding. This offers greater efficiency in the successful prosecution of illegal fishers, and also considerably reduces the distance sailed by the NPV to complete its task. The operation becomes not only far more effective per ship hour whilst at sea, but reduces the overall time necessarily spent at sea by the crew.

23.9   Detection of Illegal Imports

Persistent watch for suspicious shipping out at sea is best maintained by a MALE UAV. From this any approach to land, in other than at recognised customs ports, can be reported to the Customs and Excise Authority (C&EA). The subsequent patrol of vulnerable remote coastal areas by C&EA UAS (see Chapter 26) can detect and record evidence of the import of illegal substances or persons. This is better accomplished by a stealthy, slow-flying UAV which will not alert the criminals to the possibility that they have been detected and may remain over the scene to direct local forces to apprehend them on site or in transit.

23.10   Electronic Intelligence

Patrol in the search for enemy electronic emissions and their interception for intelligence purposes can be a time-demanding and potentially vulnerable exercise. It is therefore particularly appropriate for being carried out by UAS. The payload would include sensitive radio receiving equipment capable of sweeping through an appropriately wide range of radio frequencies, recording and transmitting the data back to the CS or other interpreting station. Such a system may be thought to warrant a self-destruct mechanism in the event of a UAV failure resulting in the system falling into enemy hands. This is a rôle requirement for all three military services with the major and most comprehensive operations usually being carried out by air forces. Therefore the operation is discussed more fully in Chapter 25.

The differences in the operations conducted by the different services may be the type of UAS used, the geographical range of the operation and possibly the range of radio frequencies of interest. For naval and air force application, a long-range, long-endurance, probably MALE or HALE, system would be employed whilst for army purposes, a medium-range UAS would probably suffice.

23.11 Maritime Surveillance

Maritime surveillance over large areas of ocean currently carried out by manned aircraft such as Lockheed P3 aircraft (US Navy) and Nimrod aircraft (UK Navy) are candidates for replacement by UAS, either MALE or HALE types carrying a range of electro-optic and infrared imaging sensors and radar. The US Navy is buying a number of Northrop-Grumman RQ-4N Global Hawks under a 'broad area maritime surveillance' (BAMS) programme.

23.12 Summary

A number of naval operations currently carried out by other means can be more efficiently, reliably and economically carried out using suitable UAV systems. The longer-range, long-endurance missions such as maritime surveillance are best carried out by land-based MALE or HALE types of fixed-wing aircraft such as Predator or Global Hawk. Off-board operations, possibly the majority of naval missions are, arguably, better carried out using VTOL UAV.

Low-level maritime or littoral operations can suffer higher winds and air turbulence than operations over land or at high altitudes. Hence, in addition to their being more readily operable from off-board where appropriate, VTOL UAV are far less gust-sensitive than most HTOL aircraft. The perceived idea that rotary-wing aircraft necessarily have an inadequate payload and range capability compared with a similarly sized HTOL aircraft for operation at low altitudes, or are necessarily less reliable, is misinformed. Developments such as the Boeing Hummingbird UAV (see Chapter 27) are showing this to be a fallacy.

24

Army Rôles

Army missions may be carried out in a wide number of areas in the world in greatly varying terrain and atmospheric conditions. These may range from the snow and ice of the poles to the heat of the tropics and may be combined with considerable height in mountainous areas. Often having to use improvised airstrips or no airstrip at all, the army UAS are likely to be subjected to more extreme environments than are those of the other services. In some areas this may be exacerbated by the need to contend with sand or dust storms.

The army UAS will usually be required to be mobile for transit from one war zone to another and also to be highly mobile within a zone to move at short notice as local military action requires. All of this will have been recognised in the early design stages of a UAS development.

24.1 Covert Reconnaissance and Surveillance

The majority of army reconnaissance and surveillance operations are over ranges of short or medium length and are intended to expose any immediate or short-term threats. They are usually required to be covert in order not to alert any enemy forces to the presence of potential opposition and will be mounted by catapult-launched or VTOL UAV in, or close to, the theatre of operations.

Some medium-range HTOL UAV systems may be deployed from improvised airstrips using UAV which are capable of a short take-off run. The longer-range missions relating to longer-term threats are usually the province of the air forces, operating from established air bases.

An example of a covert reconnaissance mission is to establish the extent of enemy positions or movements or, in another scenario, the infiltration of insurgents into friendly territory. In the past this would have been carried out by forward reconnaissance teams on foot, but under modern conditions they lack the mobility to cover sufficient territory in a timely manner and are limited in visual range. Today, the use of UAV systems offers a better solution.

Hand-launched mini-UAV systems such as Skylite B and Desert Hawk (see Chapter 4, Figure 4.23) are in regular use by the USA and other armed services. Operating at an AUM of only 3 kg, they have the merit of being relatively cheap, but their loss rate is understood to be high and they are restricted to operations at very short range (10 km) and in relatively benign conditions principally due to their very low wing-loading (10 kg/m^2).

Catapult-launched close-range fixed-wing UAV such as Luna or Sparrow of 40–45 kg AUM and higher wing-loading (20–40 kg/m^2) are better suited to operations at ranges of up to about 50 km and in moderate weather. VTOL UAV of similar mass are capable of similar range operation and under more adverse weather. They also offer the ability to be more covert than the fixed-wing UAV.

Unmanned Aircraft Systems – UAVS Design, Development and Deployment Reg Austin
© 2010 John Wiley & Sons, Ltd

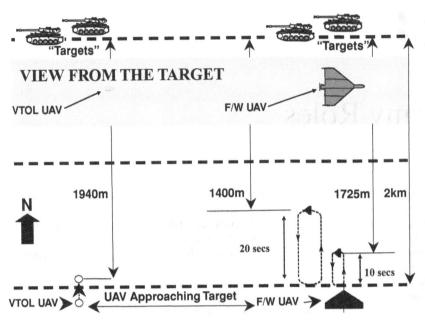

Figure 24.1   Reconnaissance mission

Figure 24.1 illustrates UAV launched from the south and proceeding northward looking for any infiltration, etc. The left-hand UAV is a VTOL type whilst that to the right is a typical catapult-launched fixed-wing type.

It is supposed that the two UAV are searching, at a typical height of 300 m and airspeed of 25 m/s. Each sees the target when 2 km away. Neither must declare their presence, and so take evasive action whilst still needing to keep the target in view.

As shown in the figure, the VTOL UAV will immediately decelerate and halt only 60 m nearer to the target, with the option of retreating backwards if desired, but still keeping its sensors trained onto the target, and possibly also zooming the optics onto it. The fixed-wing UAV, on the other hand must turn away though also with the need to maintain observation. Unless it is fitted with a rotatable, stabilised sensor turret, it will lose sight of the target until it returns onto the same, or a similar, track some time later.

It has a number of options if it has only an essentially forward-looking sensor. One will be to continue towards the target for the accepted regard time of 10 s before making a 2g banked turn onto a parallel return track and turning again onto the original line of sight to the target. This will give a 10 s view followed by a 24 s 'blind' period; this cycle repeating every 34 s.

Another option is to effect a continuous turn banked at 30° on a radius of about 50 m, at which it is estimated that it will pass through its forward line of sight to the target once every 12 s and obtain a snapshot each time. If the UAV is fitted with a rotatable turret, which implies a larger aircraft, it may perform the same 30° continuously banked turn, but retain a continuous view of the target.

There is obviously an infinite number of possible variations on these themes, including effecting turns at lower rates, but no matter which tactic is chosen, at some time the fixed-wing UAV will fly on a course across the target's view and present to an observer a projection of its wing area calculated at 50% of its plan area, depending on the tactic chosen. The projected image of the dark lower-surface of the wing will remain in view for about 3 s once every 24 s in the former tactic and once every 12 s in the latter (continuous turn) tactic. This can present an image subtending of the order of 1.3 mrad to the observers at the target. On the other hand, the VTOL type UAV will present a maximum of only 0.25 mrad. It is generally accepted that, from experiment, the human eye, without optical assistance, cannot accept an image subtending less than 0.75 mrad, assuming normal contrast.

However, this is an oversimplification. In practice, at distance the observer sees only the basal shadow formed under the airborne object. Assuming that the sun is high in the sky, say ± 3 hr from noon, a typical 0.8 m² projection of the fixed-wing UAV wing will be in shadow.

In contrast, the basal shadow of, for example, the Sprite VTOL UAV was found to be only about 400 mm in diameter, (0.125 m²) in full plan view and less than 0.02 m² when hovering at 300 m and viewed from 2 km away. That is 1/40th of that of the fixed-wing UAV. The view from the target, with the observer's eye aided by optics of magnification of about ×5, reproduced in Figure 24.1, is as accurate a representation as can be made using this medium to show the difference in detectability between the two UAV types.

Another difference, borne out in practice, is that the eye is alerted by movement across its line of sight rather than by stationary objects and more by straight lines than by amorphous 'blobs'.

24.2   Fall-of-shot Plotting

Conventional artillery batteries can range their guns onto a distant target by predictive means, but may lack accuracy due to variable wind effects and to some degree the calibration of individual guns. In the past observers, originally on geographic vantage points and more latterly in air observation post (AOP) helicopters, have provided information to allow correction of range and bearing. Manned helicopters have become very vulnerable, especially when hovering for this task. The stealthy UAV again can provide a viable answer.

Figure 24.2 shows a photograph of a Sprite UAV carrying out this task but with the UAV descended far below normal operating height for 'photo-shoot' purposes only! A hover capability for the UAV is a virtual necessity for this task. With the cross-hairs of the imaging payload stabilised onto the target, and with knowledge of the sight-line depression angle and UAV height, it is a simple trigonometric task for the CS computer to calculate any elevation and azimuth correction required and to feed that correction immediately to the gun fire-control.

24.3   Target Designation by Laser

The more sophisticated armament of laser-guided shells and missiles require that the target(s) be illuminated by a laser beam. This is reflected back from the target and the artillery shell or missile homes in onto the reflected light.

The original intent was that the laser system was to be trained onto the target(s) by advanced ground parties, with the shell or missile launched from along-side or nearby. This may have been reasonably effective, but for longer-range operation it is unlikely to be so.

For the longer-range operation where the shells may be fired from several kilometres away, not only are the ground crews very vulnerable to counter-measures, and are not very mobile if required to move quickly as the situation changes, but they are often limited in range due to intervening terrain or obstacles. It is also difficult for them to maintain the laser onto the target for an adequate period of time and the crews are not able, generally, to bring the laser beam to bear onto the target from the optimum angle.

In order to achieve the highest probability of the incoming missile striking the target, it is necessary for the greatest intensity of the reflected illumination to be predominantly back along the flight path of the missile. Most targets will reflect a large proportion of the illumination back towards the source of the lased light.

Figure 24.2   Fall-of-shot plotting

If, for example, the shell or missile is approaching from the north at 45° to the horizontal, then that is the direction from which, ideally, the laser light should originate. A laser aimed at the target from, perhaps, the east and at ground level will not have much of its illumination seen by the shell or missile. It is therefore more appropriate for the target to be illuminated from an airborne laser positioned more closely to the incoming flight-path of the missile.

Although this task has been done from an orbiting manned fixed-wing aircraft with a rotatable turret, the aircraft must stand off some distance from the target in order for the laser direction to remain at an approximately optimum angle. This can require a more powerful laser and with the greater possibility of cloud or smoke obscuration. The aircraft is also quite vulnerable to counter attack.

The task is more appropriately carried out by a hovering aircraft which can maintain the laser at the correct angle and illuminate it only at the moment in which it is required. A manned rotorcraft, however, is also vulnerable to counter-attack.

A stealthy VTOL UAV can be positioned, undetected, at the optimum point in space. Figure 24.3 shows a comparison of the visual size of a small VTOL UAV, characterised by a Sprite, with that of an attack helicopter such as an Apache. Their actual visual, radar, infrared and acoustic signatures differ even more greatly with a difference in the region of three orders of magnitude.

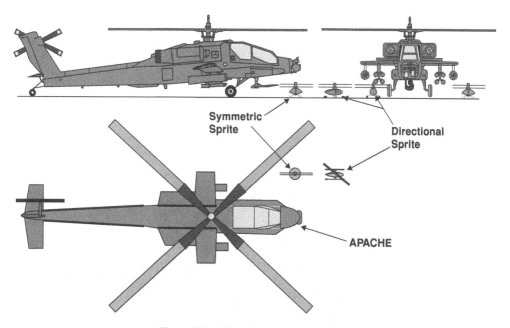

Figure 24.3   Size and signature comparison

Figure 24.4 portrays the use of the UAS to illuminate targets for missiles launched from otherwise vulnerable attack helicopters hovering in cover several kilometres away. This is one example of potential teaming between a UAV system and a manned system.

Another scenario is for VTOL UAV, as portrayed in Figure 24.5, to illuminate targets for a Copperhead battery. The United States Army's Copperhead Artillery System comprises a gun battery which fires shells which are guided onto a target by locking on to the reflected light of a laser beam trained onto that target. The UAV payload operator merely maintains the cross-hairs of the laser onto the target(s) in succession whilst the laser need not be discharged until the approach of the incoming shell.

Indeed, the optimum arrangement is for the UAV CS to be linked to the gun fire-control and the laser is discharged by the gun controller as the shell approaches the vicinity of the UAV. The gun controller will know when that occurs as calculated from the moment of the shell being fired. Thus the laser need be 'on' for only a few seconds, which considerably reduces the likelihood of it being detected by the target until it is too late, and also reduces the amount of electrical power required from the UAV generators.

24.4   NBC Contamination Monitoring

In any future conflicts, an army, whether advancing on foot or in vehicles, may encounter nuclear fall-out and/or biological or chemical gases or substances. These are all at least debilitating and usually life-threatening. An army on the move, whether it constitutes only a patrol or a complete battalion, must be assured that it will not find itself subjected to these threats. It is therefore necessary to assess the environment ahead before committing to enter a new zone.

Currently this is done by an advance patrol measuring the atmosphere close to and at ground level for contaminants. That is achieved using a hand-held device which 'inhales' and analyses samples of

Figure 24.4   Laser target designation

Figure 24.5   Artillery designation

Figure 24.6 NBC contamination monitoring

the atmosphere and displays a simplified summary of the findings on a small screen on the device. The assessing crew members must wear protective masks and clothing whilst conducting the operation and report back by radio to the main body of troops. It is a slow and clumsy procedure.

A better alternative is the use of a UAV carrying a payload which includes the NBC assessing device, but without its displays. The UAV will be sent out ahead of the proposed line of advance of troops, further if necessary than is possible by a ground-patrol. The UAV will descend to ground level at a number of points to 'inhale and analyse' the contents of the local atmosphere and transmit the data back to the unit command.

This scenario is portrayed in Figure 24.6, in this case with the UAV being controlled from an attack helicopter which receives the NBC data back from the UAV and relays it to headquarters. This is another example of the teaming of manned and unmanned systems.

An alternative scenario is for the UAS to operate alone. The options in this case are:

a) For the UAV to proceed in radio LOS to the points for investigation above which it will hover. A pre-set program will then cause it to descend to a suitable near-ground hover to inhale the pollution and then automatically climb to regain radio-link and transmit the data back to its CS.

b) A second option is for the UAV to fly out in radio LOS but trailing a fibre-optic cable. The cable is very light and of only about 1 mm diameter. It is spooled out from the UAV (not from the ground) so that it is laid statically onto the over-flown terrain. The UAV is then able to descend to ground level and remain in communication via the cable. The UAV will then climb, translate to the next point, descend and repeat the detection process as many times as is required and within the limits of the cable length.At the completion of the survey operation, the UAV is commanded to climb to LOS height and jettison the cable to return to base. Such a system under development for the Sprite UAS offered a potential cable length of 10 km wound onto a spool of a volume compatible with its mounting on the UAV.

The several advantages of UAV use in this rôle are:

a) No personnel are put at risk.
b) The data sent back can be far more comprehensive and detailed in its content than that which can be obtained by foot patrols using hand-held equipment. Data from the UAV is displayed on screens in the GCS which are overlaid with area mapping.
c) The UAV is more mobile than the ground patrol and can cover a greater range of positions more quickly. It will also be more able to respond quickly to assess conditions at other points as the situation is seen to change.

24.5   IED and Landmine Detection and Destruction

Landmines laid in the path of advancing forces can be the cause of many casualties unless discovered and destroyed. Furthermore, there are many thousands of these and other devices such as cluster-bombs and anti-personnel mines remaining in many parts of the world lying unseen to maim or kill the unwary civilian. To attempt to discover and disarm them by foot-patrols is a lengthy and dangerous business and so slow progress has been made to rid the world of such evils. There is the possibility of an important and effective rôle here for UAS.

In the operation of Sprite UAS, it became obvious that buried items, as well as surface items hidden by vegetation, could readily be detected by an appropriate, high-definition, thermal imager (TI). Detection is easier, of course, at certain times of the day, i.e. after dawn and before sunset, when the surface temperatures are changing. Different elements absorb or lose heat at different rates, resulting in greater thermal contrasts at those times.

Experiments with acoustic beams indicated the possibility of exploding the explosive items by directing an acoustic beam, pulsed at the optimum frequency, onto them. A payload consisting of the TI and acoustic beam generator could be carried best by a VTOL UAV which would hover at a safe distance and height above a discovered item to explode it. Therefore, in addition to such systems being of importance in the armoury of military forces, they might well be the solution awaited by world-wide civilian organisations.

24.6   Electronic Intelligence

As noted in the previous chapter, all three services have requirements for electronic intelligence and this is discussed more fully in Chapter 25.

24.7   Teaming of Manned and Unmanned Systems

Two illustrations have been given in this chapter of potential teaming between manned and unmanned systems. There is also great potential for teaming between different types of unmanned systems.

Proposed ground-based unmanned vehicles are limited in their visual horizons. Whether it be unmanned tanks vulnerable to unseen tank-traps or ground-reconnaissance vehicles operating in an urban environment, a stealthy UAV, preferably with a hover capability, can be of great benefit in providing an 'eye-in-the-sky' enhancing the forward vision of the unmanned ground-based system (UGS). The UAV may also be needed to relay the communications link between the UGS and its control station.

24.8   System Mobility

As noted above, ease of mobility is a necessary characteristic, particularly of the short- and close-range army UAS. Whether it be for instant mobility on the battlefield or for ease of transporting between theatres, mobility of the system is a virtue.

Figure 24.7   Phoenix and Sprite systems

(a) A back-packed system controlled via a laptop computer is the ultimate in mobility, but will inevitably lack the performance of heavier systems.

(b) A compact close-range VTOL UAS such as the Sprite system, with an aircraft of 36 kg AUM, requires only one small all-terrain vehicle which serves as GCS, UAV and support equipment storage.

(c) A close-range fixed-wing UAS such as the Cobham Raven used only a small all-terrain vehicle with a small trailer. The aircraft, of 45 kg AUM, was catapult-launched and recovered on foot. The trailer mounted the catapult launcher on its roof and contained one aircraft inside.

(d) A slightly heavier, close/medium-range fixed-wing UAS such as Phoenix with an aircraft of 177 kg, also catapult-launched, may require, in addition, a dedicated launch vehicle and a separate recovery vehicle. The increased number of vehicles, also requires extra personnel, so can considerably reduce the system's mobility and increases its response time.

As a graphical illustration of differences in mobility, operational troops of a Phoenix system and of a Sprite system are contrasted in Figure 24.7. The difference in covertness in deployment may also be compared.

24.9   Persistent Urban Surveillance

This is a very important rôle that has yet to be fully satisfied. The need is to insert a means of covert surveillance into an urban environment amongst buildings of various types from high-rise tower blocks down to single-storey units in various states of repair.

Proposals include the proposition for a UAV (in this case most appropriately a MAV or NAV) to 'perch and stare'. This implies a UAV that can hover, at least briefly, in order to alight on a vantage point and, to save power and extend mission endurance, close down the system except for its surveillance sensor(s) and communication links. The problems are:

1) The UAV must be small and inconspicuous or it will be seen and destroyed.
2) It must be able to fly under precise control in the windy, gusty urban environment in a usefully wide range of weather conditions to navigate to a vantage point.
3) It must be able to hover in order to alight onto a spot.
4) It must be able to sit securely on the spot with the sensor aimed in the required direction(s) without being disturbed by winds channelled at high speed between buildings.
5) It must be possible to achieve and maintain radio communication between the control station and the UAV whilst it is flying between buildings and then 'perched'. The radio-links are vulnerable to complete blockage or multi-path effects and also to positive jamming.

Solutions to all of these problem areas must be found before such a rôle can become feasible.

25

Air Force Rôles

25.1 Long-range Reconnaissance and Strike

Possibly the most reported of current air force rôles is that conducted by the long-range and long endurance MALE and HALE UAS such as the Predator series and Global Hawk introduced and discussed in Chapter 4.

The aircraft of these systems fly at altitudes of up to 30 000 m over ranges of order 5000 km and remaining airborne for more than 30 hr before landing to refuel. They carry high-resolution EO and TI sensor payloads in rotatable turrets and synthetic aperture radar systems (see Chapter 8). Operation is from fixed bases and the long-range communication is via satellites.

Initially the systems were limited to carrying out reconnaissance missions. However in military operations by the United States' Air Force, in the more remote areas of Iraq and Afghanistan, it was found that following the detection of targets, the delay in the arrival of airborne or land strike forces allowed the targets to disperse before effective action could be taken. This realisation led to strike weapons being fitted to upgraded marks of the Predator, subsequently called Reaper. This now allows immediate strike action to be taken. Several other air forces have purchased this class of UAS.

25.2 Airborne Early Warning

Wide-area search to look, principally with radar, for early signs of enemy movements is a long-endurance task suited to fixed-wing UAS.

25.3 Electronic Intelligence

This is also known as Signals Intelligence (SIGINT) and includes communications intelligence (COMINT). The latter refers to the task of intercepting enemy radio communications in order to be forewarned of his intent whilst included in the former is the task of listening to and recording enemy air defence radar transmissions. With that data it is then possible to determine their frequency and to develop the ability to jam them with electronic noise when required.

The aircraft must carry an array of antennae and operate for long periods over potentially hostile territory. It is therefore a dangerous as well as a dull and fatiguing job for aircrew. An unmanned system does not, of course, risk the lives of aircrew and the weight saved in the aircraft with the removal of the

Figure 25.1 IAI Heron MALE UAV equipped with SIGINT antennae

crew and operating equipment and space can be used to carry extra fuel for extending the system's range. A IAI Heron MALE UAV equipped with SIGINT antennae is shown in Figure 25.1.

25.4 Pre-strike Radar and Anti-aircraft Systems Counter

A number of UAV, possibly in swarms, could be sent ahead of a manned strike force to attract the attention of air defence systems so that their positions are disclosed. If the UAV are armed (UCAV) then the air defence systems could be attacked in order to neutralise them and remove their threat to the main manned force.

Various types of UCAV are envisaged. They would be stealthy, some smaller medium-range aircraft and some larger long-range aircraft, such as the Northrop-Grumman X-47B and the BAE Systems Taranis (see Chapter 4, Figure 4.26). This class of vehicle, currently under development, will be capable of high subsonic cruise and long range with internally carried weapons to attack targets.

25.5 Interception

Future UCAV may also be capable of intercepting opposing manned or unmanned aircraft. They could be more manoeuvrable than the equivalent manned aircraft being lighter and not being limited to rates of acceleration acceptable by manned crews.

25.6 Airfield Security

The attack of airfields may be by two means. The first is by airborne attack from which the airfield will be protected by indigenous anti-air missiles. The other is by infiltration from the ground.

(a) Airborne Attack

Should an airborne attack survive the defensive missiles and cause damage to the airfield, including disruption of the runways, it is necessary for the airfield command to quickly determine the extent of damage and which runways, and taxi-way access to them, are still operable in order for the airfield to regain operation.

It may be necessary to effect repairs should there be no, or insufficient, usable runways. In that case it would be required to determine the areas for repair which would best enable the earliest return to operability. The requirement is to carry out two phases of survey:

i) an immediate overall picture of the airfield to assess the damage and the optimum paths to repair;
ii) a lower-level survey to find if any of those paths are inhibited by timed, unexploded bombs which would greatly delay and hazard repair and if the paths are sprinkled with small anti-personnel weapons.

These tasks are best carried out by small UAV mounting high-resolution EO and IR sensors. The UAV and control station would be housed in small bomb-proof shelters, probably underground. The UAV would transmit a picture in phase (i) onto a screen in Airfield Command onto which the possible paths for repair would be plotted and further assessed in phase (ii).

This operation would benefit from the use of a small VTOL UAV, both in its ease of accommodation and launch and in its ability to fly slowly at low altitude to carry out phase (ii).

Such a UAV could also be used to explode the small anti-personnel weapons which might otherwise prevent access to the runway for defusing of the main weapons and for repair activities.

A third phase of its use might be to provide an overall view to the Command of the repair process and possibly act as a radio-relay between the Command and the repair teams if required.

(b) Infiltration

Surreptitious infiltration into an airfield to plant hidden time-fused weapons, or to make an immediate attack, is possible unless a high level of security is maintained. The usual security alarm systems would still be employed, but the maintainance of an aerial survey in periods of high alert would enable a complete watch of the whole airfield boundary to be made. The same UAS would be used as that available for post-airborne-attack operation.

26

Civilian, Paramilitary and Commercial Rôles

Few of the roles listed below have yet been adopted. The following is therefore largely a discussion of the possibilities. Rôles in which UAV systems are currently used are indicated by an asterisk * in the relevant heading.

A sense-and-avoid facility is likely to be required for the majority of rôles, whilst for some a stealthy UAV is desirable for covert operation. It will be seen that a hover ability is often required of the UAV.

Some potential operators may only require occasional or intermittent use of a UAS, or for other reasons may not feel justified in owning such a system or systems. Therefore it is likely that UAS rental companies may be set up to hire-out systems to customers to provide or to augment a facility on demand.

26.1 Aerial Photography*

Using a combination of video and high-resolution still cameras (film or digital) pictures may be obtained of geographic and constructed features, such as historic houses, castles, bridges, etc. The UAV is cheaper, and less intrusive in its use, than manned aircraft. Hover flight is advantageous for positioning. Small operating companies have been set up to do this work, but the scope is currently limited by regulatory approval and insurance acceptance to small UAV and rural areas.

26.2 Agriculture

(a) Crop Monitoring

Survey of crops is feasible using infrared and colour cameras to detect the onset of disease through changes in crop colour.

(b) Crop Sowing and Spraying*: Fungicides, Insecticides, Herbicides

UAV systems are extensively used in Japan for sowing rice and subsequently spraying the crop to maintain its health. Helicopter types are used as they can be positioned very accurately, especially where small fields are involved, and reduce the risk of spray impinging on other crops which could be damaged. Aircrew are not subjected to the noxious chemicals. There is a possibility of automatic flight of the UAV

Figure 26.1   Crop spraying

using perimeter fences for electronic positioning. The UAS is also cheaper to operate than is a manned aircraft system.

Figure 26.1 shows a Yamaha Rmax UAV spraying a crop of rice. There are reportedly some 1500 of these UAS operating in this rôle in Japan. It would appear that the regulatory authorities in Japan are more helpful than in those of some other countries.

(c) Herd Monitoring and Driving

Manned helicopters are used in some countries, notably South Africa and Australia, for this purpose, especially where herds of cattle and sheep are allowed to forage during summer months in relatively inaccessible areas. They are monitored for position and numbers – especially if tagged – and rounded up in the autumn.

It may be an appropriate use of unmanned rotorcraft systems for this task, although the manned method would be of advantage in the event of a need to attend to an animal. It may also be necessary for the UAV to carry sound emitter equipment to motivate the herd. A further complication for the UAV could be the maintenance of radio LOS. Thus such a rôle may not become a suitable application for UAS.

26.3   Coastguard and Lifeboat Institutions

A number of these operations could be carried out to economic and operator advantage by UAS.

(a) The monitoring of traffic in narrow sea-lanes is a coastguard activity required to ensure that traffic proceeds in allocated lanes. Any vessel out of position is warned and required to correct its direction. The name and registration of the vessel is recorded so that legal action can proceed if deemed warranted. The payload requirement is an electro-optic and thermal imaging reconnaissance sensor with a SAR as an option. Sortie times required could be of order 12 hr and carried out by a MALE-type UAS, but having the ability to descend to lower altitude to identify any vessel.

(b) The monitoring of coastlines for debris, etc. is another coastguard activity. This may include the results of shipwreck or the discharge of illegal and/or polluting substances such as oil. The UAS requirements would be similar to that of (a).

(c) A UAV could be the eyes of a lifeboat in its search for vessels in distress or for persons in life-rafts or in the water. This would be of particular advantage in high seas where the lifeboat crew's range of vision could be limited. The UAV could be launched from the lifeboat or, more probably, from the lifeboat station where it would be permanently based. In the latter case the UAV might be launched concurrently with, or in advance of, the lifeboat to search ahead of the vessel. It would carry electro-optical and thermal video, and possibly also miniature SAR, equipment. On finding the vessel/persons in distress, the UAV would illuminate the scene to enable the lifeboat to home-in and to assist in the rescue operation. A rotorcraft UAV would be desirable for its ease of launch and recovery, its hover capability over the scene and its lower response to turbulence compared with a fixed-wing UAV.

26.4   Customs and Excise

The surveillance of coastline, ports, bridges and other access points for the import of illegal substances is the remit of the customs and excise authority but, as discussed in Chapter 23, this is usually undertaken with assistance from naval forces. The illegal activity is more usually pursued at night, using small boats landing at remote locations, but 24 hr surveillance is still necessary.

An electro-optic and thermal-imaging payload will be required, with the UAV having the ability to descend to low level when required to obtain high-definition video of small objects and for the recognition of persons. A hover capability could be necessary, with the possible use of a mobile phone network for communication. The surveillance operation may need to be carried out covertly but a 'night-sun' may be a required option to illuminate the scene should the occasion demand.

26.5   Conservation

(a) Monitoring for pollution. UAVs have, in the past, been used to fly over suspect emissions to monitor levels of unacceptable effluent from, for example, power stations and other manufacturing facilities. Their future use in this rôle, as in many others, awaits the development of sense-and-avoid systems and the approval of the regulating authorities.

(b) UAS are being developed for high-altitude research into, for example, the ozone layer.

(c) Another proposed rôle is for the monitoring of land use.

26.6   Electricity Companies

Traditionally, power-lines have been inspected for encroaching trees, damage to structure and deterioration of insulators by employees traversing the lines on foot and climbing the pylons. This is time-consuming and arduous with a considerable element of risk.

More recently the task is often carried out by crews in manned helicopters using binoculars and thermal imagers to detect the breakdown of insulators. It is not without hazard. Trials have been carried out using UAS, with considerable success, for the inspection of power lines (Figure 26.2). Apart from the advantage of lower costs the UAS does not hazard aircrews, can operate in more adverse weather conditions and is less obtrusive to neighbouring communities or animals.

Manned helicopters inspecting power lines have caused animals to have premature offspring or to stampede, resulting in damage to farm property. Both cause financial loss to the farmers who have sought recompense from the power companies.

Hover flight is essential for the inspection task with the UAV carrying an electro-optic and thermal imaging payload, the data from which is available in real-time to the operator and also recorded. The

ADVANTAGES COMPARED WITH ALTERNATIVE METHODS

Compared with Manual Inspection a UAS offers:

Lower labour cost
More inspections per day
Less risk to personnel
Easier access

Compared with Helicopter Inspection a UAS offers:

Less Environmental Disturbance
Lower capital cost
Lower labour cost
No risk to aircrew

Figure 26.2   Powerline inspection

UAV is automatically guided along the power-line within a limited volume of airspace close to the lines using a range of distance measuring devices.

An important requirement for UAV deployed in this role is that they can be flown close to high-voltage power lines, that is within their electromagnetic fields, without adverse effects upon the control system or payload performance of the UAV. Means of countering these effects were demonstrated by the Sprite UAS.

A detailed description and cost analysis of the task, using the results of trials, is carried in Reference 26.1 and shows that using a UAS for the power line inspection task reduces the cost of the operation to about one-third of that of the other two current methods.

Although the operation would be virtually in dedicated airspace, an application to operate has not been approved by the regulating authorities in the UK.

26.7   Fire Services

UAS using small UAV may be used to patrol woods and vulnerable crops to look for hot-spots when weather conditions are conducive to the outbreak of fire. They may also be used to assist, as an eye in the sky, in directing the application of firefighting materials onto any fire, whether rural or urban. For this latter use a hover capability is essential.

The payload would be an electro-optical and thermal video sensor.

Larger UAV could be used as 'water-bombers' to replace manned aircraft in this rôle to eliminate risk to aircrew. A similar sensor suite would be employed and a hover capability would be advantageous in both loading and accurately releasing the water.

26.8   Fisheries

Fisheries protection, or the prevention of illegal fishing is carried out in many parts of the world by patrolling vessels, sometimes aided by patrolling fixed-wing aircraft.

The latter are only of limited effectiveness unless they are constantly on patrol, when their use becomes very costly. They are also very visible to pirating ships. In any case it is necessary for firm evidence of actual fishing, and probably with nets of a certain size, before any prosecution of the ship's owners could take place. The mere presence of a ship in the fishing area does not constitute an offence.

Whether this type of operation is carried out by civilian authorities or by naval forces is a decision made by individual countries. In the UK this function is the province of naval forces and so the use of UAS for this operation has been described in Chapter 23.

26.9   Gas and Oil Supply Companies

Two uses are proposed by these companies. First, the UAV offers a less expensive means of carrying out a survey of the land where the pipe lines may be installed. Second, after installation, it offers a means of patrolling the pipes to look for disruption or leaks by accidents such as a landslip or lightning strike and also damage by vehicles or falling trees. In certain areas of the world, sabotage is not uncommon.

For detailed inspection, a hover capability is required with an electro-optical and thermal imaging sensor payload.

26.10   Information Services

This can cover a multitude of activities. News organisations, television companies, newspaper publishers could have the means of covering events, whether planned or accidental. Sports events could be covered in real-time for transmission on television using high-resolution stabilised video camera payloads. Feature pictures of wildlife could also be obtained. A hover capability would be advantageous.

The UAV system is cheaper and less intrusive in its use than manned aircraft, but regulatory approval is not yet available and will most likely await the development of a reliable sense-and-avoid system. The system will most likely be required to be fully fail-safe.

26.11   Local Civic Authorities

UAS could be used to survey sites for new building work, roads, culverts, housing, etc. and also for the condition of infrastructure. In times of disaster such as floods, the UAV is a cheaper and readily available means of determining the extent of the problem and monitoring aid services within the area.

The usual electro-optical (EO) and thermal-imaging (TI) sensor payload would be required plus a night-sun system for some operations.

26.12   Meteorological Services*

UAVs such as the Aerosonde operated by the Australian meteorological office are used to sample the atmosphere over wide areas. Obviously an atmospheric sampling payload is carried in addition to the usual EO sensor.

26.13   Traffic Agencies

UAS could well be used in the future to monitor and assist in the control of road traffic. In addition to being less expensive to operate than manned aircraft, they could be covert to avoid distracting drivers. A hover capability is desirable with an EO and possibly TI payload.

26.14   Ordnance Survey

UAS could replace manned aircraft on mapping tasks which are generally dull and lengthy. The UAV would carry out a pre-programmed raster flight pattern with arrangement with Air Traffic Control. Specialised camera equipment may be required.

26.15   Police Authorities*

The main use of UAS by the police authorities would be the detection and prevention of crime and to search for missing persons. The UAV ability to remain undetected during observation activities, as well as its rapid availability compared with other systems, would be its main advantages for police operation.

A number of police forces are experimenting with the use of mini- or micro-UAV for reasons of limiting cost and operating under the regulations pertaining to model aircraft. However these regulations severely restrict the use of the system and the very lightweight UAV are able to carry only payloads of limited performance and to operate only in fair weather. Reports to date indicate that this approach is unlikely to result in a viable system being adopted. In reality the rôle awaits the availability of a more robust, all-weather, fail-safe UAS with a UAV capable of carrying a higher resolution EO and IR payload and probably a sense-and-avoid system.

Approval of the appropriate regulatory authority would be necessary for general operation. A UAV with stealth characteristics is highly desirable and a hover capability is essential.

The proposed use of MAV/NAV for Urban Surveillance in a 'perch and stare' rôle is applicable to civilian policing operations as well as for military use, as discussed in Chapter 24.

26.16   Rivers Authorities and Water Boards

Rivers Authorities have used UAS successfully in monitoring watercourse flow and water levels. They have been used to assist in the control of flooding and monitoring of pollution (see Reference 26.2).

UAS could also be used for the monitoring of reservoirs for pollution or damage and also for pipeline monitoring for security purposes. However such operations have not been reported in more recent years and their future operation must depend upon the approval of the regulatory authorities.

26.17   Survey Organisations

Although this use may be seen to overlap with other organisations, such as the Ordnance Survey Authorities, survey organisations have used UAS for geological purposes in searching for mineral deposits, etc, and for archaeological survey where buried remains of ancient buildings and roads have been discovered using colour-sensitive film. However, this is likely to be limited to an occasional requirement, and the UAS would be more likely to be hired from other operators or hire contractors.

26.18   Communications Relay

Although it is not clear that any future UAS may be dedicated for civilian communications relay purposes, it is obvious that such a facility might be added to UAS for use whilst engaged in other primary activities. For example, if a UAV is assisting police or fire services in monitoring or illuminating the scenes of incidents, a radio relay function could assist communication between officers on the ground and their headquarters.

26.19   Landmine Detection and Destruction

This function is largely the province of the Army and is discussed in Chapter 24.

26.20   Other Applications

In the author's experience, once a reliable and cost-effective UAS is available and in operation, its operators will be approached by other potential users with enquiries regarding their possible applications. Sometimes these are quite unexpected as, for example, an enquiry from the Austrian Police regarding the possibility of accurately dropping hand-grenades to trigger incipient avalanches. Once UAS are in regular civilian use their range of use may well exceed anything envisaged today.

There is a cautionary note to this prophecy. In the history of mankind, almost any new technology has proven to be as much a tool for evil as for good if it reaches criminal hands. A number of 'doubtful' enquiries were received for sales of the Sprite System. Safeguards must be put into place to prevent the criminal use of this new technology.

References

26.1.   R. G. Austin and G. Earp. 'Powerline Inspection by Unmanned Aircraft – a Business Case.' Nineteenth Bristol International UAV Systems Conference, Bristol University, 2003.

26.2.   O. T. Addyman. 'Possible Application of RPV in the British Water Industry.' Third Bristol International UAV Systems Conference, Bristol University, 1982.

Part Four

UAS Future

27

Future Prospects and Challenges

27.1 Introduction

The author is unaware as to who coined the phrase 'the future is now', but it most certainly applies to UAV systems.

In several cycles over many decades it was thought that the UAV had at last arrived, only for it to subside into a marginal activity. Now, however, their use seems to be rapidly proliferating, at least in military operation.

Even in the period during which it has taken to prepare this book, many new developments have taken place. Several systems have moved up the capability scale, for example MALE systems taking on the mantle of HALE systems aided by technology developments in power-plants, materials and sensors.

Whereas, in the past, UAV systems gained through the technology spin-off from other endeavours, now much research into new technology is being addressed purely to advance UAV systems. The spin-off from this is likely to be adopted to assist advances in other fields such as bioengineering as well as other aeronautical applications. This is particularly true of the developments in miniaturisation and material integration driven by the needs of UAV systems.

The author's experience with, in particular, the Sprite system was that would-be customers in both military and civil fields, approached us to ask of the possibility in using the UAV in their area of activity. It was not necessary for us to think of possible applications. The customers came to us!

So it is difficult to attempt to predict the possible future applications. We will probably be surprised as to how wide the opportunities may be. The limiting case, however, is more likely to be legislation or public concern rather than technical constraint.

One area already often mooted is that of replacing aircrew in passenger-carrying airliners. The author remains very sceptical about this ever happening since analysis readily shows the cost saving of replacing aircrew and their accommodation by the systems required for autonomous control is quite small within the overall costs of running an airline. The latter includes, of course, insurance against loss of life in the air and on the ground as well as loss of the aircraft and damage to real estate. The certainty of a large initial premium placed upon UAV airline operation could greatly increase the overall operating costs.

Any unlikely saving must also be balanced against the psychological reaction of passengers who would prefer to think of an aircrew up-front who also want to arrive home safely that night! Passengers have been deterred from flying with particular airlines on much less tenuous grounds. It would be a courageous – or foolish – airline that first launched such a service.

The author accepts that time may yet prove him to be wrong, but suspects that, if it occurs at all, he will not then be here to be called to task! There are, however, some more immediate and obvious issues that warrant discussion.

Unmanned Aircraft Systems – UAVS Design, Development and Deployment   Reg Austin
© 2010 John Wiley & Sons, Ltd

27.2   Operation in Civilian Airspace

Current operation of UAS is, with few exceptions, limited to military airspace – i.e. on ranges owned and controlled by the military or in zones of military conflict. In previous chapters some of the many possible civilian and commercial operations in which UAS may prove to be more effective and economic than manned systems have been outlined. The major hurdle to the use of UAS in these operations is the limitation currently imposed by the regulatory authorities.

Public concern with the prospect of unmanned aircraft flying around the skies and possibly crashing onto people and property or colliding with other aircraft is perfectly understandable. It is necessary that regulations are in place to prevent cavalier and ill-considered use of non-airworthy and unreliable systems causing loss of life and damage to property. This is in the interest of responsible manufacturers and users of UAS who would wish to see public confidence building in the responsible deployment of well-conceived systems to the economic and environmental advantage of all.

As previously discussed, the airworthiness of a properly developed and proven system should not be a barrier. The proving of such systems would be expected to take, and have taken, a route to achieving and demonstrating their airworthiness similar to that long established for manned aircraft systems. In fact it may be argued that the airworthiness of a UAS could be engineered to exceed that of a manned system since those accidents caused by human error or infirmity would be largely eliminated.

The difficulty currently arises in obtaining sanction to use air-space. This comes from the concern that unsighted UAV may collide with other aircraft. Some authorities see the onus of avoiding any collision between UAV and manned aircraft or UAV and UAV falling solely upon the UAS since it is the 'new kid on the block' and that there will be no airspace dedicated to UAV. The UAV may therefore be expected to have full responsibility to sense other aircraft and to take effective avoiding action.

27.2.1   Be Seen, Sense and Avoid

The twin mantras of current collision avoidance, especially outside of controlled airspace where the majority of UAS operations are relevant, are 'see and be seen' and 'see and avoid'. For UAV the "See" may be interpreted as "Sense".

First we look at the requirement to *be seen*. As any pilot knows, the unfortunate truth is that when two aircraft are on a collision course their visual images of each other are at their lowest, with the ultimate case being when they are approaching head-on. In that condition their closing speed is also at a maximum. There is no apparent relative movement of the approaching aircraft which makes it even more difficult to detect (see Chapter 7 on Stealth).

A UAV will usually be much smaller than a manned aircraft making it that much more difficult for it to be visually detected. The UAV infrared and radar images may also be smaller than those of manned aircraft but, since few manned aircraft, especially light aircraft, are equipped with those detectors, the visual image is of the greater concern.

It may be necessary for the UAV to be equipped with a distinctive visual enhancement such as a pulsed beacon light or lights. This has been done and is not particularly difficult except for mini, and smaller, UAV in which electrical power may also be a problem.

Turning now to *sense and avoid*, several regulatory authorities require that UAV must be equipped with a sense-and-avoid system which will detect other aircraft and take avoiding action before they can be allowed to operate in civilian airspace and that the onus to avoid a collision rests with the UAV. This is also referred to as 'detect, sense and avoid' or DSA.

The ASTM F-38 Committee (Reference 27.1), one of several organisations attempting to quantify a standard, published a proposed standard for DSA collision avoidance (F2411-04 DSA Collision Avoidance) that requires a UAV to be able to detect and avoid another airborne object within a range of $\pm 15°$ degrees in elevation and $\pm 110°$ in azimuth and to be able to respond so that a collision is avoided by at least 500 ft. The 500 ft safety bubble is derived from the commonly accepted definition of what constitutes a near mid-air collision. This gives airframe and avionic/DSA electronics manufacturers a target for certification. Another authority recommends that a UAV be able to detect and avoid another airborne object approaching from any azimuth direction, i.e. Through ± 180°.

It is also suggested (References 27.2 and 27.3 for example) that:

a) the UAV sense-and-avoid system must 'be equal or better than the theoretical see and avoid capability of a human pilot'; and
b) it must have the reliability of a flight-critical system.

The question then is – what level of efficiency is required of a DSA system to be comparable with that of a human pilot? In order to answer that it is necessary to know how efficient is a human pilot in achieving DSA. It may also be questioned whether the current DSA ability of a 'human pilot' is adequate. Also, being human, the ability will inevitably differ from one pilot to another and is probably as much a function of attention-span as of eyesight. This makes it all the more difficult to define a scientific standard.

As an exercise in viewing the problem, we might consider the worst-case scenario where a 2–4 seat light aircraft is approaching head-on to another of the same type. With an effective frontal area of about 16 ft^2 or 1.5 m^2, using the theory presented in Chapter 7, the other aircraft might become visible under ideal, maximum contrast conditions at a range of 2000 m. However that is a highly improbable situation and reflected light illuminating the other aircraft, thus reducing the contrast with its background, or misty or polluted atmosphere will significantly reduce the distance of detection to a fraction of that. If we make the practical, but not particularly pessimistic, assumption that the distance is halved to 1000 m we can construct a probable scenario.

If the two aircraft are both flying at about 100 kt (50 m/s) then they have a closing speed of 100 m/s. Even if the pilot is looking in the direction of the other aircraft at the moment of its emergence, i.e. ignoring the possibility he might be head-down, map-reading or checking instruments, there will inevitably be a 'reaction-time' before he takes avoiding action to turn to the right, away from his original flight path. This reaction time is customarily accepted as being 2 s. Although this may be seen as very optimistic, it has been taken for the following calculations.

The statutory distance quantified as a 'near-miss' is 500 ft (150 m) – Reference 27.2, and we can assume that the turn is applied to achieve a lateral displacement of that order from the original flight path. Using this input, calculation can be made to estimate the rate of turn that the avoiding aircraft (or UAV), flying at 50 m/s, must make in order to have a 150 m lateral distance apart from the oncoming aircraft when that aircraft passes it.

In Figure 27.1, line A shows the degree of banked turn that the aircraft/UAV must make to achieve the separation distance for a range of flight speeds of the oncoming light aircraft. It can be seen that, for example, if the speed of the light aircraft is 100 kt, the aircraft/UAV must enter and maintain a banked turn of 24° (1.1g) or if the light aircraft is flying at 200 kt then the avoiding aircraft/UAV must make a banked turn of over 50°, pulling in excess of 1.6g.

Line B follows from similar calculations with the avoiding aircraft flying at 100 kt, but with the oncoming aircraft a medium-sized commercial jet aircraft with a fuselage diameter of 4.2 m. The ideal-condition detection distance for that aircraft is calculated to be 4 km with a more realistic distance of 2 km. The example indicated is with the airliner flying at 450 kt. To avoid this aircraft, the avoiding aircraft must turn banked at 48° (1.5g).

The author believes that a human pilot flying a typical light aircraft would be unlikely to achieve this performance. Therefore it would seem that, in reality, the regulating authorities will require the sensing and reaction components of a UAV sense-and-avoid system to exceed the capability of a human pilot. Further, the avoid component may require the UAV to be more manoeuvrable than a typical light aircraft.

Provided that detection is achieved at or earlier than the ranges used above, the UAV should be able to meet the necessary level of manoeuvrability. However, the picture is changed if the UAV flies at a greater speed, for example 200 kt.

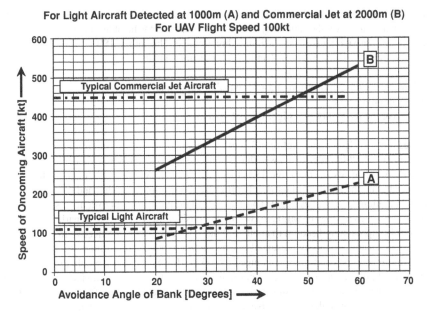

Figure 27.1   Bank angle to avoid

In the same scenario with the light aircraft, if the oncoming aircraft is flying at 100 kt, the UAV must bank at 53° (1.66*g*) to avoid it and at 60° (2*g*) if the oncoming aircraft is flying at 130 kt. With the light aircraft flying at speeds greater than 130 kt, and for all cases with the jet airliner, it is unreasonable to expect the UAV to achieve the necessary manoeuvrability.

It may be concluded therefore that to meet the proposed requirements of the regulating authorities, the UAV should fly at low speed in uncontrolled airspace and have a sensing system which can 'see' further than can a human pilot and scan almost instantaneously over the whole specified range in azimuth.

In *controlled* airspace the UAV would have to be fitted with a transponder system so that the Air Traffic Controllers can monitor its progress and other aircraft made aware of its type and position. In that case, the UAV would be required to fly in a flight-lane in the same direction as other aircraft and at a speed as compatible as possible with other traffic. The UAV might therefore not be required to have an all-around field of vision but 'only' comply with the ASTM suggested requirements of ± 110° in azimuth and ± 15° in elevation.

At least one transponder system, the T2000UAV-S by Micronair Avionics of Australia, is being developed for UAV fitment; it has a mass of about 500 g. This extra mass and power consumption could probably be acceptable in UAV as low in AUM as close-range systems. Mini- and micro-UAV will not be able to carry them unless future development reduces their mass penalty by two orders of magnitude, but there is unlikely to be a requirement to operate such systems within controlled airspace.

There remains the major problem of the safe operation of UAS in *uncontrolled* airspace.

Sense-and-avoid Systems in Development

A number of different companies and organisations are attempting the development of sense-and-avoid systems, even though the specification for such systems has yet to be established and agreed. The US government and industry consensus group, RTCA 203 has set a target for a specification to be agreed by 2013 for a system to enable UAV to operate in uncontrolled airspace with contingencies for UAV to enter controlled airspace. Not an easy task.

It may be that in order to arrive at this specification the analysis, possibly conducted for the first time, will show that in certain circumstances the proposed requirement for the sense-and-avoid system to be 'equal or better than the human pilot' will indeed need to be 'better'. This is especially since a mid-air collision between manned aircraft resulting in loss of life is seen as a catastrophe, but loss of life caused by a collision with a UAV would be untenable.

Indeed, with ever-increasing density of air traffic it may be concluded that the capability of a human pilot is inadequate and that manned aircraft, also, should carry a sense-and-avoid system. In that case a spin-off from UAS technology, if affordable, would benefit general aviation.

Some opinions currently believe that a fusion of EO and IR sensor images will be required to detect other aircraft, whilst others favour a radar system which will offer detection at longer range. A problem with relying upon radar returns is that many UAV will have very low radar signatures, in fact some will have been designed to minimise their signatures. Hence it may be a requirement that UAV operating in civilian airspace carry radar enhancement equipment.

It then follows that military stealth UAV transiting civilian airspace will need to have radar enhancement which can be turned on and off whether it be emitters or reflectors. Some organisations, developing sense and avoid systems, propose a combination of EO, IR and radar sensors.

The down-side of all of these proposals is the likely cost, mass, dimensions and power requirement of the future systems, and although electronic development is rapidly advancing it is unlikely that it will reduce those features to a level, within the foreseeable future, which can be borne by other than the largest and most expensive UAS.

Potential future civilian operations would largely be carried out by smaller UAV systems with the UAV probably having a DGM in the range 30–60 kg with a UAV unit cost probably in the range £20 000–60 000.

One of the significant economic advantages of a UAS operation arises from the first cost of the system being lower than alternative means. Even if it were light and small enough to be carried by the UAV, a DSA system costing as much or more than the cost of the UAV would almost certainly destroy its economic advantage. Until this issue can be satisfactorily resolved, the outlook for civilian UAS looks bleak.

27.2.2 'Primary' Civilian Operations

There are some operations in civilian airspace which might rationally be exempted from carrying potentially heavy and expensive sense-and-avoid systems. If these 'primary' operations were allowed, carried out at first in remote sparsely populated areas, it might be the answer to the better understanding of the UAS integration in more open airspace and build confidence in their reliability whilst still performing a useful service for the operator.

One, already in operation for some years, is the agricultural crop seeding and spraying of rice crops in Japan by the Yamaha R MAX UAS (see Chapter 4, text and Figure 4.20). It is understood that some 1500 R MAX VTOL UAV are in regular use for this purpose. The R MAX system is regulated by the Japan Agriculture Aviation Association (JAAA), who regard the system as agricultural equipment, and it is certified as such for reliability in operation. The Association require that the distance between the unmanned helicopter and its operator should not exceed 150 m.

This is a departure from the current regulations in Europe where it would not be authorised under model aircraft 'sport flying' rules as the AUM of the UAV greatly exceeds 20 kg and would be disallowed exemption in a heavier category since it would be seen as operating for 'hire and reward'. Perhaps the Japanese authorities have different guidelines from the European ones, or operation over rice fields may be seen as a special case.

Another primary operation which might be allowed in the future, but so far has been forbidden, at least in the UK, is that of powerline inspection. From powerline trials carried out in the USA and detailed operational and cost analysis by the UK EATech Organisation (see Chapter 26 and Reference 26.1), the cost of carrying out the operation with a suitable VTOL UAS would be of order one-third of that of the current methods which use either manned helicopters or foot patrols. The UAS operation has other benefits such as being more environmentally friendly and potentially causing less damage to the powerline environs.

The UAV would fly at levels no higher than the height of the pylons and be closely 'locked-on' to the line of the conductors by three different and independent types of sensor, and so remain in otherwise unusable airspace. It would continuously transmit a video picture to the operator (and elsewhere if required) of its surrounding scene so that any obstacle or unlikely deviation from its track would immediately be seen and acted upon. The UAV could also mount a high-visibility strobe light. Initial operations would be carried out in the more remote, scarcely populated areas of the country and only progress to other areas with the prior and express agreement of the authorities.

The current rule in the UK, however, asserts that the UAV must stay within the sight of the operator. To do so totally removes the economic advantage of the UAV operation and so prevents the use of that operation as a 'building block' for future UAS applications.

It is probably not within the remit of the regulating authorities to allow flight of UAV beyond the sight of the operator in uncontrolled airspace even in the above circumstances, without the inclusion of a DSA System which, as previously observed, may not be possible at an acceptable price and mass within the foreseeable future.

A change in legislation would probably be required to allow the authorities the power to permit the powerline inspection operation without the DSA system. It is therefore only with greater foresight on the part of the national and international governments to change the legislation that such operations could ever be allowed. If this were done, the demonstration of the safety and efficiency of these primary operations could then sensibly lead the way, in time, to acceptance of other civilian UAS applications.

27.3 Power-plant Development

As a general approximation, the mass of the power-plant in most aircraft of moderately high performance is about 10% of the design gross mass of the aircraft. Typically, the fuel carried is about 10–15% of the DGM for light aircraft of medium range. The payload of these aircraft usually is of order 40–50% of DGM.

In the case of ISTAR UAV, where the payload may be of imaging sensors or other light electronic systems and so of a lower fraction of DGM, more fuel may be carried to extend their range. The fuel load in a UAV may therefore be 20–25% of the DGM, raising the proportion of total power-plant and fuel mass to one-third or more of the DGM. Any reduction, therefore in the mass of the power-plant or in its specific fuel consumption can have a very significant effect on the reduction in size or increase in range of a UAV.

To that end, developments are taking place to advance the technology of power-plants not only in improving current forms of internal combustion (IC) engines, but also in introducing new forms of energy production which include electrical power and fuel-cell systems.

Improvements in IC Engines: Piston Units

The current level of performance of two-cycle and four-cycle engines is presented below in Figure 27.3 in terms of their effective mass per unit power without fuel, and the increase incurred by providing fuel and tankage to allow a range of endurances. The mass of the tankage that has been added was taken as

10% of that of the fuel to be accommodated. Over the power ranges indicated, there appears to be no quantifiable change of the parameter with size of unit (see Chapter 6, Figure 6.6).

Steady reduction in the mass of these types of engines is being made through the use of lighter and stronger materials and reduction in fuel consumption by improved fuel/air mixing, distribution, turbo-charging and better ignition. There is a limit to increasing power output through higher compression ratios due to the onset of pre-ignition with volatile fuels.

On the other hand several operators, especially naval forces, are demanding the use of 'heavy' fuels, (diesel oils and kerosene) rather than petroleum, for reasons of safety. These fuels require a high compression ratio and have higher calorific values, and so may be the way forward. The challenge is to find means of achieving the necessary compression ratios without increasing the engine weight and of developing lightweight, compact, reliable and affordable injection systems and turbo-chargers. For the smaller engines the injection systems have, in the past, dwarfed the engines in size and mass. If this challenge can be met, it carries the added advantage of higher power output and lower specific fuel consumption achieved with fuels of higher calorific value and combustion at higher temperatures and pressures.

Another line of approach has been the 'stepped piston engine' pioneered by Bernard Hooper Engineering of the UK. The principle of the operation of this type of engine is illustrated in Figure 27.2.

A stepped piston engine uses piston movement to provide suction and then compression to feed the charge into another cylinder. A flange, or step, around the base of the piston creates a secondary chamber which draws the fuel/air mixture in on the piston's downstroke. On the upstroke, the mixture in this chamber is passed into the upper chamber of an adjacent cylinder where the mixture is compressed and then burnt to power the downstroke and subsequently is exhausted. The piston of each cylinder operates 180° out-of-phase with the other. For a detailed description see Reference 27.4.

It is claimed that the stepped piston engine offers a high durability solution for low mass two-cycle power-plants with high efficiency and low emissions. The engine allows operation on a wide range of fuels without the need for pre-mixed oil since the engine has its own integral four-stroke type recirculatory sump lubrication system.

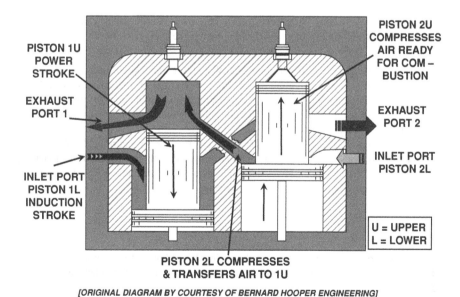

Figure 27.2 Stepped piston engine principle (Reproduced by permission of Bernard Hooper Engines)

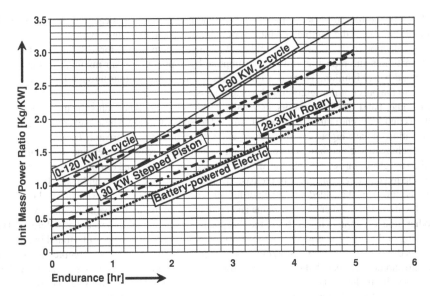

Figure 27.3   Power-plant performance

Other advantages that are claimed for this system are that it has the simplicity and smoother torque delivery of a two-cycle engine with the fuel efficiency of a four-cycle engine:

a) it has no valve gear;
b) the ingoing mixture is not contaminated by the remains of the exhaust gases;
c) the pistons have low thermal loading;
d) the pistons are more effectively lubricated from a wet sump. Plain bearings can be used, as with a four-stroke engine resulting in a more durable unit than is the average two-cycle engine.

Prototypes have been built of a unit producing 30 kW of power and its performance is shown in Figure 27.3. The result indicates a basic mass/power ratio lower than either of the two- or four-cycle engines, but with a fuel consumption between the two. At this point, with a single unit, there is no indication of any size effect within this type. Although research is continuing on the configuration there are no known plans for production of this engine type.

Further shown in Figure 27.3 is the same performance parameter for a current 28 kW rotary engine. This shows a marked improvement in performance compared with the more conventional engines. However, the limited data on rotary engines does indicate a strong scale effect where smaller units are far less efficient than equivalent two- or four-cycle engines.

IC Engines: Turbo-shaft

No shaft-drive gas-turbine engines of less than about 200 kW are known to be currently used in aircraft. The performance of a 200 kW unit, the Rolls-Royce Allison C250, is virtually identical, in the terms of Figure 27.3, with the battery-powered electrical motor shown.

At lower powers, which would be relevant to most UAV other than MALE or HALE UAS, none of this type of engine is available for serious use. In smaller sizes they have, due to scale effects, a higher fuel consumption and mass/power ratio than piston engines. They are also expensive. This makes them uncompetitive, in spite of their smoother power output and ability to use a range of 'heavy' fuels, with the alternative engine types. This is a pity since their other attributes, especially in the free-power-turbine configuration for VTOL aircraft, are very desirable features.

Perhaps it is time that a research programme is launched to find means of improving the efficiency of small turbine engines. If this is successful, the engines would be much in demand for UAV and for light aircraft, and production in significant numbers would reduce their unit cost.

Electric Power

Recent developments in lightweight electric motor and battery design are making feasible the use of electric propulsion from storage batteries. In particular the improvement in rechargeable battery performance through use of Li–S (lithium sulphur) technology has reduced the mass/energy ratio to about one-quarter that of other battery types. The resulting performance of a high-technology battery-motor combination may be seen in Figure 27.3 to be marginally better than the rotary engine shown.

Unfortunately a down-side of the system is the large volume of the batteries. For a given energy storage, even Li–S batteries occupy four times the volume of that of fossil fuels, presenting a problem for all other than short-range UAV.

Fuel Cell Technology

A fuel cell works by catalysis, separating the component electrons and protons of the reactant fuel (electrolyte), and forcing the electrons to travel through a circuit, hence converting them to electrical power. The catalyst typically comprises a platinum group metal or alloy. Another catalytic process takes the electrons back in, combining them with the protons and the oxidant to form water as a waste product if hydrogen is the fuel used. Some electrical power is required to operate the system but the majority (95%?) is available to drive an electric motor instead of using batteries.

Recent developments by companies such as Protonex (Reference 27.5) indicate a mass/power ratio for a hydrogen-powered fuel cell to be of order 1 kg/kW. To this must be added the mass of an electric motor raising the ratio for the installation to 1.2 kg/kW, making the installation heavier than any of the other systems. It is claimed, however, that the process is very efficient with energy conversion being in the region of 95% compared with that of about 35% at best for most IC engines. Although there is little information available on the fuel consumption, the hydrogen must be contained in pressure vessels which must weigh more per mass of fuel compared with the tanks for fossil fuels. Therefore no attempt has been made to illustrate fuel-cell system performance in Figure 27.3.

It is believed, however that, although the mass of a fuel cell–electric motor power system installation is currently heavier than other alternatives, its fuel-tank mass per hour may be less than for other systems. This characteristic may therefore make it particularly appropriate for use in future long-endurance systems such as MALE and HALE UAS which operate from fixed air-bases and have high levels of utilisation rather than for use in shorter-range and battlefield systems.

As an illustration of the potentially low specific fuel consumption of the fuel cell, a Puma UAV equipped with a Protonex fuel cell–electric motor combination has flown for 9 hr compared with the 2 hr endurance of a Puma normally equipped with a battery-powered motor. Claimed advantages for the fuel cell based system are:

- higher efficiency than any other fossil-fuel-based technology,
- modular and easy to install,
- in most cases fuel cells are zero-emission devices,
- zero or very low noise, except for occasional vibrations.

Perceived disadvantages are:

- highly expensive due to exotic materials and complicated design and assembly,
- highly sensitive to fuel contamination adding expense for filters and cleaners,
- skilled personnel needed for maintenance and overhaul.

This is a simple overview comparison of the systems which does not take into account smaller differences incurred by different gearbox or control subsystems appropriate to the alternative powerplants. A more detailed analysis may be found in Reference 27.6.

27.4 Developments in Airframe Configurations

ISTAR UAV systems seem to be polarising around the HALE and MALE systems on the one hand and close-range systems on the other. The configuration of the longer-endurance systems seem to be well established and little change may be expected other than an ongoing refinement brought about by lighter material and more efficient power-plant development. More rapid development may be seen in the sensors carried and in the deployment of weapons.

Greater changes may be forecast in the smaller, close-range and mini systems, not only through miniaturisation of their payloads and sub-systems, but in the airframe configurations employed. Generally operating in more adverse situations at low altitude in turbulent and polluted air, more versatile systems are required. In particular, systems able to operate more successfully in high winds and turbulent air into and out of small spaces are seen to be necessary. Greater emphasis is being placed on VTOL systems.

Other upcoming UAS applications which do require a different approach to the airframe configuration are the extra-high-altitude, pseudo-satellite systems, and UCAV systems.

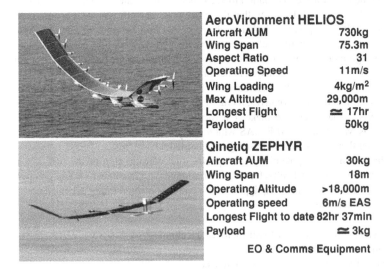

AeroVironment HELIOS

Aircraft AUM	730kg
Wing Span	75.3m
Aspect Ratio	31
Operating Speed	11m/s
Wing Loading	4kg/m^2
Max Altitude	29,000m
Longest Flight	$\cong$ 17hr
Payload	50kg

Qinetiq ZEPHYR

Aircraft AUM	30kg
Wing Span	18m
Operating Altitude	>18,000m
Operating speed	6m/s EAS
Longest Flight to date	82hr 37min
Payload	$\cong$ 3kg

EO & Comms Equipment

Figure 27.4 Extra-high-altitude UAV (Reproduced by permission of AeroVironment Inc. and Qinetiq plc)

27.4.1   Extra-high-altitude UAV

The two leading UAV in this category, Helios and Zephyr, are shown in Figure 27.4. The Helios Prototype was an electrically powered experimental lightweight flying wing developed by AeroVironment, Inc., under NASA's Environmental Research Aircraft and Sensor Technology (ERAST) programme. Designed to use energy derived from the sun by day and from fuel cells at night, the Helios Prototype was expected to be the forerunner of high-altitude unmanned aerial vehicles that could fly on ultra-long duration environmental science or telecommunications relay missions lasting for weeks or months without using consumable fuels or emitting airborne pollutants.

Helios reached an unofficial record altitude for non-rocket-powered aircraft of 96 863 ft (29 441m) on 13 August 2001. Unfortunately, the program suffered a major setback when the Helios experienced control difficulties while on a checkout flight on 26 June 2003, and the aircraft suffered some structural failures and was subsequently destroyed when it fell into the Pacific Ocean about 10 miles west of the Hawaiian island of Kauai. The experimental fuel cell system carried aboard the Helios Prototype on that flight was lost.

The control loss has been reported as resulting from large wing deflections on encountering air turbulence. This increased wing dihedral, and caused the aircraft to become unstable. The following large-amplitude oscillations caused structural failure. No replacement for Helios has been built.

By contrast, Zephyr is a fraction of the size of Helios, having only $\frac{1}{4}$ of its wing-span and 1/24 of its AUM. It, like Helios, converts solar energy to electrical energy to power the motors/propellers during daylight hours but, instead of relying upon fuel-cell energy for night-time operation, it stores remaining day-time energy in highly efficient silicon–sulphur batteries to be used during the night.

Qinetiq has concentrated on miniaturising and raising the efficiency of systems and general saving of weight in the airframe. The record flight of more 82 hr made at heights of over 18 000 m, indicates the success of that formula. Qinetiq believe that Zephyr is very close to an operational system and has one more step of improvements to design a robust and reliable system that will remain airborne for months.

Qinetiq is also working with the American aerospace company, Boeing, on a US defence project codenamed Vulture. It is predicted that this would be the aircraft with the largest wing-span in history and capable of carrying a 450 kg payload. US commanders say the design must be able to maintain its position over a particular spot on the Earth's surface uninterrupted for five years and to act as a pseudo-satellite.

27.4.2   VTOL UAV

As has been observed elsewhere in this book, the operational advantages of a UAV having a vertical flight capability, are belatedly being understood. Therefore more attention is now being paid to the development of a number of different systems to achieve this. As discussed in Chapter 4, fixed-wing aircraft, whether manned or unmanned, over the years have reduced to a limited range of proven configurations.

This is not so with VTOL aircraft in which there is a number of different configurations, each with its own particular characteristics. Some of these configurations were illustrated in Chapter 4, but more of each main type is shown in Figures 27.5 and 27.6 as an aid in addressing those characteristics.

The most ubiquitous of VTOL types, both in manned and UAV versions, are the single-main-rotor (SMR) type of helicopter. The reason for their proliferation in manned systems is that they best suit the small to medium size of helicopter for which there is most demand. Scale effects tell against larger-sized rotors and so the largest helicopters use two smaller rotors (tandem rotor machines) rather than one large one. Until a VTOL UAV of greater than about 20 000 kg is required, it is unlikely to take the form of a tandem rotor helicopter.

Single Main Rotor Helicopter

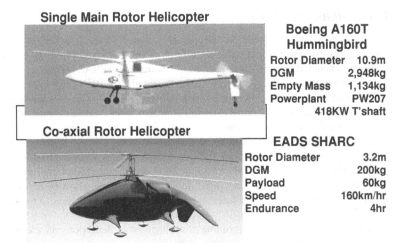

Boeing A160T Hummingbird

Rotor Diameter	10.9m
DGM	2,948kg
Empty Mass	1,134kg
Powerplant	PW207
	418KW T'shaft

Co-axial Rotor Helicopter

EADS SHARC

Rotor Diameter	3.2m
DGM	200kg
Payload	60kg
Speed	160km/hr
Endurance	4hr

Figure 27.5 Advanced rotorcraft configurations. *Source*: EADS Defence and Security; Boeing - Frontier Systems

Air Flow into T-Hawk Duct in Hover Mode

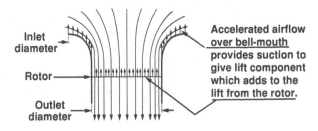

Inlet diameter

Rotor

Outlet diameter

Accelerated airflow over bell-mouth provides suction to give lift component which adds to the lift from the rotor.

Honeywell T-Hawk MAV

FENSTAR MAV

Figure 27.6 Enclosed-rotor VTOL UAV. *Source*: GFD Projects UK

The Firescout (see Chapter 4, Figure 4.13) was chosen for its task as an airframe appropriate to the 270 kg payload, already existed as a 4–5 seat manned helicopter (Schweizer 333) with well-proven systems. Converting the manned aircraft to a UAV with a slimmer and more streamlined fuselage was presumably seen to have less risk than designing and developing a completely new airframe and new dynamic systems from scratch for the job. The lighter and more streamlined fuselage gives the UAV a speed of 220 km/hr and 6 hr endurance at an AUM of 1432 kg.

The UAV has a similar AUM and speed performance to its manned predecessor, but three times the endurance, largely because it carries extra fuel instead of passengers. The Firescout is being evaluated by the US Navy for off-board operations, the result of which is being viewed with interest by navies of other countries.

The Schiebel S-100, on the other hand, is required to carry a smaller payload mass of 50 kg. No equivalent manned machine existed and so it was developed 'from scratch' as a UAV and would have had to undergo the discipline of fundamental design and testing, but gaining by starting with a 'clean sheet of paper'.

The S-100, also shown in Figure 4.13, offers a speed and endurance similar to that of the Firescout, even though it operates at the lower AUM of 200 kg. Apart from being acquired in significant numbers by Saudi Arabia, it is being evaluated by both French and German navies for off-board operation.

Another SMR helicopter designed from the start as a UAV is the Boeing A160T Hummingbird, shown in Figure 27.5. This machine was conceived originally by Frontier Systems Inc. and later acquired by the Boeing Company. The concept was to develop a MALE category VTOL UAV, with an eventual aim of achieving a capability of operating up to 10 000 m, a speed of 260 km/hr and with a 24 hr endurance. Simple calculations will show that this is no pipe-dream for a helicopter UAV and the Hummingbird completed a 19 hr flight in May 2008, well on its way to the design target.

By taking advantage of designing it originally as a UAV, dispensing with the mass and volume requirements for aircrew and their support systems, a light, streamlined structure can result. In the case of the Hummingbird a disposable load in excess of 60% of the aircraft AUM is realised and with a useful 200 kg payload, 90% of that is available to carry fuel.

The disc loading is average for a helicopter of that AUM and results in a power requirement of about 400 kW to hover out of ground effect (OGE) at design gross mass and sea level (SL) ISA conditions. The aircraft, however, will only require about 155kW to hover at the end of a sortie when most of its fuel will have been consumed and its AUM is approximately halved. This lower power requirement is, of course, reflected in a lower fuel consumption as the sortie progresses, and together with the large fuel capacity, enables the projected long endurance and long range to be realised. This is especially so as, with the low-drag fuselage, the power required in forward-flight can reduce to less than two-thirds of the hover power.

However, Boeing/Frontier has incorporated in the Hummingbird a difference from other helicopters. Helicopter rotors are customarily governed to operate within a quite narrow speed (rpm) band principally because, especially for SMR helicopters, the main and tail rotors input oscillatory forces into the airframe at a range of different frequencies. The airframe stiffness is designed to avoid those frequencies in order to obviate damaging resonances.

The Hummingbird includes a transmission which enables the rotor speed to be varied according to the operating condition of the aircraft. i.e. lower rotor speed at lower AUM, altitude and forward speed, with increased rotor speed at greater mass, altitude and forward speed. This must present a design challenge to avoid a clash of frequencies somewhere in the spectrum.

The author suspects that the saving in power brought about by the rotor speed change over the sortie may be little more than 20 kW, compared with more than 200 kW resulting from the reduction in airframe weight and aerodynamic drag compared with the equivalent manned helicopters, and questions whether the complication is worthwhile.

Also, as the power-plant is a Pratt and Whitney PW 207 free-power-turbine turbo-shaft engine, the output can be governed to run over the range of speeds proposed for Hummingbird independently of the gasifier spool. This should lose at most 5% in fuel efficiency at the extremities of the range and it might be thought to be a simpler, more reliable and lighter solution than a variable ratio transmission which might be thought to be hardly necessary.

So why was a SMR configuration chosen for the S-100 and Hummingbird UAV instead of the alternative option of a coaxial rotor system? The decision was probably not an easy one, but perhaps rested upon the background experience of the company's engineers and possibly also a perceived view that that was how most manned and model helicopters looked.

There are a few manned coaxial rotor helicopters in operation, but mostly military, largely because their overall height is greater than that of the SMR and they require taller hangars. This characteristic is less restrictive on the smaller UAV so they are beginning to be adopted as a configuration appropriate to that use. Two examples were previously shown, the 300 kg AUM Beijing Seagull in Chapter 4, Figure 4.14, and the 36 kg AUM MLA Sprite in Figure 4.19. A more recent development, the EADS SHARC of 200 kg AUM is shown in Figure 27.5. (An interesting feature of the Seagull is that, though designed as a UAV, the fuselage was subsequently modified to carry a pilot).

How do the SMR and coaxial rotor configurations compare? The coaxial configuration offers the following advantages:

a) The power-plant, transmission and rotor system can form a compact module for servicing or replacement. The same module can readily be used in a range of different airframe types.
b) The aircraft is much more compact than the equivalent SMR, enabling it to be operated more readily in confined spaces, for example on a ship's deck.
c) It has no tail rotor vulnerable to striking ground objects or personnel.
d) Oscillatory force inputs are limited to main rotor frequencies so that the dynamic design of the airframe is easier, and it has no relatively flexible tail boom to suffer fatigue problems.
e) Less power is required to fly, especially in the hover. A tail rotor usually consumes of the order of 10–15% additional to the main rotor and the coaxial rotor system itself is more efficient than an SMR by about 4–6% within the disc loading range of most current or projected VTOL UAV.
f) Due to the symmetry of the configuration, there is no cross-coupling between the control modes. In a manned version this makes it much easier to pilot. In the UAV, the design and construction of the AFCS is far simpler.
g) The moment of inertia in pitch and yaw is less than for a SMR with its tail boom and tail rotor mass. Thus its response to control in those modes is quicker.
h) There is no side-wind limitation as with a tail rotor, hence it can operate in higher winds and irrespective of the wind direction.
i) Again due to its symmetry, its response to air turbulence is less than with the SMR, some responses being zero.

The disadvantages of the coaxial configuration are few, but a potential concern is that in autorotation the yaw control moment first reduces and then, at greater rates of descent, reverses in direction. The author has not found this characteristic to be of great significance as the aircraft does not naturally suffer disturbances in the yaw mode, especially with a plan-symmetric body. Lateral manoeuvres do not require the aircraft to turn into a new direction. However, if this characteristic is seen to be of concern, an addition to the AFCS programming could readily adjust the yaw control input to change in amplitude and direction as a function of main rotor blade pitch.

The greater height of the configuration in the larger sizes may require a taller ground transport vehicle to accommodate it, but a heavier UAV will need to be transported in a larger (and therefore taller) vehicle anyway.

The aerodynamic drag of the two (smaller) hubs and the transmission between them will be somewhat greater than will a single (larger) main rotor hub. This is offset by the added drag of the tail rotor hub and blades and mounting boom of the SMR. In practice, which type has the lower overall drag is down to the detail in the individual design.

The response of the aircraft in roll may be slower than that of an equivalent SMR due to its greater height and therefore roll inertia. However the difference will be slight.

Enclosed-rotor VTOL UAV

Concern is sometimes expressed as to the danger presented to operators of a spinning, exposed rotor. This may have some validity in the larger sizes of UAV where the rotor mass is significant. The author's experience with the Sprite system was that in many hundreds of hours of operation in varied conditions, it was never a problem.

The Sprite blades were designed to be frangible on impact, but were never required to stand the test other than in the laboratory. On start-up, the rotor system remained stationary, with the engines running at idle, until the aircraft was ready to take off. Not until then were the rotors engaged.

However, the concern still remains with some authorities and this has given rise to a number of enclosed-rotor configurations. Another reason for an enclosed-rotor configuration is that MAV types intended to enter buildings may 'self-destruct' if the machine inadvertently strikes walls or other objects.

Two types of enclosed rotor UAV are shown in Figure 27.6. One, the Honeywell T-Hawk MAV (previously briefly discussed in Chapter 4) is of a ducted-fan configuration, the other, Fenstar by GFS Ltd. uses a top-mounted centripetal fan that blows air outwards over the outer surface of the UAV.

Honeywell T-Hawk

The T-Hawk reportedly carries a very small payload of about 0.5 kg, has an AUM of 7.7 kg and a fan diameter of about 0.33 m. Little more data is generally available other than that a 3.2 kW heavy-fuel engine is in development to replace the current aero-model engine. The thrust is thought to be generated by the fan thrust plus negative or positive pressures generated on the walls of the duct.

In hover-flight the fan sucks air downward, thus producing a negative pressure (relative to the ambient atmospheric pressure) above the fan and a positive pressure below it. The duct shape above the fan has an entry diameter a little larger than that at the fan so providing a vertical suction vector. The exit diameter is also slightly larger than the fan diameter, acting as a diffuser, and therefore limited in diameter by airflow separation. With the positive pressure, this may generate a small upward thrust vector. However since the shape of the T-Hawk duct implies that there is very little bell-mouth form in the intake, the lift from the duct will be small. Air-friction on the duct walls and on the control vanes at the base of the duct must exert a downward force. An area of fan, and therefore thrust, at its centre must be lost to the drive shafts and support.

Without the test data, it is impossible to know what is the sum of all of these different forces. If the assumption is made that the duct thrust balances the air-friction download and the loss of thrust from the fan hub, then the resulting thrust, equal to the aircraft mass, may be considered to act over the whole diameter of the fan.

If standard actuator theory is used and with the disc loading exceeding 1000 kN/m², assuming no fan tip loss, the air-stream velocity through the fan is derived to be about 20 m/s. The power required from the engine for the aircraft to hover at SL ISA conditions must be at least 1.8 kW. With extra power required to cover transmission losses, drive ancillary systems and to allow manoeuvring and operation in hot and high conditions, it is not surprising that an engine to develop over 3 kW is being sought for such a lightweight machine.

With a conventional value of disc loading, an engine of that power would adequately support an open-rotor machine of 18 kg DGM and a payload mass of at least ten times that of T-Hawk. (Note that the disc loading of the T-Hawk is far off the scale of even large transport helicopters, see Chapter 6,

Figure 6.13) This is a measure of the cost of providing an enclosed-rotor machine in the form chosen by Honeywell. However, it does carry its small payload with a compactness allowing the system to be back-packed.

The Company are to be congratulated on their achievement, especially in paring the weight of the aircraft down so low. A mere 10% increase in the weight of the aircraft components would have resulted in a machine with zero payload. That the advantages offered by the T-Hawk to the US Navy are recognised is reflected in the Navy's order of 90 systems which include 180 UAV. No doubt a challenge for Honeywell (Reference 27.7) in the future is to increase the payload and extend the endurance of the T-Hawk type.

GFS Fenstar

The Fenstar UAV is an entirely different approach to providing an enclosed-rotor VTOL UAV. An in-flight photo of a prototype Fenstar, which is aimed at operation in urban surroundings, is shown in Figure 27.6. The intent is to provide a UAV that can 'perch and Stare' and to be able to recover from collisions with obstacles both outside and inside buildings.

Unlike T-Hawk, it does not obtain its lift by accelerating a downward airflow inside a duct, but instead around the outside of its saucer-shaped body. This utilises the 'Coanda' effect whereby an airflow follows the curved surface of a body, provided that the curve is designed so that the pressure gradients are not such as to cause the flow to separate from the surface.

The airflow is generated by means of a centripetal fan mounted on top of the UAV body which draws in air from above and centrifuges it out radially and evenly across the upper surface of the body. The Coanda effect then keeps the airflow attached to the UAV surface until it flows downward around the body periphery to create a lifting force.

The technology is by no means as mature as that of T-Hawk. Whilst some hover tests have been made in order to assure pitch, roll and yaw control in hover mode, full wind-tunnel testing has yet to be made to ascertain its characteristics in forward flight, climb and descent. However, an electrically powered prototype of the Fenstar has demonstrated that it will fly. The task now remaining for GFS (Reference 27.8) is to obtain a better understanding of the aerodynamics of the Fenstar and to ensure that the configuration can be produced as a practical UAV able to carry a usefully positioned payload and to sustain flight over a meaningful flight envelope in adverse weather conditions.

Other VTOL UAV Configurations

Of the two other VTOL configurations discussed in Chapter 4, the Aerovironment Sky Tote (Figure 4.27) and the Selex Damselfly (Figure 4.28), at the time of writing no future programmes have been announced.

27.4.3   UCAS

The two foremost UCAS programmes are the Northrop-Grumman X47B and the BAE Systems Taranis. Both of these are illustrated in Figure 4.26.

The X47B programme is sponsored by the US Navy for operation from aircraft carriers as a strike weapon, eventually carrying small diameter ordnance internally. It is anticipated that the first ground tests will have been made in 2009 with operational tests from carriers to begin at the end of 2011. The Taranis is a smaller and lighter UAV and is seen more as a research system in the development of a future UCAS. The Taranis programme is scheduled for completion in 2010.

The challenge for both programmes is the development of:

a) a stealthy, high subsonic yet manoeuvrable, UAV capable of long-range operation but from short- to medium-length runways,
b) an on-board control system that has a high degree of autonomy.

To address some aspects of the former of these two challenges, a five-year programme called FLAVIIR (flapless air vehicle integrated industrial research) was set up by BAES and the Physical Sciences Research Council with several UK universities. The programme is managed by Cranfield University. The programme is developing a small flapless aircraft using fluidic thrust vectoring by air jets. It is expected that the results will be brought together in a flying demonstrator in about 2010.

For the latter challenge, BAE Systems have developed a research UAV, HERTI, (high-endurance rapid technology insertion, see Chapter 28, Figure 28.8) as part of an overall risk reduction programme for Taranis. The UK Government has funded a Defence Technology Centre (DTC) covering the systems engineering for autonomous system (SEAS) to study the technologies and systems integration aspects of advanced autonomous unmanned systems.

27.5 Autonomy and Artificial Intelligence

A number of manufacturers of UAS already claim that their systems are capable of autonomous operation. However, the understanding of the subtle differences between what constitutes automation and autonomy appears to be a 'woolly' area. To quote a BAE Systems press release on Taranis, for example:

> The brains of Taranis are now designed and coherent. What we have designed is a system that can autonomously control the aircraft to taxi, take off, and navigate its way to a search area while reacting to any threats or other events. It will then route its way around the search area in whichever way it wants to, locate the target, and then use its sensor system to transmit a series of images and views back to the operator to confirm it is the target to be attacked. Then, once it has been authorised to do so, it autonomously attacks that target, routes its way back home, lands and taxies back.

Chambers's Twentieth Century Dictionary defines autonomy as 'The doctrine that the human will carries its doctrine within itself'. This may be interpreted as 'Thinks for itself'.

Encarta Online English Dictionary defines automation as: 'An activity carried out by prior arrangement when certain conditions are fulfilled without the need for a decision', and autonomy as: 'Having the ability to make decisions and act on them as a free and independent agent' or 'Having the freedom to determine one's own actions'.

The author has underlined the words in the above press release which imply that the Taranis will have brains to 'think for itself'. That requires a measure of intelligence.

The reason given for the desire to imbue a UAS with 'autonomy' is that high-level 'automation' requires a complex program to be written into a UAV before flight. This could delay the operation and impose an element of inflexibility in the event of a change to the mission, On the other hand a UAV having 'autonomy' would merely require an instruction of 'what to do' to be entered and not the more complex sequence of instruction of 'how to do it'.

In the author's view another definition of autonomy in a UAS might be: 'Having the intelligence and experience quickly to react in a reasoned manner to an unplanned situation'. An experienced human pilot is able to do that, but no programming yet can cover the immediate recognition and decision-making required to avert an *unexpected* adverse situation.

It may therefore be postulated that for a UAS to operate with autonomy, a measure of artificial intelligence (AI) must be present within it plus a readily available source of experience. Somehow the latter might have to be extracted from the memory of an experienced pilot and enshrined in the processing elements of the system.

This is a tall order and, as observed in Chapter 10, Section 10.6, is a fully autonomous UAS actually desirable? Questions of its monitoring arise since, given intelligence, it might refuse to complete a mission or, even worse, exhibit behaviours which are undesirable for the operators. In the final analysis, the human operator has to take full responsibility in law for the result of actions of his machine.

Automation, however, is more understandable, predictable and certifiable for reliability. Automatic systems built-in to the UAV's control system offer several advantages. Such systems might, for example, monitor engine condition, take action automatically in the event of a power failure and alert the operator to the situation.

Another automatic action might be through the monitoring of fuel state and distance from base, warning the operator should the fuel state approach critical in terms of ability to return to base and initiate that return. The operator should always have the option to over-ride that action if, for example, the mission circumstances were such that it was necessary to accept the UAV being expended.

Very high level automation, where a number of subsequent decision branches are programmed into the on-board computer, so that progressive decisions may be made by the UAV without reference to the operator, is often referred to as autonomy.

Several authorities have attempted to define levels of autonomy rising from the lower levels, which are obviously automation, up to higher levels which may be classified as autonomy. The following table is derived from an example offered by the authors of Reference 27.9.

Locus of authority	UAV computer authority	Computer level of authority	Levels of human–machine interface
Computer monitored by human	Full	5b	Computer does everything autonomously
Computer monitored by human	Full	5a	Computer chooses action, performs it and informs human
Computer backed-up by human	Action unless revoked	4b	Computer chooses action and performs it unless human disapproves
Computer backed-up by human	Action unless revoked	4a	Computer chooses action and performs it if human approves
Human backed-up by computer	Advice and, if authorised, action	3	Computer suggests options and proposes one of them
Human assisted by computer	Advice	2	Computer suggests options to human
Computer assists human if requested	Advice only if requested	1	Human asks computer to suggest options and human selects
Human	None	0	Whole task done by human except actual operation

Advantages of 'on-board' automation/autonomy include:

a) reduction of operator workload,
b) reduction of possible operator errors,
c) reduction in the use of radio frequency bandwidth.

Artificial Intelligence

The study of artificial intelligence probably began in the 1930s and has 'enjoyed' a roller-coaster ride in a similar fashion to that of UAS development. If the author understands correctly, there are, in general, two approaches made. One is a bottom-up approach which attempts to develop neural networks akin to the operation of a human brain. The other, known as top-down, attempts to simulate the performance of a human brain by using high-speed computer algorithms.

Seemingly, neither has as yet been successful in achieving the level of human understanding or creativeness. It is recognised that today's supercomputers are still short by a factor of ten in achieving the equivalent of human brain in terms of speed and capacity. Although parity may be achieved at some future time, the cost of it is likely to be prohibitive.

27.6   Improvement in Communication Systems

As already discussed in Chapter 9, this is one of the principal challenges in expanding the use of UAS.

References

27.1. ASTM International, or American Society for Testing and Materials (ASTM). ASTM F2411 - 07 Standard Specification for Design and Performance of an Airborne Sense-and-Avoid System.
27.2. 'See and Avoid'. T. Mildenburger, European Cockpit Association – UAV 2007 Paris Conference.
27.3. NATO Document PFP (NNAG-JCGUAV) D (2008)0002.
27.4. 'Stepped Piston Engines for Multi-Fuel UAV Application'. P. R. Hooper, IMechE Conference on Propulsion Systems for Unmanned Aircraft, Bristol, 14 April 2005.
27.5. Protonex Technology Corporation,153 Northboro Road, Southborough, MA 01772-1034,USA.
27.6. NASA Tech Memo 2005-213800.
27.7. Honeywell Aerospace, 1944 E Sky Harbor Circle, Phoenix, Arizona 85034, USA.
27.8. GFS Ltd, 3 St David's Square, Peterborough, PE1 5QA, UK.
27.9. 'Effective Operator Engagement with Variable Autonomy', A.F. Hill, F. Cayzer and P.R. Wilkinson, *Proceedings of the 2nd SEAS DTC Technical Conference*, Edinburgh University, UK, 2007.

28

UAV Systems Continuing Evolution

28.1 Introduction

The impact that evolving technology has had on human affairs makes for an interesting study. In military endeavours, throughout the centuries, technology has offered new weapons with an imperative to find a counter. Technology usually provides that counter.

In civilian life, evolving technology, most usually handed down from military initiatives, may be used by entrepreneurs to adapt it to new uses or leapfrog competitors' products. A somewhat similar pattern is seen to be taking place with UAS.

They were first seen as a possible counter to both the offensive and defensive weapons of opponents. Subsequently military UAS capability was expanded in turn as enabled by advancements in technology, principally that of the electronics and logistics of guidance and control. The availabilities of steadily evolving technologies which paced the evolution of UAS were principally:

a) gyroscope systems and their increasing accuracy,
b) air data systems and their increasing accuracy,
c) radio command systems with increasing range and protection,
d) radio and radar tracking systems,
e) imaging sensor development in several wavelengths and fusion,
f) radio down-link performance,
g) radio, laser and acoustic altimeters,
h) GPS systems,
i) high-speed computers and net-work centric communications,
j) sense-and-avoid technologies.

The early steps towards UAS, as known today, were by pioneers who may not have visualised where their development might eventually lead.

In the following 'guide through time' a number of systems, represented by pictures of the aircraft component, are shown in each era or decade. These are not to be taken too objectively since some systems overlap successive decades, being progressively developed with evolving technologies and for new rôles. At a given point in time some systems adopt available advanced technologies whilst others do not. The types shown for each decade are merely chosen to represent a trend and are not intended to be all

Unmanned Aircraft Systems – UAVS Design, Development and Deployment   Reg Austin
© 2010 John Wiley & Sons, Ltd

Figure 28.1   Early cruise missiles

inclusive. For a more detailed account of the early history of UAS, the following book is recommended: *Unmanned Aviation* by Laurence R. 'Nuke' Newcombe, published by AIAA; ISBN 1-56347-644-4.

28.2   Cruise Missiles

The earliest systems were developed as long-range armament (the forerunners of today's cruise missiles) in such devices as the US Navy's 'aerial torpedo' of 1917, the US Army's 'Kettering bug' of 1918 and the British Army's 'aerial Target' begun in 1914 (a name designed to mislead from its real purpose, see Figure 28.1).

The aerial torpedo was an unpiloted biplane bomber made of wood, weighing just 270 kg, including a 136 kg explosive charge as payload, and was powered by a 40 horsepower Ford engine. The method of guiding the aerial torpedo to its target was primitive but ingenious. Once wind speed, wind direction, and target distance had been determined, the number of revolutions the engine needed to take the missile to its target was calculated. A cam was then set to drop automatically into position when the right number of engine revolutions had occurred. Once airborne it was controlled by a small gyroscope, its altitude measured by an aneroid barometer. When the engine had completed the prescribed number of revs, the cam dropped into position, causing bolts that fastened the wings to the fuselage to be pulled in. The wings then detached, and the bomb-carrying fuselage simply fell onto its target.

The Kettering bug was a lighter biplane designed to carry an 82 kg explosive charge. It worked on a similar principle to the Navy machine. A replica can be seen in the US Air Force Museum in Dayton, Ohio.

The aerial target (A.T.) was a radio-controlled pilotless monoplane powered by a 35 horsepower engine. The A.T. concept vehicles achieved their purpose which was to prove the feasibility of using radio signals to guide a flying bomb to its target.

None of these devices were developed satisfactorily enough to be used by the military before the end of World War I. They signalled, however, the beginning of a new technology, though the guidance methods were very crude and unreliable.

Queen Bee Gunnery Target

Wing Span	8.9m
AUM	830kg
Speed	170km/hr
Engine	104KW Gipsy Major
Guidance	Radio Control

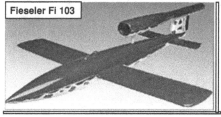

V1 "Vengeance Weapon"

Wing Span	5.3m
AUM	2,150kg
Speed	640km/hr
Range	250km
Propulsion	Pulse-jet
Warhead Mass	850kg
Guidance	Autopilot

Figure 28.2   WW2 unmanned aircraft

In the mid-1920s interest in pilotless systems was revived in the UK and, in particular for the Royal Navy. A monoplane capable of carrying a warhead of 114 kg, over a range of 480 km, was built at the Royal Aircraft Establishment, Farnborough, and first flown in 1927. It was powered by a 200 bhp Armstrong-Siddeley Lynx engine. It was given the name 'long-range gun with Lynx engine'(!) which was subsequently shortened to 'LARYNX'. It is shown, in Figure 28.1, on a launch ramp aboard a Royal Navy destroyer.

Unlike the WWI machines, it was fitted with radio control for the launch mode, after which the autopilot restrained it to fly on a pre-set course at a pre-set height to a pre-set range. Of the twelve built, only the last five were fitted with a warhead and tested in the Iraqi desert. A modicum of success was achieved, although some flights were plagued by engine failures. It did, however, introduce a measure of radio control.

28.3   World War II Systems

At this point, in Great Britain, it was decided to abandon the 'cruise missile' concept and to concentrate on target aircraft with full-mission radio control although its range was limited. This began with the installation of radio control into some Fairey naval aircraft and designated 'Fairey Queen'. The success of these led to the construction, between the years 1934 and 1943, of no less than 420 radio-controlled target aircraft for Naval and Army use. The aircraft were versions of the De Havilland Tiger Moth constructed in wood and were named 'Queen Bee'. They served to give gunners of both services very necessary training in WWII. Figure 28.2 shows a Queen Bee being operated by Royal Air Force personnel.

A smaller, lighter, model-aircraft-sized UAV, RP4, was developed by the Radioplane Company of the USA, and several thousand of these were produced during WWII for use as gunnery practice targets by US forces. Through these aircraft, the technology and use of an early form of radio control was developed.

The cruise missile concept, however, was revisited by Nazi Germany with the 'V1 Vengeance Weapon' that was aimed at southern England from the Low Countries. It was the first cruise missile to use jet propulsion, albeit with a pulse-jet.

**Northrop Falconer or Shelduck ,
later designated BTT**

Crossbow Decoy

Figure 28.3   1950s systems

The guidance system was only a little more sophisticated than earlier devices. It used a barometric system to regulate speed and height and a vane anemometer to determine distance travelled. The latter operated a counter, pre-set depending upon the anticipated head-wind vector. This initiated the aircraft into its terminal dive after a number of anemometer turns indicated the aircraft to be in the vicinity of the target. Heading was held from the input of a directional gyro and a magnetic compass. It was a very inaccurate weapon due to gyro drift and errors in the barometric and anemometer sensors compounded by inadequate wind-speed data, but it contributed to the early development of a flight-control system.

The United States Navy also experimented with 'assault drones' during this period. The 'Project Fox' drones were fitted with an early television camera in the nose and were radio-controlled from a manned aircraft fitted with a television screen. Some successes were achieved against target ships. These experiments gave early experience in 'hands-on' guidance and control.

28.4   The 1950s

In the post-war period the US Company Radioplane, later to become Northrop, developed a successful series of unmanned target aircraft variously named 'Falconer' or 'Shelduck' (Figure 28.3). These continued in production until the 1980s, adopting evolving radio and control system technologies on the way. They were finally given the designation BTT (basic training target) and were produced in greater numbers than any other piston-engine-powered target aircraft.

Another development was the use of UAV as anti-radar detection decoy systems. A number of 'Crossbows' (Figure 28.3) were carried by intruding bombers, such as the B47, and released to confuse the opposing radar systems. Released from parent aircraft, they could be radio-controlled and directed from on-board video images.

28.5   The 1960s

With the arrival of fast jet-propelled military aircraft, this period saw the further development of faster, longer-range target aircraft such as the 'Ryan Firebee' (later Teledyne-Ryan) which had been initiated in the late 1950s (Figure 28.4). These were later modified to carry bombs for release onto ground targets.

Figure 28.4   1960s systems

Subsequently the Firebee, and other target UAV, were adapted to carry still cameras for reconnaissance purposes over enemy territory. They operated at high altitudes, where control via direct radio line-of sight from a GCS was possible, or at lower altitudes controlled from a stand-off manned aircraft. The photographs were developed at base after the return of the UAV. Such intruders were less easy to detect and less easy to shoot down than manned reconnaissance aircraft, also they would not give rise to the diplomatic incidents attendant upon the capture of a human pilot.

The Firebee could be ground-launched with a rocket-assisted take-off system or air-launched from a manned aircraft such as the Lockheed C130 Hercules transport. They were recovered by a deployed parachute on returning to a suitable area for landing. Other Firebee variants were adapted for dispensing chaff to confuse radar systems. There was significant evolution of Firebee over many years.

The various types, even one with supersonic speed capability, were given a plethora of different designations and some may even remain in service today, having been fitted with GPS navigation systems and other advanced sensors. Firebee systems were manufactured in considerable numbers and, although not initially configured as true UAS, as known today, contributed considerably to UAS technology.

The Northrop 'Chuckar' (Figure 28.4) was also a jet-propelled target begun in this decade, but it was much smaller and lighter than the Firebee. It used control methods similar to the Firebee, but a Mk2 version in the 1970s had later technology, more accurate autopilot system for out-of-sight control, and a few were fitted with a radiation-seeker and warhead to attack enemy radar systems.

The Gyrodyne DASH (drone anti-submarine helicopter) was probably the first 'battlefield UAV' fielded by the United States (Figure 28.4). It was a dedicated design for the purpose and not a development of a target system. Its task was to fly from US Navy frigates and to carry torpedoes or nuclear depth charges to attack enemy submarines that were out of range of the ship's other weapons. In terms of control systems, it could be considered a backward step since it carried no autopilot or sensors, being merely radio controlled and, presumably radio or radar-tracked for positioning. It did, however, introduce a different rôle and for the first time the use of a rotorcraft UAV. A large number just fell into the sea and it became known as the 'splash-a-DASH-a-day' programme. Possibly these losses might have been avoided if it had been fitted with a contemporary autopilot system.

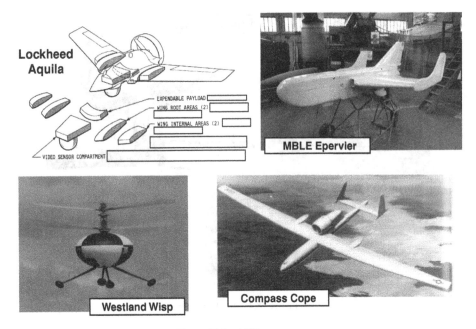

Figure 28.5   1970s systems

28.6   The 1970s

This decade saw the introduction of several UAS designed from the outset for reconnaissance and surveillance, both for the shorter-range, but ground-mobile 'battlefield' systems and the longer-range higher-altitude systems. With the heightening pressures of the Cold War, these systems were becoming more sophisticated, both in terms of their mission requirements and the security of their communications. Understandably, the details of these were wrapped in much secrecy and reports of them could be misleading. Therefore in some respects the author has to resort to an element of supposition. One system designed for the shorter- range use was the Lockheed Aquila (Figure 28.5).

The mission capabilities of Aquila would have provided a very valuable asset if achieved, but it was probably too ambitious for the then current technology. The UAV was to be a small, piston-engine-powered, propeller-driven aircraft to be portable by four soldiers and of modular construction. It would provide ground commanders with real-time battlefield information, by video down-link, about enemy forces located beyond the line of sight of ground observers.

Aquila was expected to fly by autopilot, carry sensors to locate and identify enemy point targets in day or night, use a laser to designate the targets for the Copperhead artillery projectile, provide conventional artillery fall-of-shot adjustment, and survive against Soviet air defences.

Achieving the survivability expectation required development of a jam-resistant, secure communications link. Methods of achieving this were discussed in Chapter 9. Navigation would have been by radio-tracking of the UAV from the GCS. This would have been operated only occasionally in short bursts to update the position estimated by the on-board autopilot computer.

The problem was that, although the airframe could be built within available technology, the implications of assembling it and operating it as an entire system were not fully understood. The sub-system technologies required for satisfactory integration within the specified size of airframe, had not been developed. In the attempt to develop them belatedly it was found that the sub-systems were too heavy, too large, demanded excessive power or suffered from mutual interference. For example, use of the secure link, which probably used the spread-spectrum technique still in its infancy, degraded the video quality, which interfered with the ability to do targeting.

The aircraft was launched from a hydraulically operated catapult and this was successful. However, another problem was brought about by the decision to recover the aircraft by catching it in a net strung between two vertical posts. The accuracy of guidance required to achieve this reliably was not available and many aircraft were destroyed during the test programme. In an in-flight emergency, a parachute was deployed which lowered the aircraft inverted, but this was not the prime method of recovery as it inevitably caused damage to the airframe, but not to the payload.

Although it is understood that some successful deployments were achieved to demonstrate its surveillance and laser designation capability it did not achieve this with an acceptable level of reliability. Work to achieve a reliable system solution proceeded on through the 1980s at considerable expense, but the programme sadly was eventually cancelled, it is understood, because of unacceptable cost.

A less demanding programme might have seen a system initially in useful service limited in range and to reconnaissance and surveillance missions. Other mission equipments might then have been added, and the range extended, as the technologies became available.

A less ambitious practical battlefield system, designed more within the available technologies, was the MBLE (Belgium) Epervier (Figure 28.5). Development began in the 1960s and the system was in operational service with Belgian forces by the mid 1970s. The airframe was powered by a small turbo-jet engine and launched from a truck-mounted ramp by JATO bottle rocket motor. The recovery was by parachute.

The flight profile was achieved through a pre-programmed autopilot assisted by radio guidance. Payloads consisted of a daylight camera or infrared linescan camera, the products from both of which were processed on return at the GCS.

The Boeing Gull HALE UAS was the winner, in 1971, of the Compass Cope competition organised by the US Air Force for the first high-altitude long-endurance surveillance UAS (Figure 28.5). The vehicle aim, which was achieved, was to demonstrate flight at 16 770 m, and an endurance of at least 20 hr. Payloads of up to 680 kg were to allow operational versions to carry out photo-reconnaissance, communications relay, and SIGINT over a range of 300 km by day and by night in all weather conditions.

The prototype was flown manually under direct radio control, but production systems were designed to have automatic take-off and landing with autopilot-based mission control.

Sadly, the programme was terminated in 1977, reportedly on the basis that the necessary payloads could not be developed within an acceptable time. It appears that yet another programme was committed to and lost through over ambitious expectation of technology. More than ten years elapsed before another attempt was made to develop this type of system with the availability of much-improved technology, particularly GPS for autonomous navigation.

In previous eras, it had become a concern that attrition of UAV was largely the result of accidents on launch and, more so, on recovery. Westland and others were considering the use of vertical take-off UAV for the shorter range operations. The Westland Wisp system (Figure 28.5) was designed in response to a UK MOD requirement for a special-purpose very short range system. It used a simple gyroscope autopilot for stability and was operated manually by line-of sight radio control. It carried a daylight TV camera which sent back real-time images to the operator. It was also noteworthy in adopting for the first time the new 'plan-symmetric' (PS) rotorcraft configuration. Three of these aircraft were flown and, in addition to solving the problems of launch and recovery, they gave insight into the advantages of a hover capability during surveillance missions.

28.7   The 1980s

A surveillance system which spanned a number of decades, beginning its development in the 1960s, operated in an early form in the 1970s and with further development during the 1980s, entered service in

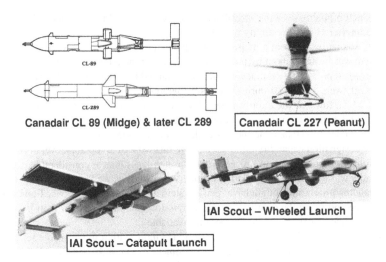

Figure 28.6   1980s systems

a more advanced form in the 1990s is the Canadair CL-89 and CL-289 series (Figure 28.6). It remained in service into the 2000s. The final development was carried out under the prime contractorship of Bombardier/EADS.

The original CL-89 was developed to provide non-real-time visual intelligence of enemy territory within an operating radius of 70 km. The UAV was launched by a solid propellant rocket from a zero-length ramp and recovered by a parachute and airbag. The UAV carried a fixed film camera or a linescan infrared camera by day. For night surveillance the linescan infrared camera only was used. The images from both cameras were processed and interpreted after recovery.

A major requirement of the UAV was that it be highly survivable in the face of the defensive systems of a sophisticated enemy. With its small diameter (0.33 m), stub wings of 0.94 m span and high speed (740 km/hr) it was difficult to detect and destroy.

Guidance was by a preset programme controlling an autopilot supported by vertical and directional gyros and barometric sensing of speed and altitude. Navigation was by computer-based dead-reckoning.

The later, CL-289, had an enlarged airframe (diameter 0.38 m and wing-span of 1.32 m and AUM increased from 108 to 240 kg) to extend its operating radius to 200 km. It also carried a video data transmitter which could send real-time infrared images to the GCS when the UAV was within 70 km of its base. Further additions for service in the 2000s were a GPS system for more accurate navigation and terrain-following and a SAR/MTI radar for all-weather operation.

The 1980s also saw continuing rotorcraft UAV development activity for close-range reconnaissance and surveillance duties. The Canadair CL-227 Sentinel VTUAV (Figure 28.6) adopted a prolate plan-symmetric body as had the Westland Wideye of the previous decade. The first prototypes of the CL-227 were powered by Wankel rotary-piston engines and, like Wideye, had payloads consisting of daylight video cameras stabilised to look only vertically downward. This was soon recognised by Canadair as unsatisfactory and phase II aircraft were equipped with TV cameras capable of forward as well as downward vision. Phase II aircraft, flying in 1981, also abandoned the piston engines for a Williams turboshaft unit. Phase III aircraft, first flown at the end of 1987 introduced an IR payload.

A still later version, flying in 1996 abandoned the plan-symmetric configuration, having streamlined fairings added to the body above and below the rotors. This was renamed 'Puma'. It had an up-rated engine and added a GPS navigation system.

Further developments by Canadair were of two configurations, the CL-327 Guardian which reverted to the PS configuration and the CL-427 which retained a streamlined shape. Both types were of greater AUM (350 kg) with larger rotors and more powerful engines than the Sentinel's in order to increase their operating range and endurance for naval operation. Although successful shipboard launches and recoveries were demonstrated, development had ceased by the end of 1999.

Appearing virtually parallel in time with the Canadair VTUAV, was the M L Aviation Sprite VTUAV system. This was a much smaller (36 kg AUM) multi-service UAV and of oblate-shape body. (see Chapter 4, Figure 4.19). This successfully demonstrated off-board operation at sea with the US Navy earlier in the same year as did the CL-227. It also demonstrated the advantage of the oblate PS rotorcraft, with its lower centre of mass and lower aerodynamic drag, compared with the prolate. Successful field trials in various rôles were carried out by Sprite UAS in several different countries and for different customers, both military and civilian but, in spite of military orders, production was short-lived due to external factors.

Armies were by now looking to extend UAS real-time ISTAR operations to somewhat greater range (of order 100 km). This was made possible by the increased accuracy and reliability of flight control systems. A number of close/medium-range HTOL UAS were developed. Typical of these, and probably one of the first in the field, was the IAI Scout (Figure 28.6) which introduced the now ubiquitous twin tail boom, pusher propeller configuration. This system, together with the similarly configured Tadiran Mastiff led to the IAI (Mazlat) Pioneer which, with upgrades, remained in service with Israeli and US Forces until the mid 2000s.

All of these types employed for guidance and control, a full three-angular attitude gyro-stabilised computer-based AFCS and automatic tracking secure two-way data link. The up-link could be used to direct the flight profile in real-time or to transmit occasional position updates and select a flight program for the next phase of flight, thus minimising the time in which radio transmission was active.

Parallel development of payload sensors opened up the possibility of a greater range of mission payloads to include electronic warfare (EW), ECM, laser target designator, decoy and communications relay in addition to optical and IR TV imaging payloads. Increased guidance accuracy also enabled more reliable recovery into ground-based nets, although recovery onto ships still presented a problem for HTOL aircraft.

28.8   The 1990s

The more general availability of the Global Positioning System (GPS) and satellite communications freed the UAV from the limitation of operating within radio tracking range or of relying on relatively inaccurate onboard navigation systems based upon computerised dead-reckoning systems with their gyroscope and air data inputs.

These navigation systems, now with digital flight control systems (DFCS), enabled UAV to operate out to much greater ranges with positional accuracy. The result was the development of medium- and long-range systems, the former characterised by the Denel Seeker and similar systems presented in Chapter 4 and the latter by the General Atomics Gnat (Figure 28.7).

The General Atomics Gnat, powered by a piston engine, may be considered to be the precursor of a new breed of MALE and HALE long-endurance/long-range systems. The Gnat systems saw operational service in the mid-1990s over Bosnia and Croatia in a reconnaissance role using EO and IR sensors and, later, signals intelligence (SIGINT) equipment. Several upgrades of the UAV took place through A, B and C models. The data shown in Figure 28.7 is for the A model. Experience with the Gnat systems paved the way by the end of the decade for a larger development, the Predator MALE UAS, and then the Northrop-Grumman Global Hawk HALE UAS in the following decade.

Operations with the Gnat and early mark of Predator enabled reconnaissance missions to be carried out from higher altitudes than previously, which offered some protection from detection and ground

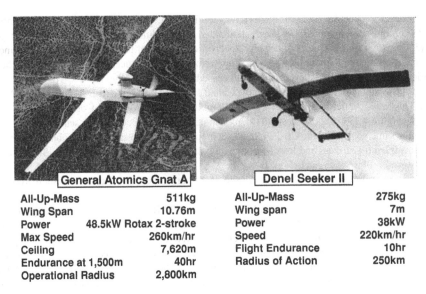

General Atomics Gnat A		Denel Seeker II	
All-Up-Mass	511kg	All-Up-Mass	275kg
Wing Span	10.76m	Wing span	7m
Power	48.5kW Rotax 2-stroke	Power	38kW
Max Speed	260km/hr	Speed	220km/hr
Ceiling	7,620m	Flight Endurance	10hr
Endurance at 1,500m	40hr	Radius of Action	250km
Operational Radius	2,800km		

Figure 28.7   1990s systems

fire. However the EO and IR payloads could not see through cloud and this forced the UAV to descend to altitudes which then made it more vulnerable. The answer to this was the development of synthetic aperture radar (SAR) systems for UAV which enabled reconnaissance to be carried out with images obtainable through cloud.

Also of note in this decade was the development and introduction in Japan of the first large-scale production and use of a rotary-wing UAV the Yamaha R50, and a slightly enlarged model, the R.Max. This system, employed in agriculture for rice sowing and crop-spraying (see Chapter 4, Figure 4.20), was successful in finding a niche market within which regulatory authorities allow it to operate. It is understood that some 1500 units have been produced to date.

28.9   The 2000s

Initial SAR payloads were heavy and this, with the desire to extend the range and endurance of the UAS still further, required the development of larger, heavier and more capable systems. These appeared in the form of the Predator B model, now powered by a turbo-propeller engine, and the still larger and higher-altitude Global Hawk UAV powered by a turbo-fan engine (see Chapter 4 for details).

This decade saw a much increased use of UAS in military roles. Some systems, notably the General Atomics Predator, Northrop Global Hawk and the Boeing/Insitu Scan Eagle have amassed operating hours measured in hundreds of thousands compared with just thousands in earlier decades.

Civilian operations, although potentially more extensive than military, have not come to fruition due to the perceived difficulty in ensuring separation between manned and unmanned systems.

Another development in this decade was the realisation that when military reconnaissance missions showed the existence of enemy forces against which it was necessary to strike, by the time a ground attack could be mounted, the enemy force had moved to a less assailable position. Thus some medium- and long-range UAV were modified to carry armament for an immediate response. The later, enlarged and more heavily armed, version of the Predator B became known as the 'Reaper' and is shown in Figure 28.8.

AUM   500kg Endurance   24hr
Wingspan   12.6m Ceiling   >6000m
Engine   Rotax 86kW Max Speed   220km/hr

BUAA M-22
Co-axial Rotor RPH
AUM   60kg
Payload mass   10kg
Endurance   1hr 30min
Mission   Surveillance

Figure 28.8   2000s systems (Reproduced by permission of BAE Systems)

As a further example of the potential of UAS to be developed with advances in technology, General Atomics have recently announced the C version of Predator which elevates the former MALE system to a HALE system, possibly competing with the Global Hawk system, but also offering a degree of stealth.

This evolution is shown graphically in Figure 28.9 which shows merely three steps in the development of Predator from the A version of 1994 to the latest C version which, in addition to carrying its predecessor's search sensors, offers higher speed, operation at HALE-type altitudes, and carries its munitions internally to maintain its stealth characteristics.

There is a proposal for NASA to acquire Global Hawks for world-wide upper-atmosphere sampling and analysis, together with information for use in weather forecasting. In the longer term this activity is expected to be carried out by developments of the Zephyr UAS, early versions of which have demonstrated very high altitude and long endurance flights in this decade. Current plans of NASA and Qinetiq are to extend its operating altitude even further and its endurance to a continuous flight for 6 months, descending only for maintenance. It would also have its payload capability increased to allow carriage of atmosphere sampling equipment.

Automation, as previously discussed in Section 27.5, reduces operator workload, can reduce operator error and reduce the radio bandwidth required for control and data transmission. Work is continuing in this field in several countries. BAE Systems, for example, is using the Herti (high-endurance rapid technology insertion) UAS to develop and demonstrate this capability.

The high endurance of Herti, 24 hr, results from a combination of a fuel-efficient engine and the low propulsive power requirement obtained at low speed through use of low-drag, long-span wings derived from a sailplane. Given adequate power, the UAV has the potential to considerably exceed the claimed 6000 m altitude since it operates at maximum AUM at a very low span-loading of 0.39 kN/m. (This compares with the 3.23 kN/m of Global Hawk and the extremely low value of 0.018 kN/m of the stratospheric Zephyr).

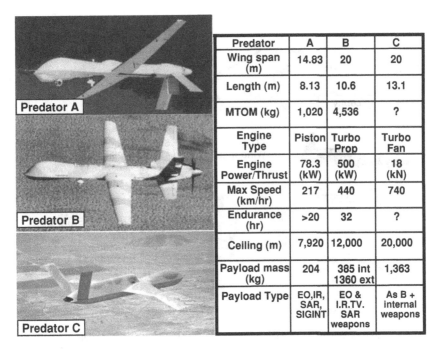

Predator	A	B	C
Wing span (m)	14.83	20	20
Length (m)	8.13	10.6	13.1
MTOM (kg)	1,020	4,536	?
Engine Type	Piston	Turbo Prop	Turbo Fan
Engine Power/Thrust	78.3 (kW)	500 (kW)	18 (kN)
Max Speed (km/hr)	217	440	740
Endurance (hr)	>20	32	?
Ceiling (m)	7,920	12,000	20,000
Payload mass (kg)	204	385 int 1360 ext	1,363
Payload Type	EO,IR, SAR, SIGINT	EO & I.R.TV. SAR weapons	As B + internal weapons

Figure 28.9   Predator evolution. (Reproduced by permission of General Aeronautical Systems Inc)

Herti, therefore, offers a very wide-ranging capability as a research tool and with its payload of 150 kg can carry a range of equipment to assess approaches to issues such as automation, sense-and-avoid, improvements in sensors and communications, etc. There is also the possibility that it might enter service as a persistent, low-speed surveillance facility.

A return of interest in VTOL UAS has been growing in this decade. It has been occasioned by the awakened realisation that such UAV offer the following advantages, especially for shorter-range, lower-altitude operation, compared with fixed-wing UAV:

a) lower response to air turbulence,
b) compatibility with stealth design,
c) ease of positioning in surveillance, designation and other missions,
d) simplicity of launch and recovery, resulting in less support equipment and lower loss rates.

Statistics of HTOL UAV operation reveal that, of 98 incidents causing significant damage or total loss to a fleet of Global Hawk, Hunter, Pioneer, Predator and Shadow UAV between July1991 and December 2001, about 40 occurred during launch or recovery. The acquisition by Saudi-Arabian forces of a considerable number of the German Schiebel Camcopter S-100 is an indicator of this realisation. In addition to the VTOL UAV shown and discussed in Chapters 4 and 27, they are also known now to be under development in other countries as exampled by the Chinese BUAA M-22 shown in Figure 28.8 which also adopts the plan-symmetric configuration.

Another interesting development in this period, however, was that of the Boeing/Insitu Scan Eagle UAV. This, as described in Chapter 12, is catapult-launched, but recovered by locating and hooking onto a vertically suspended section of rope. It is reported that to date some 100 000 successful recoveries have been made by this method and it goes someway to alleviating the recovery problems of HTOL UAV.

However the technique is probably limited to lightweight UAV and, although tested by the US Navy, that organisation is continuing to favour the introduction of the Northrop-Grumman Firescout VTOL UAS for more general use.

During this period, also, aided by the availability of satellite systems, improved radio communications and faster computer systems, Network Centric Operations have been developing to enable a greater and more timely range of information to be acquired and distributed to military operators at all levels.

28.10   The 2010s

A number of issues are expected to be addressed in this coming decade. As previously noted, apart from the exceptional application of the RMax UAS in Japan, there has been no significant use of UAS for civilian tasks. This has been primarily due to the several airworthiness authorities, including the EASA and FAA denying the use of suitable airspace to UAS until UAV are equipped with a reliable sense-and-avoid system (see Chapters 5 and 27).

Although a number of organisations are addressing the problem, the earliest prediction of a sense-and-avoid system being available has been made by ASTRAEA. This is a body (Autonomous Systems Technology Related Airborne Evaluation and Assessment) set up by the UK government with UK industry participants to investigate bringing together the technologies required to achieve an appropriate sense-and-avoid system. ASTRAEA has also been discussing applicable regulations with the UK CAA and predicts that a system could be trialled by 2012.

In the USA, the multi-vehicle unmanned aircraft systems sense-and-avoid (MUSAA) programme is expected, in 2010, to award a $2.4 million, three-year contract to develop a sense-and-avoid system. However, as discussed in Chapter 27, such a system would need to be small, lightweight and affordable enough to be installed in civilian UAV without destroying their economic and operational *raison-d'être*. An encouraging development was the landing and take-off of a Predator UAV at the Oshkosh General Aviation show in July 2009. Although special arrangements were undoubtedly made, it may be seen as a 'light at the end of the tunnel'.

As noted in Chapter 9, another desirable improvement, to further the use of UAS in civilian operations, is the availability of a radio bandwidth dedicated to them. It is hoped that this will also be achieved in this time-scale.

This decade will inevitably see a continuation of developments in electronic technology which will improve the efficiency of many current systems and possibly enable the development of aspects of UAS currently not even foreseen.

Current systems which can be expected to be improved include the fusing of imaging sensors in the optical, IR and radar frequencies to form a single powerful image which offers an improved conspectuity (ability to see), combining the advantages of all the frequencies with none of their faults, and the reduction in bandwidth requirement of UAS radio-links.

With the expansion of the use of military UAS, especially by the United States forces, a shortage of suitably qualified UAV 'pilots' or, more appropriately, 'operators' is becoming critical according to the US Air Force Chief of Staff, General Schwartz. This shortage may be encountered by the forces of other countries and also by the civilian industry if civilian operations take off.

It is questionable as to whether UAV operators should be drawn from people with piloting qualifications as is currently generally the case. If this were to remain a requirement then, with the proliferation of UAS, a severe shortage will become inevitable. Hence it is seen that educational, experience and fitness standards for intake to training courses must be agreed and work is ongoing to determine this.

Training courses and methods suitable for future UAS operators must also be agreed and courses set up for them. Simulation of operations will obviously pay a large part in this, but a final element of involvement in physical operation may also be required and the flying facilities for this made available.

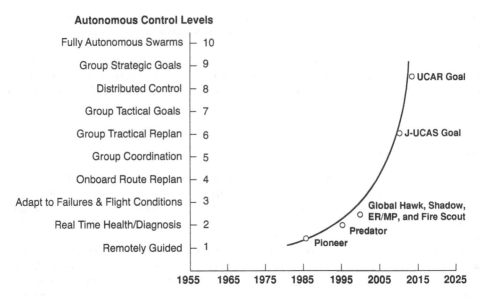

Figure 28.10   Predicted trends in autonomy (DOD Roadmap)

Subsequent certification and regular appropriate physical and ability checks to maintain the approved standards will almost certainly be required.

As discussed in Chapter 27, there will be continuing effort to raise the level of UAS automation to counter the growing shortage of radio bandwidth availability and to reduce the manpower requirements of growing UAS deployments.

Figure 28.10 reproduces a graph, courtesy of the US Department of Defence, which predicts the future rate of development of UAS autonomy. To the author's knowledge, there is, as yet, no universally agreed definition of the various levels of UAS autonomy. Hence the levels shown in Figure 28.10 do not accord exactly with those shown in Chapter 27, Reference 27.9, but the general expectation is clear. Unmanned systems are expected to proliferate and the rôle of the human will increasingly be that of a user and operator more than a controller.

28.11   Into the Future

Men, my brothers, men the workers, ever reaping something new;
That which they have done but earnest of the things that they shall do.

For I dipt into the future, far as human eye could see,
Saw the Vision of the world, and all the wonder that would be;

Saw the heavens fill with commerce, argosies of magic sails,
Pilots of the purple twilight, dropping down with costly bales;

Heard the heavens fill with shouting, and there rain'd a ghastly dew
From the nations' airy navies grappling in the central blue;

Far along the world-wide whisper of the south-wind rushing warm,
With the standards of the peoples plunging thro' the thunder-storm;

Till the war-drum throbb'd no longer, and the battle-flags were furl'd
In the Parliament of man, the Federation of the world.

There the common sense of most shall hold a fretful realm in awe,
And the kindly earth shall slumber, lapt in universal law.

From 'Locksley Hall', written by Alfred, Lord Tennyson in the 1830s.

Unlike Tennyson, the author has no crystal ball, but holds the hope that the near future will see legislators and engineers work together to enable the safe application of unmanned aircraft systems to civilian uses for the benefit, rather than the destruction, of humanity.

Appendix A

UAS Organisations

This listing cannot be considered to be all-inclusive as, for example, conferences continue to be set up and many turn out to be 'one-offs'. New associations and sites are established as the industry grows.

A.1 Conferences

Industry Associations and several press organisations such as the Shephard Press and Flight International mount conferences (see Section A.3). UAS Conferences are put on by the following learned bodies, the first of which to be established were the Bristol Conferences.

American Institute of Aeronautics and Aerospace (AIAA),
1801 Alexander Bell Drive, Suite 500, Reston, VA 20191-4344, USA
Tel. +1 703 264 7500 or 800 639 AIAA; website: www.aiaa.org; email: custserv@aiaa.org

The Bristol-International UAV Systems Conferences,
University of Bristol, Department of Aerospace Engineering, Bristol BS8 1TR, UK
Tel. +44 (0) 117 928 9764; website: www.bris.ac.uk/aerospace/uavs;
email: aero-uav@bristol.ac.uk

The Royal Aeronautical Society,
4 Hamilton Place, London, W1J 7BQ, UK
Website www.aerosociety.com

The Royal Melbourne Institute of Technology (RMIT),
Australasian Conference on Unmanned Aerial Vehicles (Affliated to the Bristol UAV Conference Series) Chair Person: Dr Arvind Kumar Sinha
Australian Aerospace and Defence Innovations, 506 Lorimer Street, Fishermans Bend, Melbourne, Australia, Vic 3207
Tel. + 61 438 578 060; email: arvind.sinha@aadi.com.au

A.2 Industry Associations

Euro UVS,
88 rue Michel-Ange, 75016 Paris, France
Tel. +33 1 46 51 88 65; Fax. +33 1 46 51 05 22; website: www.euro-uvs.org; email: info@ euro-uvs.org

Unmanned Aircraft Systems – UAVS Design, Development and Deployment   Reg Austin
© 2010 John Wiley & Sons, Ltd

The Association for Unmanned Vehicle Systems International (AUVSI),
2700 S. Quincy Street, Suite 400, Arlington, VA 22206. USA
Tel. +1 703 845 9671; email: info@auvsi.org; website: www.auvsi.org

Unmanned Air Vehicle Systems Association Ltd, The Mille, 1000 Great West Road, Brentford,
Middlesex TW8 9HH, UK
Tel. +44 (0) 208 568 1116; Fax. +44 (0) 208 568 1116; email: secretary@uavs.org; website:
www.uavs.org

UAVNET – A networking organisation funded by the European Community to advance the
development of UAV in civilian operations
website: www.uavnet.com

European Unmanned Systems Centre, Unit 4a, ParcAberporth, Aberporth, Ceredigion,
SA43 2DW, UK
Tel. +44 (0) 1239 814 814; website: www.eurousc.com

A.3   Press Organisations

The Shephard Press Ltd., 268 Bath Road, Slough, SL1 4DX, UK
website: www.uvonline.com
(Unmanned Vehicles Magazine and annual handbook)

Janes Unmanned Aerial Vehicles and Targets, Jane's Information Group
website: www.janes.com

Flight International, Quadrant House, The Quadrant, Sutton, Surrey, SM2 5AS, UK
website: www.flightglobal.com; email: international@flightglobal.com

Unmanned Systems Magazine, published bi-monthly by The Association for Unmanned Vehicle
Systems International, UAVSI (see Section A.2)

A.4   Useful Websites

Barnard Microsystems Ltd: www.barnardmicrosystems.com

Federation of American Scientists: www.fas.org

Flight International: www.flightglobal.com

Google UAV Alerts

Shephard: www.uvonline.com

UAV Forum: www.uavforum.com

A.5   Test Site Facilities

Finland

The Finland-based Robonic Arctic Test UAV Flight Centre is proposing the development of a common
European unmanned air vehicle flight training centre, around its UAV flight test centre
website: www.robonic.fi

Sweden

North European Aerospace Test range (NEAT). The site offers an air-restriction area of 360×100 km, and a 1650 km² restricted land area sometimes extended to 3000 km². Contact: NEAT, Esrange, PO Box 802, SE-981 28 Kiruna, Sweden
Tel. +46 980 72399; website: www.neat.se; email: neat@neat.se

United Kingdom

ParcAberporth, Ceredigion (Cardigan), West Wales. The site covers an area of 22 km diameter with a runway of 1200 m and unrestricted airspace up to 1500 m. Contact: Welsh Development Agency, Mid Wales Division, Ladywell House, Newtown, Powys SY16 1JB, or contact Alex Bricknell on +44 (0)1686 613 143; Fax: 01686 627889; website: www.wales.gov.uk; email: ein.@Wales.GST.gov.uk

United States of America

Great Plains Test Center, Salina, Kansas. The site covers about 140 km². Contact: Sharon Watson, Director, Public Affairs Office. Tel.: (785) 274-1192; website: www.Kansas.gov/ksadjutantgeneral

Idaho National Laboratory. The site covers an area of about 2200 km² and is at an average height of 1500 m above sea level. Airspace is available up to 300 m, but the airstrip for HTOL aircraft is of only 300 m long and so is only suitable for smaller HTOL UAV or for VTOL systems. Contact Scott Bauer, +1 208 526 8967 or Tom Harper, +1 208 526 0113; website: www.inl.gov; email: Scott.Bauer@inl.gov or Thomas.Harper@inl.gov

White Sands Missile Range, New Mexico, USA. The site covers an area of 3565 square miles at altitudes ranging from 4000 ft AMSL to 9000 ft AMSL. Contact: Commander, U.S. Army White Sands Missile Range, CSTE-DTC-WS-TC, Building 100, New Mexico, 88002-5000, USA. website; www.wsmr.army.mil; email: TeamWhiteSands@wsmr.army.mil

A.6   Regulators

Federal Airworthiness Administration (FAA),
800 Independence Avenue, SW Washington, DC 20591. USA
Tel. +1 866 835 5322); website: www.faa.gov

European Aviation Safety Agency (EASA)
Postal address: Postfach 10 12 53, D-50452 Cologne, Germany
Visiting address: Ottoplatz 1, D-50679 Cologne, Germany
Tel.: +49 (0)221 8999 0000; website: www.easa.eu.int; email: info@easa.eu.int

Civil Aviation Authority (CAA)
CAA House, 45-59 Kingsway, London, WC2B 6TE, UK
Tel: 020 7379 7311; website: www.caa.co.uk; email: infoservices@caa.co.uk

Index

Printed in the USA/Agawam, MA
May 23, 2019

703786.002

120674

THE HUMAN DIMENSIONS OF NATION MAKING

Essays on

Colonial and Revolutionary

America

Edited by

JAMES KIRBY MARTIN

973.3
H918

The State Historical Society of Wisconsin

Madison   1976

Alverno College
Library Media Center
Milwaukee, Wisconsin

Copyright © 1976 by
THE STATE HISTORICAL SOCIETY OF WISCONSIN

All rights reserved

Manufactured in the United States of America by
NAPCO Graphic Arts, Inc., New Berlin, Wisconsin

Library of Congress Cataloging in Publication Data
Main entry under title:
The Human dimensions of nation making.

  Essays written in honor of Merrill Jensen upon his
retirement from the University of Wisconsin.
  "A listing of Merrill Jensen's publications": p. 362.
  Includes bibliographical references.
  CONTENTS: Ferguson, E. J. Merrill Jensen, a personal
comment.—Martin, J. K. The human dimensions of nation
making, Merrill Jensen's scholarship and the American
Revolution.—Webb, S. S. The trials of Sir Edmund Andros. [etc.]
    1. United States—History—Revolution, 1775–1783—
Addresses, essays, lectures. 2. United States—History—
Colonial period, ca. 1600–1775—Addresses, essays, lectures.
3. Jensen, Merrill. 4. Jensen, Merrill—Bibliography.
I. Martin, James Kirby, 1943– . II. Jensen, Merrill.
E208.H84   973.3   75–30821
ISBN 0–87020–158–1

PREFACE

THROUGHOUT his long and distinguished career Merrill Jensen has served the craft of history in multiple ways. His commitment to the highest standards in scholarship and teaching has affected the thinking and writing of untold numbers of persons. Even his scholarly critics admit that his voluminous writings have had and will continue to have a profound influence in shaping our conceptions of the critical era of the American Revolution. His many Ph.D. students well remember Professor Jensen's uncompromising insistence upon meticulous, detailed research and careful, exacting analysis of sources as the cardinal principles of solid scholarship. Thirteen of his former students have come together through this volume to honor their graduate program mentor as Merrill Jensen retires from the University of Wisconsin after three decades of faculty service. All of his friends and students wish him well as he pursues his many scholarly interests in the years ahead.

The research essays that follow have not resulted after the imposition of some artificial theme upon the authors. Rather, each article represents an individual scholarly entity, fully capable of standing by itself. Each advances our knowledge and understanding of some aspect of the early years of struggle and nation making in the American experience. It is only appropriate that the essays flow out of original research and that they are not inhibited by an artificial theme because Merrill Jensen has always insisted that historical scholarship must begin with the sources.

Many individuals have supported efforts to make this volume a reality. Three in particular made especially valuable contributions. James Morton Smith and Paul Hass of the State Historical Society of Wisconsin and Jackson Turner Main of the State University of New York at Stony Brook deserve much more than the usual acknowledgments. The editor sincerely appreciates their commitment, their support, their time, and their guidance.

JAMES KIRBY MARTIN

Rutgers University
April, 1975

CONTENTS